The Effects of Constant Light on Visual Processes

The Effects of Constant Light on Visual Processes

Edited by

Theodore P. Williams

and

B. N. Baker

Florida State University
Tallahassee, Florida

PLENUM PRESS · NEW YORK AND LONDON

Library of Congress Cataloging in Publication Data

Symposium on the Effects of Constant Light on Visual Processes, 2d, Florida State University, 1979.

The effects of constant light on visual processes.

Includes index.

1. Retina–Congresses. 2. Light–Physiological effect–Congresses. 3. Photoreceptors–Congresses. 4. Retina–Wounds and injuries–Congresses. I. Williams, Theodore P. II. Baker, Barbara N. III. Title.

QP479.S95 1979 612'.843 79-26293

ISBN 0-306-40328-5

Proceedings of a symposium on the Effects of Constant Light on Visual Processes, held at Florida State University, Tallahassee, Florida, and sponsored by the Psychobiology Research Center, April 25–27, 1979.

A Division of Plenum Publishing Corporation
227 West 17th Street, New York, N.Y. 10011

Printed in the United States of America

Preface

"...And the evening and the morning were the third day...And God made two great lights; the greater light to rule the day and the lesser light to rule the night...And the evening and the morning were the fourth day." The First Book of Moses, called Genesis (1: 13, 16, 19).

There was daytime and nighttime before there was a sun or a moon. An interesting thought! How long were each of those first three days? Without a sun to reckon its length, a day could have been longer or shorter than 24 hours. Animals, says Genesis, appeared on the fifth day and by that time the sun and moon were illuminating the earth, presumably in cyclic fashion and with a period of 24 hours. A good thing, too, as readers of this monograph will ascertain.

The papers collected into this volume are written versions of 45 minute talks given at a symposium on "The Effects of Constant Light on Visual Processes", held at The Florida State University in Tallahassee on April 25-27, 1979. The conference was supported by the Psychobiology Program and handled, logistically, by the Center for Professional Development and Public Services. It was recognized that limitations on time and funds made prohibitive the invitation of others who may be doing interesting and related work. But, our earnest hope is that what is compiled here is a good blend of "true" light-damage and "relevant related" work.

The symposium began with a full day devoted to characterization of light-damage in mammals and a delineation of some of its manifestations.

The ensuing sessions dealt with diverse topics including constant-light effects on invertebrates: insects, lobsters, spiders, and, of course, the horeshoe crab, Limulus. What monograph on visual processes would be complete without reference to Limulus?

In one short but interesting session we heard papers dealing with aspects of human retinal light damage and one on the possible light damage of the visual pigment protein, opsin. The latter raises the question: does the visual pigment "wear out" and, if so, is that the reason rod outer segments are renewed?

Because light-damage is expressed primarily in the photoreceptor cells and because it is sometimes reversible, it seemed logical to us to include a session in molecular renewal of the photoreceptors. In this context we heard papers on rod-disk shedding, protein synthesis and lipid metabolism.

Apart from these sessions, there was time for extended discussion at meals and at evening socials. Our only regret about this monograph is that the reader of it will not be able to enjoy any of the conviviality and only part of the scientific exchange of those after-hour meetings. We enjoyed sponsoring this conference and want to thank especially the Psychobiology Program for the opportunity of presenting it and Beverly Tranchand who made all the wheels turn smoothly.

T. P. Williams
B. N. Baker

Tallahassee

Contents

PART II: INVERTEBRATES

PART III: SELECTED TOPICS

PART IV: MOLECULAR ASPECTS OF PHOTORECEPTOR PHYSIOLOGY

Vertebrates

THERE ARE DIFFERENT KINDS OF RETINAL LIGHT DAMAGE IN THE RAT

Werner K. Noell

Department of Physiology
State University of New York School of Medicine
Buffalo, New York 14214

In the following, I will review experimental results of our laboratory during the last 15 years on the damaging effect of light. My purpose will be to distinguish between at least two different kinds of retinal pathology caused by light. The question whether light exerts its damaging effect by one or several mechanisms has plagued us from the very beginning of our study, i.e. once it had been ascertained that the visual cells and the pigment epithelium of common breeds of albino or pigmented rats are irreversibly affected and destroyed by prolonged exposure to visible light of moderate or even weak intensity. The question inhibited the planning of experiments for the analysis of the mechanism of the effect and for the evaluation of physical and biological factors that influence the magnitude of the effect for which a reliable quantitative measure had to be found.

We, therefore, searched for a condition of exposure under which light produced an irreversible effect with such a consistency and rapidity that the assumption of essentially one mode of action of light for the initiation of the non-physiological reaction (over the whole retina) seemed a reasonable assumption. We found this condition by exposing the freely moving rat or one eye of the anesthetized rat to green light above 100 Ft-C at an artificially elevated eye temperature (1). Visible light then was maximally damaging within less than 4 hours of exposure, and fractionally effective in less than ½ hour, as quantitatively assessed by the surviving electroretinogram, ERG.

The uncertainties, however, remained for all other conditions of exposure, and especially for the quantitative evaluation of the striking dependence of the retinal vulnerability on the "light history" of the animal, i.e. on the long-term adaptation of the retina to the environmental lighting conditions in which the animal had lived prior to exposure (2-5). The study of differences between retinas adapted to a cyclic light condition and those adapted to darkness could yield clues as to the mechanisms of damage, provided the retinal vulnerability could be assessed reliably.

The following brief review is not intended to be exhaustive: it will deal exclusively with our own experimental results. Instead, reference is made to the extensive and thoughtful review by J. Lanum (6) on the damaging effect of light.

The rhodopsin-mediated light damage of the first kind.

This sort of damage is characterized histologically, at 1-2 weeks after exposure, by the absence of the outer retinal layers including the pigment epithelium in an area which varies in size from very small to almost the whole retina (1). The variations in the size of the lesion relate to the vulnerability of the animal's retina, the animal's motor activity during exposure and, most definitively, to the duration of exposure at a given light intensity. Our most extensively conducted experiment in which this type of damage was the almost exclusive effect of constant light will be used to illustrate the features of this damage.

The 19 male albino animals of this experiment had been purchased (as one group) from the Charles River Laboratories at the age of 6 weeks. They were shipped in light-shielded boxes and then housed, 2 rats per wire-mesh cage in special quarters illuminated from the ceiling by 60 watt tungsten light bulbs in a 12-hour per day rhythm. The light bulbs were outside the direct view of the animals. Illumination at the most exposed cage site measured less than 2 Ft-C; it was less than 0.1 Ft-C in the darkest corner of every cage. The animals generally slept during most of the light cycle in the typical posture of healthy rats in the least illuminated region of the cage.

The animals remained in this cyclic light environment for 8 weeks, before being exposed to the damaging light of our standard procedure (1). Exposure time varied between 12 and 26 hours. Exposure was in circular wire-mesh cages ("chambers") which were surrounded by 1/8 inch thick green acrylic plastic (Plexiglass #2092, transmitting in the 490 to 580 nm range). Three circular 32 watt fluorescent lamps (G.E. Cool White) were stacked around the green plastic which also was covered by 3 sheets of white translucent paper for reducing the level of chamber illumination. Depending upon position within the chamber, illumination varied between 12 and 25 Ft-C. Temperature within the chambers was adjusted to 23-25°C.

Following exposure the animals were transferred to (constant) darkness. Their ERG's were measured on the 10 or 11th day after exposure under light Nembutal® anesthesia reinforced by 1% xylocaine® to the cornea. Our routine ERG stimulus is the maximal (xenon) flash of the Grass Photostimulator given twice to each eye at 2-5 minutes interval. In albino controls the flash elicits a sharp a-wave of 600-800 μV and a b-wave which peaks at 1200-1600 μV; variability in control ERG measurements are predominately functions of age and of the "light history" of the animal, i.e. of the light environment in which it had been maintained prior to the test. Fast wavelets ("oscillatory potentials") preceed the b-wave peak and are often slightly higher in the control animals.

For histological study, the eyes were removed immediately after the ERG test and fixed in Zenker's solution. They were cut in paraffin at 5μ thickness in 15 steps of 200-250μ in a ventro-dorsal direction.

Fig. 1 shows a reconstruction of the histological appearance of the retina of a rat of this experiment which had been exposed for 24 hours. The heavily shaded area, located mainly in the dorsal half of the retina, indicates the region of complete disappearance of the layers of the visual cells and of the pigment epithelium as is illustrated in Fig. 2. Bruch's membrane (Fig. 2)

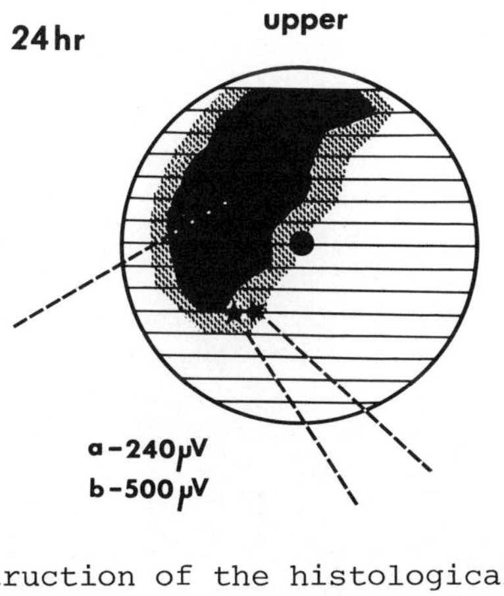

Figure 1. Reconstruction of the histological damage as measured in the eye of an albino rat (eye #2L) exposed for 24 hours to "green" light of 12-25 Ft-C (see text). The eye was removed 20 days after exposure following an ERG test. Note a-wave and b-wave values at the bottom. At 6 days after exposure, the a-wave was 330 μV and the b-wave was 460 μV.

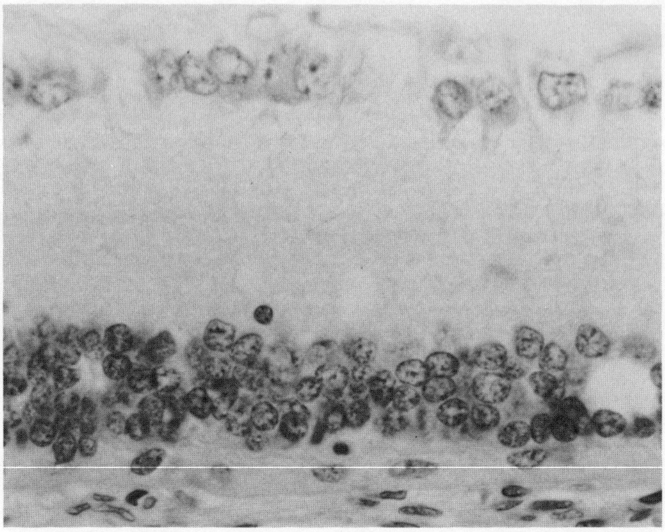

Figure 2. Photomicrograph of the retina from the center of the lesion of Fig. 1. The visual cells have disappeared; the pigment epithelium is absent. The right of the figure shows a typical cyst in the inner nuclear layer. (Reduced 20% for purposes of reproduction.)

separates a narrow, fibrous outermost part of the retina from the essentially normal choroid. This fibrous outermost part, lies directly beneath the inner nuclear layer and shows some invasion of non-retinal cells. No visual cell elements are evident including cone nuclei, which, in rats, generally have a higher survival potential than the rod cells. The number of inner nuclei and ganglion cells are not decreased at this time (1-2 weeks after exposure) except when replaced by small vacuoles in the damaged area. These vacuoles seem to increase in number with time from exposure; they are first evident 5-7 days after exposure.

The region of maximal pathology is surrounded by a narrow rim (stippled in Fig. 1) of less severely affected tissue characterized typically by the survival of several rows of outer nuclei in the absence of the pigment epithelium (Fig. 3, 4). Going to the periphery of the lesion, pigment epithelial cells reappear but the continuity of the epithelium is not the normal one.

The reappearance of the pigment epithelium is associated with the survival of (shortened) outer and inner segments and the survival of several rows of visual cell nuclei (Fig. 4).

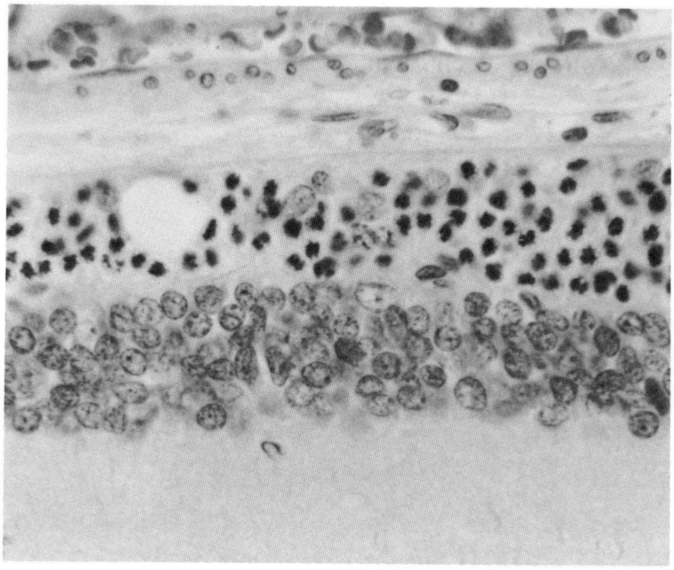

Figure 3. Photomicrograph of the retina at the rim of the lesion of Figs. 1, 2. The outer nuclei layer is partially preserved, but the pigment epithelium and all inner and outer segments have disappeared (see text). (Reduced 20% for purposes of reproduction.)

Outside the shaded and stippled areas of Fig. 1, the outer and inner segments appear light-microscopically normal. The rows of outer nuclei are no more than 1/3 reduced in the vicinity of the lesion.

The weakest manifestations of the pathology of the kind described above is illustrated in Figs. 5-8 from another retina of the same experiment. Exposure time was also 24 hours. Again pathology is circumscribed (Fig. 5) but much smaller in size than the lesion of Fig. 1. It is less severe in terms of cell loss (Fig. 6) and resembles in its histology the tissue at the rim of the severe lesion (see Fig. 4). The pigment epithelium is preserved but differs from normal by clear confluent regions indicative of degenerative vacuolization. The most discernable change is a 40-60% reduction in the number of outer nuclei, i.e. it is associated with degenerative changes in the rod and cone layer. Outside the stippled region of Fig. 5, the retina has virtually the normal appearance (Fig. 7 and Fig. 8).

The lesion s as described above have been observed in numerous experiments. They are typical for the damage produced by a short-term exposure which can be as short as one-half day and which can

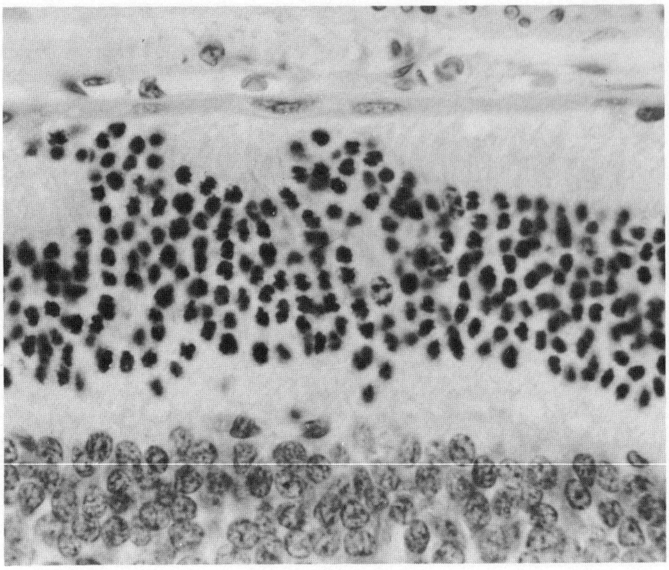

Figure 4. Photomicrogrpah from same retina as Figs. 1-3, close to the edge of the lesion. The pigment epithelium is present but is abnormal in appearance with discontinuities between neighboring cells; the outer nuclei layer is distorted as is typical for retinal regions with localized pigment epithelial damage. (Reduced 20% for purposes of reproduction.)

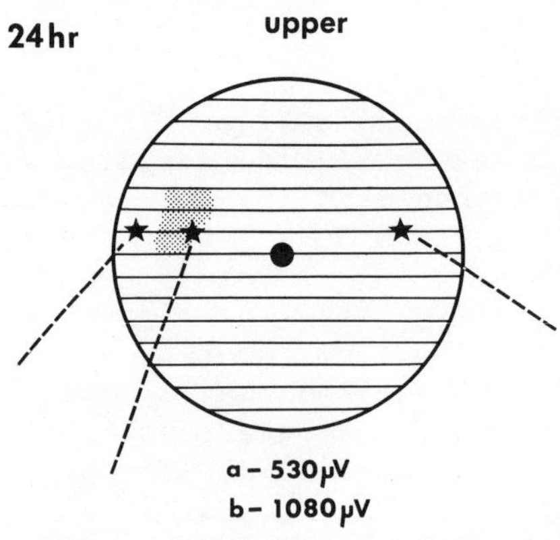

Figure 5. Reconstruction of a "minimal" lesion (eye #18R) which is characterized by the localized loss visual cells without destruction of the pigment epithelium. The eye was removed 10 days after the 24 hour exposure to 12-25 Ft-C, as was the eye of Figs. 1-4.

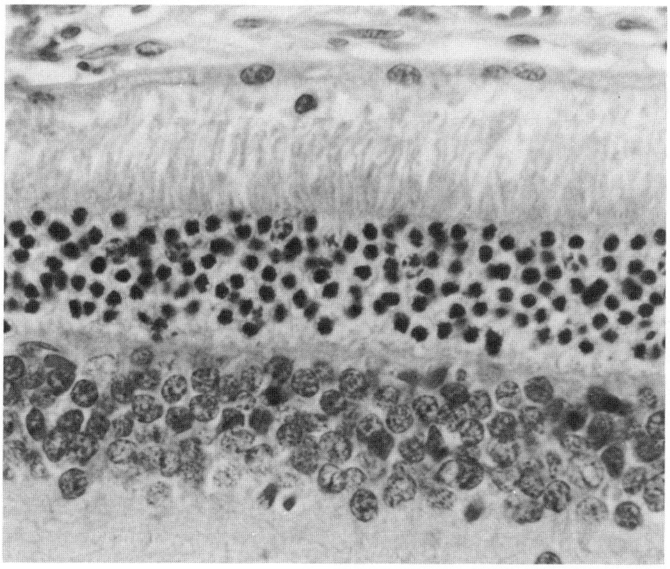

Figure 6. Photomicrograph of the lesion of Fig. 5 near the border of the lesion. About 50% of the visual cell nuclei have disappeared; the outer segments of the surviving cells are distorted and not all surviving inner segments carry an outer segment. The pigment epithelium has vacuoles; a phagocytotic cell (center) has invaded the rod layer. (Reduced 20% for purposes of reproduction.)

render a retina non-functioning in an irreversible manner. The vulnerability of the albino or pigmented rat to this kind of light damage is greatly enhanced by constant maintenance, prior to exposure, in a dark environment for several weeks and by exposure during an artificially elevated body temperature. A high body temperature during exposure to light above 100 Ft-C which bleaches rhodopsin to a level of less than 15% in 10 minutes, destroys the pigment epithelium and the visual cells over virtually the whole retina within 3 to 4 hours (1).

The ERG is one of the easiest means of quantitatively assessing the damage by light after exposure. Fig. 9 illustrates ERG's of the experiments just described. The two traces at the top of the figure are from 2 eyes of the experiment in which a histological "lesion" was not found. The 6 remaining ERG's were recorded from eyes with lesions of the kind described above. They are arranged in the order of increasing lesion size. The major abnormality is a reduction in amplitude which is virtually the same quantitatively for a-wave, b-wave and wavelets.

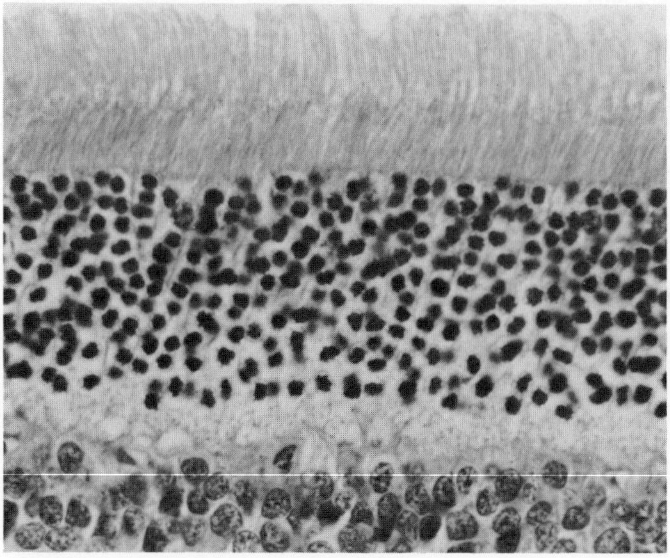

Figure 7. Photomicrograph of the non-lesioned retina from the same slide as Fig 6 near the area of the damage. The retina has separated from the pigment epithelium during the fixation procedure. (Reduced 20% for purposes of reproduction.)

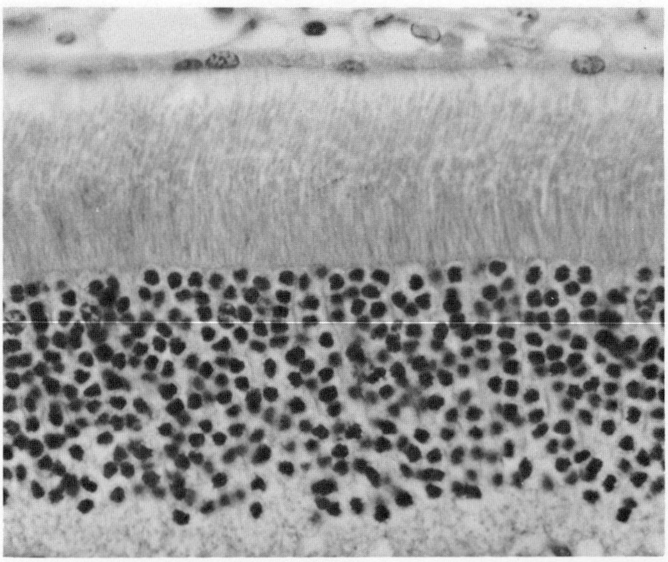

Figure 8. Photomicrograph from the same slide as Figs. 6 and 7 but from the retinal quadrant opposite to the site of damage. The retina seems "normal". (Reduced 20% for purposes of reproduction.)

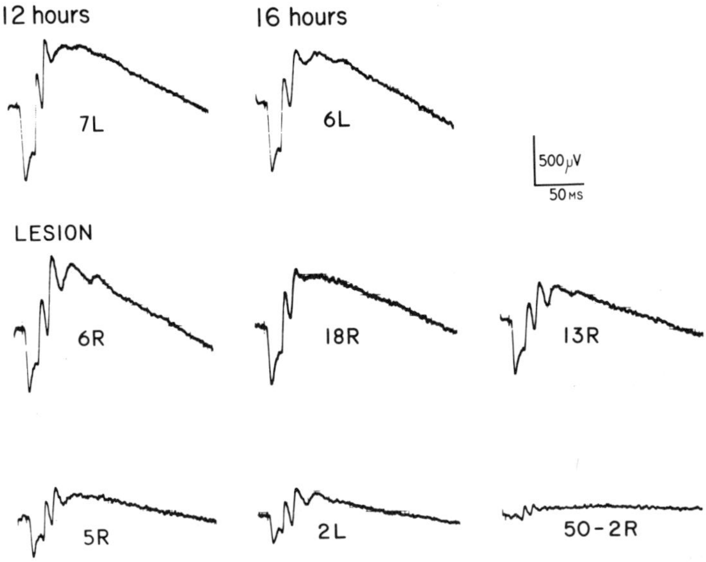

Figure 9. Selected ERG's of the experiment of Figs. 1-8 (see text). The ERG of eye #50-2R is from a different experiment with exposure for 48 hours at 10 times higher illumination than in the others. Note that eye #6R showed a lesion as defined (see text) while the fellow eye (#6L) lacked such a lesion. Eye #7L (and its fellow eye) also did not show a lesion.

Approximately 60% of the 37 eyes of the experiment (12-26 hours exposure to 12-25 Ft-C) showed a localized abnormality as described. The mean a-wave of these eyes was 459 ± 110 μV(SD). The remaining eyes had an average a-wave of 668 ± 65 μV(SD), which does not differ from our average control values for unexposed animals. The average difference between the 2 eyes of the same animal was 62 ± 47 μV(SD) for those showing a lesion; it was 48 ± 32 μV(SD) for animals in which neither eye showed a lesion. In two animals one eye had a minimal lesion, while none was found in the other eye. The two eyes with a lesion had a-waves of 555 μV and 585 μV, while the a-wave of the non-lesioned fellow eye was 585 μV and 630 μV, respectively.

In Fig. 10 the amplitude of the ERG (a-wave) has been plotted as a function of the relative size of the non-lesioned retina, using measurements of lesion size (cf. Figs. 1 and 5). Theoretically, an almost linear relationship between ERG amplitude and functional area should prevail if the non-lesioned area generates an ERG of normal form while the lesioned area has lost ERG generation by the death of the visual cells. Fig. 10 shows that the curve fitting the experimental points deviates markedly from linear as if the

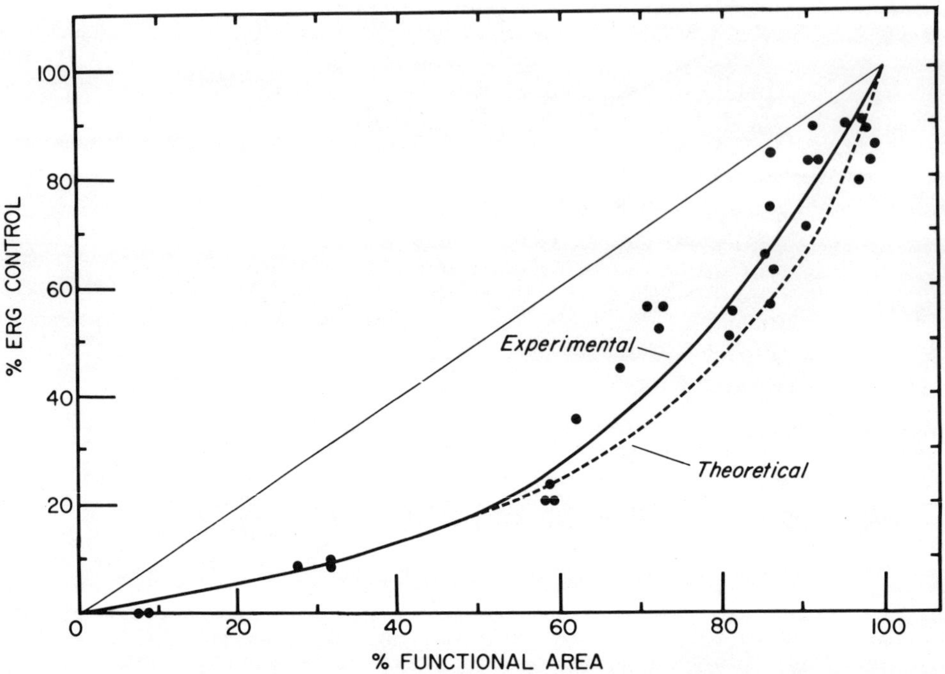

Figure 10. Correlation of the ERG (a-wave) with the retinal area outside the "lesion" considered "functional" because of the preservation of the outer segment layer. The control a-wave value was 650 µV. The measurement of the functional area was from reconstructions as illustrated in Figs. 1 and 5. A total of 30 eyes were used which include those of the experiment of Fig. 1-9. Additional eyes from other experiments were chosen in order to cover the low ERG range. The "theoretical" curve relates ERG and functional area on the assumption that the lesion, the non-functional retina, includes the complete loss of visual cells and the destruction of the transretinal resistance component given by the pigment epithelium.

ERGs were more affected than could be accounted for by the loss of the visual cells in the damaged area. In other words, the curve suggests that the electrogenic function in response to the light stimulus is abnormal in the non-lesioned retina of "normal" histology. The retinal lesion, however, is also characterized by the loss of the pigment epithelium. Because of its basal and apical membranes, and its cell to cell coupling, the pigment epithelium provides the major component of the transretinal resistance which determines ERG current flow (7). Hence disappearance of the pigment epithelium shunts the electrical current generated by the non-lesioned area to a greater degree than would the localized disappearance of only

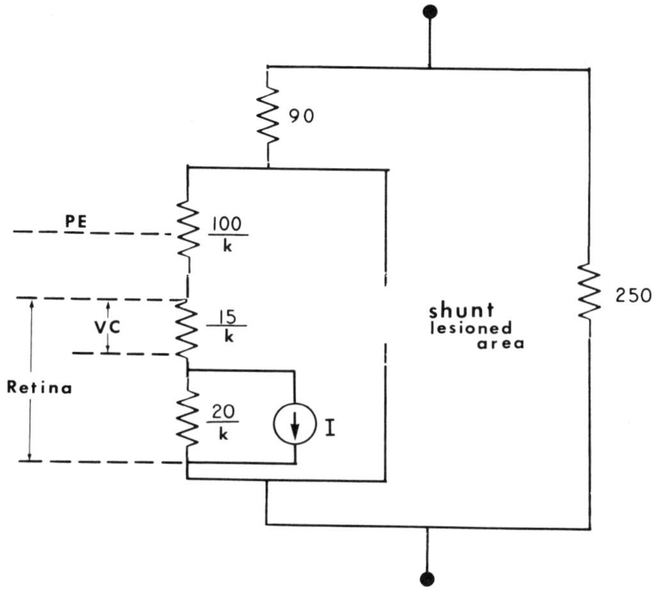

k = fraction of retina functional

	Extent of Lesion	
	VC	PE+VC
SHUNT (Ω)	$\frac{120}{1-k}$	$\frac{20}{1-k}$
ERG (% Control)	$\frac{k}{5-4k} \times 100$	$\frac{k}{1.1-0.1k} \times 100$

Figure 11. The equivalent circuit for the theoretical curve of Fig. 10. The transretinal resistance components are (in ohms): 20 for the inner layers of the retina, 15 for the visual cell layers (VC), 100 for the pigment epithelium (PE) and 90 for the extraretinal series resistance between the electrodes. A parallel resistance of 250 ohms is assumed for the flow of any current generated within the (whole) retina. The light-produced lesion is assumed to represent a shunt of the current through retina which lacks PE and VC. The resistance through this shunt then is given only by the resistance of the inner layers (20 ohms). This shunt resistance is zero (as indicated in the drawing) when all retinal layers are destroyed locally as for instance with photocogulation.

the visual cells. The equivalent circuit, shown in Fig. 11 was used to assess the shunting ability of a retinal lesion characterized by the loss of visual cells and pigment epithelium. The resistance values are as computed for the rabbit retina (7). The resulting "theoretical" curve (Fig. 10) for the relationship between ERG and non-lesioned area fits the experimental data surprisingly well.

We thus conclude that the reduction in ERG after light damage of the first kind may not be the result of abnormal function in the non-lesioned area. For comparison with the preceding figure, Fig. 12 is included which shows data for lesions in the rabbit retina. These lesions resulted from intravenous iodoacetate (IAA) and were characterized by visual cell loss and survival of the pigment epithelium.

Additional support for this conclusion of a normal function of the non-lesioned area is provided by estimates of the sensitivity of the surviving ERG. For this estimate, ERG measurements were made in affected animals for stimulus intensities over a 5 log unit-range. Data were used from 20 control rats and from 50 rats with damage after strong-light exposure for 6 to 48 hours producing the whole spectrum of abnormally low ERGs in response to the routine maximal xenon flash. Regression lines were computed for the relation between the a-wave amplitude in response to the maximal stimulus and the b-waves in response to the same flash attenuated by neutral density filters (least squares method). They are shown in Fig. 13. The coefficient of correlation was .977 for the relation between the a-wave and the b-wave of the maximal stimulus and .932 and .916

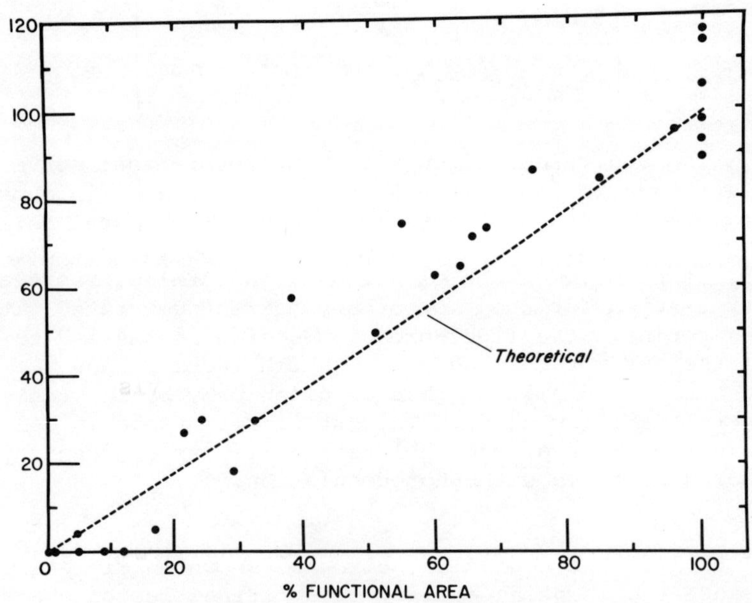

Figure 12. Data from rabbit experiments for lesions produced by iodoacetate (IAA). The lesions are characterized by the loss of the visual cells only. The theoretical line derives from the network of Fig. 11.

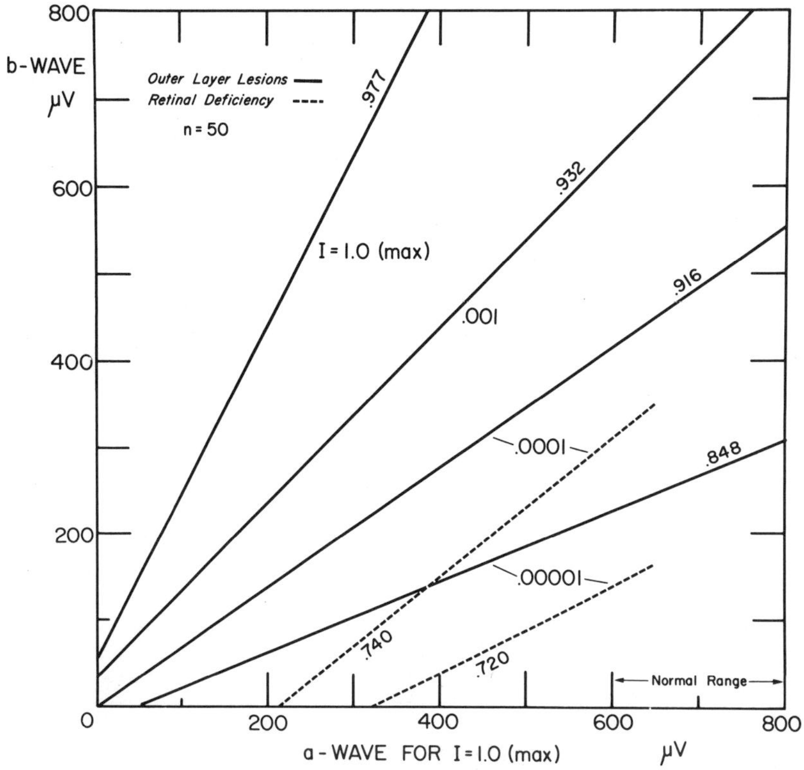

Figure 13. Computer-fitted regression lines (least squares method) for the relationship between the amplitude of the a-wave in response to a xenon flash of maximal intensity (Grass Photostimulator) and the b-wave elicited with the same kind of flash and attenuated by neutral density filter. These attenuated flashes were below the a-wave threshold of the controls. This threshold was near I = .01. The ERG's are from 50 light-damaged retinas; they cover the whole range of a-wave amplitude, from zero to the normal range. Light exposure (120-200 Ft-C) preceeded the ERG test by at least 10 days. The middle number beside each line is the intensity of the flash in arbitrary number. The upper number is the coefficient of correlation (e.g. .977). The broken lines are for the ERG's of the retinas of vitamin-A deficient animals.

for the relation with the b-wave in response to a stimulus attenuated by 10^{-3} and 10^{-4}, respectively. It is concluded that the sensitivity of the functioning retina is within the normal range for the damage by light of the first kind. Results similar to those of Fig. 13 were obtained for a correlation of the a-wave (xenon flash) and the b-waves of ERG's elicited by 50-msec (tungsten) flashes over a 4 log unit range below the threshold for a-wave generation (Fig. 14).

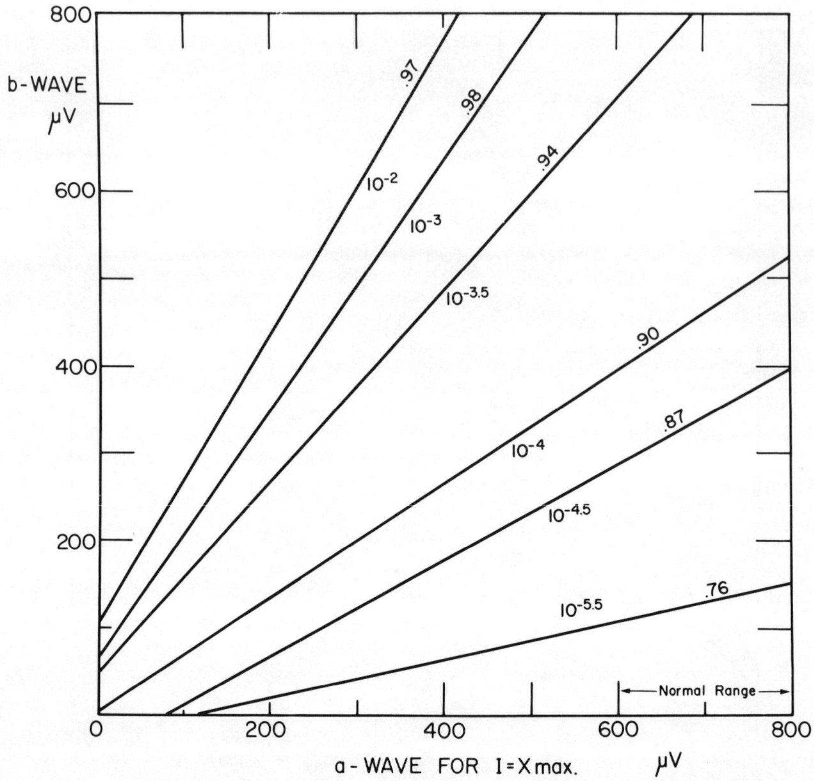

Figure 14. Regression lines as in Fig. 13. Different light-damaged eyes. The a-wave was measured from the response to the xenon flash of maximal intensity, as in Fig. 13. The weak flashes were from a tungsten source and were 50 msec. in duration. Their intensity was 10^{-2} - $10^{-5.5}$ below that of the maximal tungsten flash which was 1.3 log units above a-wave threshold. The high correlation, as shown by Figs. 13 and 14, between the a-wave in response to a maximal xenon flash and the b-wave in response to the same flash or weaker flashes below a-wave threshold suggests that (a) the function of the surviving outer retinal layer is independent from the size of the non-functioning area of the retina in light-damage of the first kind; (b) the function of the surviving visual cells is probably not different from normal, including photoreceptor sensitivity (see intercepts with x and y) and (c) the ERG elicited with the maximal xenon flash adequately assesses the electrical function of the surviving retina.

The question then arises whether the disappearance of the pigment epithelium, which characterizes the light damage of the first kind, is a primary damaging effect of light on this structure. Three arguments can be presented that it is not. First of all, it is certainly not a thermal lesion as was ascertained early in our study by a consideration of the damaging light intensities of different light sources (1). Secondly, limited measurements of the action spectrum of the damage made at the heightened retinal vulnerability by an elevated eye temperature, strongly suggests that the initial event is the absorption of light by rhodopsin. Thirdly, exposure of rats afflicted with hereditary retinal degeneration produced no significant pigment epithelial change when the exposure occurred at a stage of the disease at which the visual cells had been lost but rhodopsin-rich debris (8) was still attached and intertwined with the pigment-epithelial processes.

The hypothesis, therefore, is favored that the damage to the pigment epithelium results from a change in the chemical environment at the apical side of the epithelium. The visual cells may release a "toxic" product, or induce a change within the intercellular space which exceeds physiological limits.

The question then may be asked whether this chemical change is secondary to the "death" of the visual cell, or to a decomposition of the outer segments with which the pigment epithelium is in close contact. The answer to this question also seems to be in the negative. Histological changes of the pigment epithelium have not been observed, as a rule, with rapid, massive visual cell death following the administration of iodoacetate or following the exposure of the retina to x-irradiation or high oxygen pressure. These experiments, however, were mainly performed in the rabbit and must be repeated in the rat to exclude the possibility that the rat's pigment epithelium is especially vulnerable.

The involvement of the pigment epithelium is, however, one of the first histological signs of the light damage under discussion. It was markedly swollen and vacuolized at both the LM and EM levels (9) when the retina was immediately fixed after 3-4 hours of light exposure at an elevated body temperature. Inner and outer segments were then still preserved light-microscopically. In the EM study, it was estimated that about 70% of the outer segments were of normal appearance at the time of a severe pigment epithelial change. Similar severe changes with vacuolization were seen in the outer plexiform layer with massive swelling, especially of the Müller cell processes. Likewise, the presynaptic endings of the visual cells were severely affected. These findings favor the assumption of a detrimental chemical change in the spaces surrounding the visual cells.

The early loss of pigment epithelial function probably has a minor role for the development and extent of the visual cell damage. This is suggested by a comparison with the visual cell changes following the destruction of the pigment epithelium by the action of intravenous iodate in the rabbit (10). Electrophysiological studies very strongly suggest that the effect of iodate on the pigment epithelium is a primary one and that the visual cells are secondarily involved due to the loss of pigment epithelial function. These secondary changes include degeneration and loss of inner and outer segments with preservation of the outer nuclei. Light-induced changes in the outer nuclear layer and in the layers of rods and cones as illustrated in Figs. 3 and 4, may well result from pigment epithelial damage. They are of the same kind as observed with the iodate effect.

The loss of pigment epithelial function may also be responsible for the finding that signs of ERG recovery do not extend beyond one day after exposure resulting in light damage of the first kind. On the contrary, ERG attenuation increases somewhat between the first and tenth day after exposure.

Rhodopsin-Mediated (?) Light Damage of the Second Kind.

Our light damage of the second kind is first of all classified by a characteristic ERG abnormality indicated by a predominant reduction of the a-wave. It is classified histologically by the survival of the pigment epithelium in the presence of widespread visual cell layer abnormalities with loss of visual cells. The occurrence of this kind of damage was observed under a variety of conditions, generally those of a reduced retinal vulnerability compared to that necessary for the production of the damage of the first kind. The damage of the second kind also tended to become more prominent at relatively low levels of constant illumination. Nevertheless, we were able to demonstrate this damage using the strong light of our standard procedure as originally used for the damage of the first kind. This was achieved by studying albino animals which at the start of the exposure were 20 days old, i.e. at an age when the ERG has just reached adult dimensions. The animals were of the same breed as those described above.

Table 1 shows ERG results (mean data) from an experiment in which the initially 20 day old rats, born in the laboratory, were

Table 1

ERG (μV) after 12 d exposure			
3 d		42 d	
a-	b-	a-	b-
51	299	154	661

exposed for 12 days to the light in our standard chambers at a temperature within the chamber of 23.5-26.0°C. Illumination of the space within the cage to which the animals were restricted, varied between 100-160 Ft-C depending upon location. Until exposure the rats had been in a complete dark environment to which they were returned after the end of exposure. They were separated from their mother at the start of the experiment.

Table 1 shows the typical ERG effect. Three days after exposure the average a-wave in this group of animals is very low (50 μV), as low as in older animals after a 1 to 2 day exposure to the same light resulting in a large lesion of the first kind covering more than 75% of the retina. The b-wave, however, is relatively high, about 6 times higher than the a-wave. Typically a substantial recovery in both a- and b-wave occurs during the following weeks, so that 1-2 months after exposure the a-wave is 3 times higher than at the first test. Simultaneously, the b-wave response becomes larger. Thus, the light damage of the second kind has a reversible component from which recovery occurs slowly within several weeks. Probably, this is the same kind of recovery as studied by Kuwabara electronmicroscopically (11).

The ERG, 2-3 months after exposure relates to the irreversible fraction of the damage of the second kind. It is not simply reduced in amplitudes but generally still shows an abnormally high b/a ratio. Thus, the late ERG data suggest that the surviving function of the retina is irreversibly abnormal.

Fig. 15, lower rows illustrates the ERG tracings from 3 animals of this experiment. All had been for 12 days constantly in the strong light. In addition to the above described change in amplitude, the records show that the latency of wavelets and b-wave is abnormally long early after the exposure, indicating a slowing of transmission of signals from visual cells to the secondary neurons.

Low a-wave ERG's with a relatively high b-wave have been recorded under a great variety of experimental conditions and in clinical cases in macular degeneration and other diseases. They are the early signs of hereditary visual cell degeneration in the C3H mouse (Fig. 16) and rat (Fig. 17). They are typical for the ERG in vitamin A deficiency when exposure to weak cyclic light has measurably decreased the rhodopsin or when a deficient animal is exposed to strong light for 1 to 2 days (2, 3). As shown in Fig. 18, vitamin A deficiency protects against light damage of the first kind provided the deficiency has become manifest prior to exposure by maintenance of the rat in weak cyclic light (2). The ERG's of such animals, exposed to light, resemble the ERG's which we used above for the characterization of the light damage of the second kind. The abnormality persists when the animals are maintained on the deficient diet but reverses to normal usually within 5 days upon

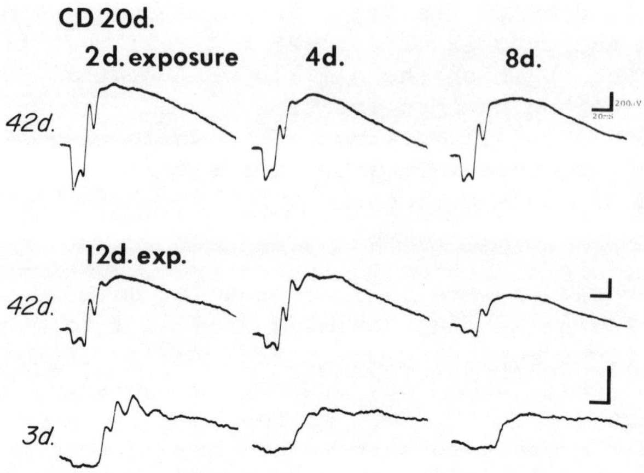

Figure 15. ERG's of CD rats exposed to strong light starting at the age of 20 days. Chamber illumination through green plastic was between 100-160 Ft-C depending on location; lowest illumination was at the bottom of the cage, the highest was 1" up from the floor of the chamber looking towards the side. Top row: 3 different animals exposed to 2, 4 or 8 days. The ERG was recorded 42 days after the end of exposure; middle row: 3 different animals, each exposed for 12 days. ERG was recorded 42 days after exposure; bottom row: same animals as in middle row above; the ERG was recorded 3 days after the end of exposure.

the relief of the deficiency by a high dose of retinol. Thus, the deficient animals demonstrate a reversible damage which may represent a sub-type of the damage of the second kind.

ERG's with a high b/a ratio are also typically recorded after iodate poisoning when the pigment epithelium has disappeared (Fig. 19). The retinas after iodate show degeneration of rod and cones in association with the survival of the majority of outer nuclei.

The histological appearance of the retinas of our young animals after exposure to strong light for 2 to 12 days showed small lesions of the damage of the first kind, characterized by the disappearance of the pigment epithelium. Fig. 20 illustrates the typical sharp edge of such a lesion in an animal which had been exposed to the strong light for 4 days, starting with the age of 20 days. In contrast to the damage of the first kind, the retinas outside these small lesions were abnormal with respect to the width of the outer nuclear layer and the appearance of outer and inner segments. This abnormality was widespread over the retina, the more so the longer the exposure time. We assume that this widespread abnormality, confined to the visual cells, would have been the only damage, had the animals been exposed to a lower light intensity.

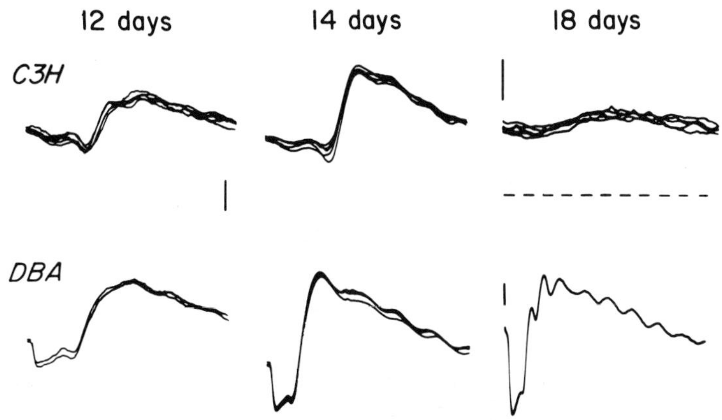

Figure 16. ERG's in response to the maximal xenon flash from the mutant mouse (rd, C3H) in comparison with the ERG of a normal mouse (Rd, DBA). Dashed line indicates 10 msec for each dash; vertical lines denote 100 μV for the nearest records. (From Ref. 10). Note that the a-wave is more affected than the b-wave as retinal development is over-taken by visual cell degeneration (visual cell death). The latency of the b-wave is relatively long at an early age in both mouse strains; it remains long in the mutant rat as degeneration progresses.

The most significant histological manifestation of the irreversible fraction of the damage of the second kind was the reduction in visual cell number, as evident by the number of surviving rows of the outer nuclear layer. Fig. 21 shows the almost maximal expression of this damage in an animal of the same group as Fig. 20. Barely one row of outer nuclei has survived, while the pigment epithelium is preserved.

The appearance of outer and inner segments correlates with the number of surviving outer nuclei. The greater their reduction, the greater is the degeneration of outer and inner segments, and their transformation to shortened and thickened stumps. Typically the outer and inner segments are only moderately reduced in length but remarkably thickened when 30-50% of the outer nuclei have survived (Figs. 22, 23). We assume that all these outer nuclei carry inner and outer segments, but we are unable to decide at the present time whether the degeneration of these organelles is mainly the result of the disappearance of neighboring units and of changes in the environment of the surviving ones or is directly related to light damage.

Conditions which favor damage of the second kind, are exposure at young age and exposure to low light intensity. In addition,

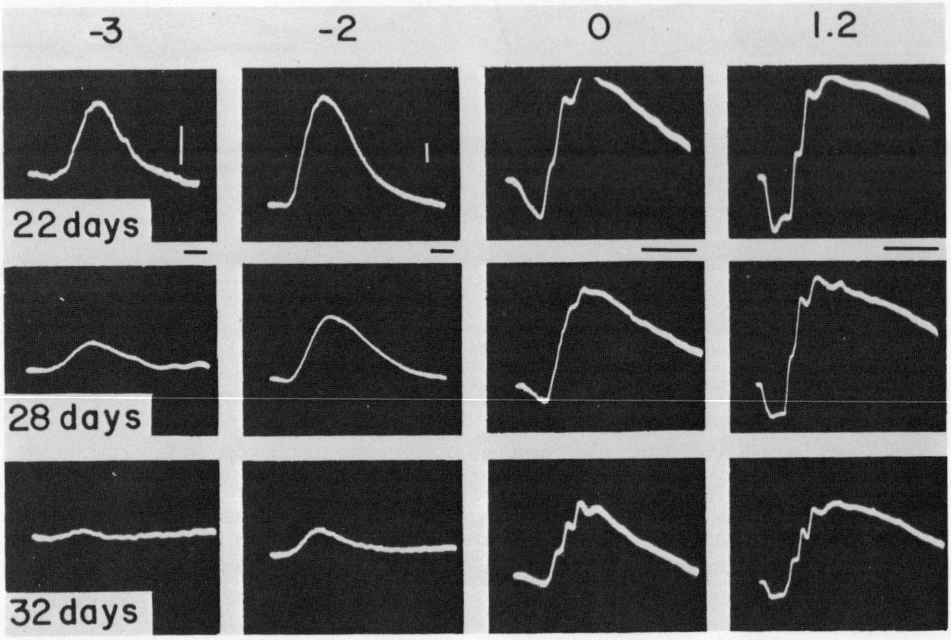

Figure 17. ERG in response to xenon flashes of different intensity for different ages of the mutant (rd) rat of a hooded strain. Stimulus strength is given in log arbitrary unit on top of figure. Time marks denote 20 msec. Calibration lines denote 200 μV, one for the left-hand column, the other for the 3 columns to the left. Note the predominant reduction in a-wave amplitude with the progress of the degeneration (visual cell death). The ERG in response to weak stimuli (e.g. log I = 3) also is more affected than the ERG to the maximal stimulus, i.e. the retinal sensitivity of an increasing fraction (with age) of the still functioning visual cell is abnormally low.

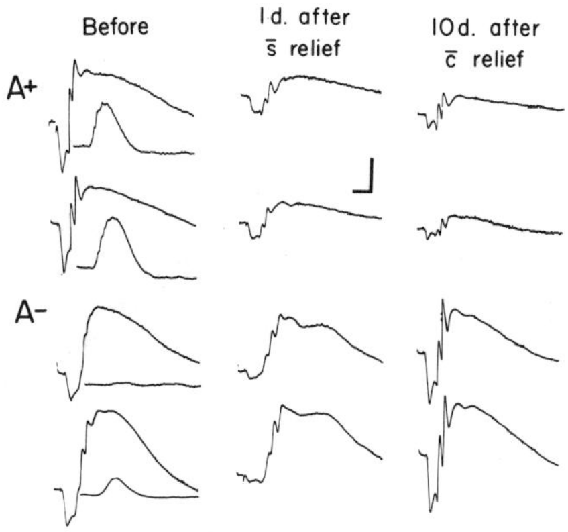

Figure 18. ERG's of vitamin-deficient rats (A-) before and 1 and 10 days after a 40-hour exposure to strong light (green, 120-180 Ft-C). Comparison with non-deficient rats (A+). Vitamin A deficiency was relieved by an I.P. injection of retinol palmitate after the first ERG test following exposure. (Adapted from Ref. 2). The ERG at 1 day after exposure shows the strongest manifestation of vitamin A deficiency: the a-wave is very low and the b-wave has a late, slow component ("hump"). This b-wave change has not been observed so far in light damage of the second kind. Threshold is increased whenever the a-wave is reduced in vitamin A deficiency (see also regression lines in Fig. 13). This seems to be true for all cases of predominant a-wave reduction, but it has not yet been tested for light damage of the second kind.

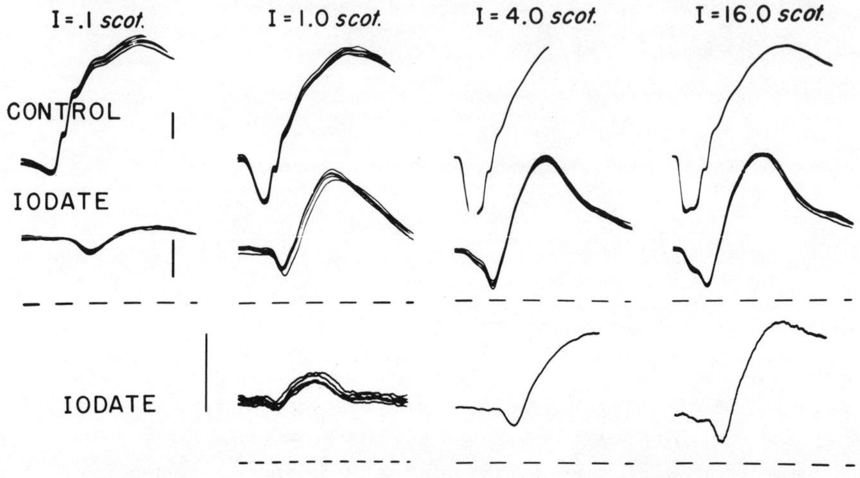

Figure 19. ERG abnormality of the rabbit resulting from iodate poisoning (I.V. injection) 2 days (middle row) and 3½ days (bottom row) prior to the ERG test by xenon flashes of different intensity (I =). The vertical line for each row denotes 100 μV; each dash of the horizontal lines indicates 10 msec. (From Ref. 12). Note in the abnormal records the marked reduction of the early negativity, the first downward stroke, which signals the appearance of PIII. The late negativity, just preceeding the b-wave, is probably not of PIII origin and may be generated by the Müller cell as is the b-wave. This negativity has been typical for the iodate effect; it has not been observed so far for ERG's of light-damaged retinas. Therefore, neither the ERG's of vitamin A deficiency nor those of iodate poisoning are of the same pattern as those recorded after light-damage of the second kind. Thus, on the basis of the ERG pattern, light damage of the second kind seems most closely related to the hereditary degeneration of the visual cells, especially as observed in the mutant rat.

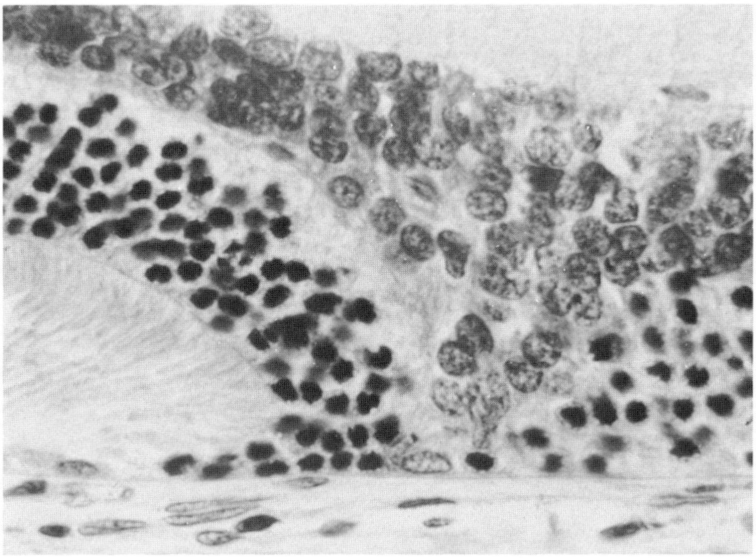

Figure 20. Retinal photomicrograph from a young rat exposed to strong light for 4 days. The eye was removed 42 days after exposure. The a-wave was 285 μV, the b-wave 850 μV.

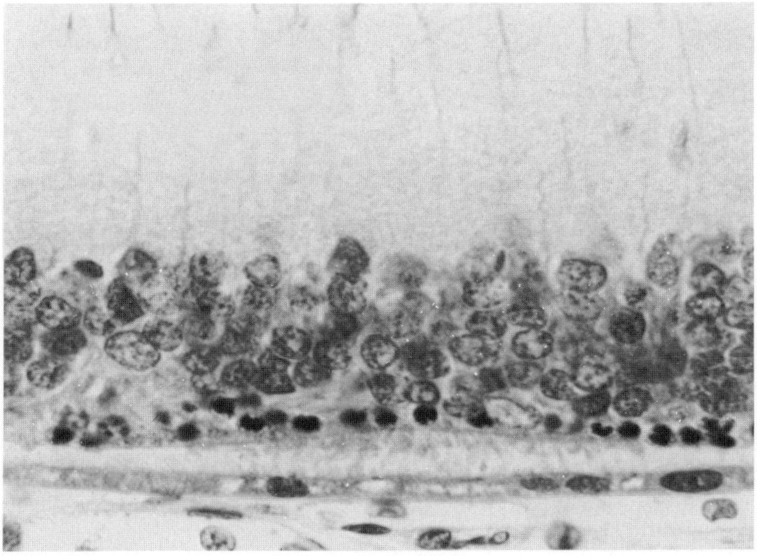

Figure 21. Photomicrograph from a young rat exposed to strong light for 8 days. The eye was removed 42 days after exposure. The a-wave was 170 μV, the b-wave 800 μV. Note the vacuolization of the PE which is not present in regions where visual cell loss in less severe.

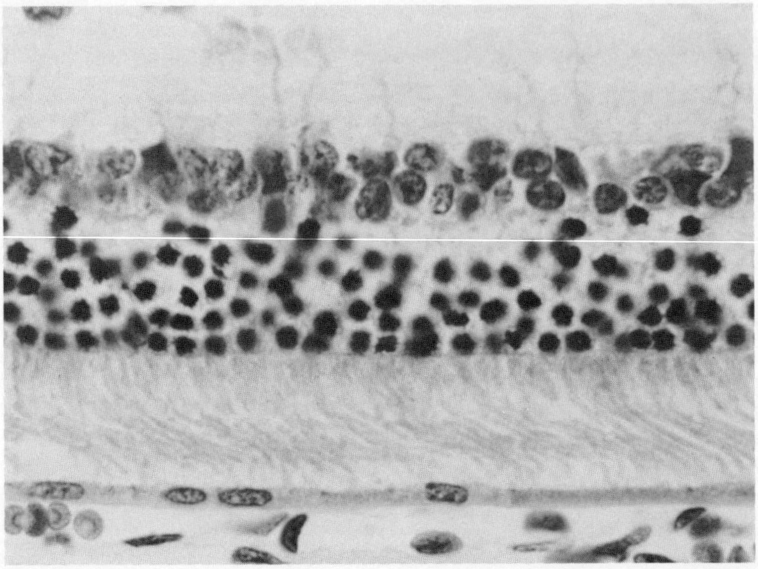

Figure 22. Photomicrograph from an adult rat exposed at the age of 200 days for 24 hours to the same strong light as were the animals of Fig. 21 and 22. The rat had been in a cyclic light environment (5 Ft-C) for 6 months prior to exposure. The eye was removed 3 months after exposure. Visual cell number is markedly reduced; the surviving inner and outer segments are well aligned but thickened; the pigment epithelium is preserved. The a-wave before eye removal was 165 μV, the b-wave 460 μV. Animals maintained continuously in darkness instead of cyclic light prior to the exposure had lost the ERG irreversibly with exposures for 12 or 24 hours. After 6½ hours of exposure of the dark-maintained animals of this experiment, average a-wave amplitude was 111 μV, and average b-wave amplitude 212 μV. The average values for the cyclic light maintained rats with a 24 hour exposure was 240 μV (a-wave) and 680 μV (b-wave).

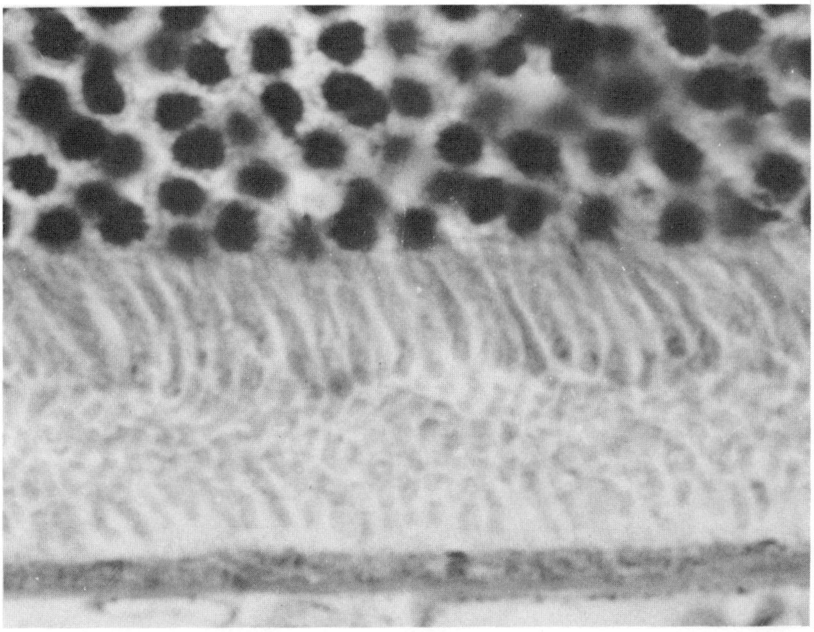

Figure 23. Enlarged photomicrograph of a retinal region similar in appearance as that of Fig. 22, from an animal of the same group as Fig. 22, demonstrating the thickening of the inner segments.

damage of the second kind predominates (Fig. 22) when the animals have been maintained in a non-damaging cyclic light environment prior to exposure. It seems that the protection against damage of the first kind which such environment conveys, enables the damage of the second kind to become evident especially with a prolongation of exposure time.

The action spectrum of the damage of the second kind has not yet been determined. We assume that damage of the second kind is rhodopsin-mediated as is, with great probability, the damage of the first kind. This assumption is justified by their similarities, and by the fact that the damage of the second kind is restricted to the visual cell while that of the first kind, despite it being rhodopsin-mediated, also involves the pigment epithelium.

It is our working hypothesis that the two kinds of light damage result from different mechanisms of the reaction to light and that their differentiation will enable one to study the mechanism of light damage in a definitive way.

This work was supported by a Public Health Service Research Grant (EY 00297) from the National Eye Institute. I am indebted to Dr. Brigitte Heiss for the histological preparations and evaluations.

REFERENCES

1. Noell, W. K., Walker, V. S., Kang, B. S. and Berman, S. (1966) *Invest. Ophthalmol.* 5, 450-473.
2. Noell, W. K. and Albrecht, R. (1971) *Science* 172, 76-80.
3. Noell, W. K., Delmelle, M. C. and Albrecht, R. (1971) *Science* 172, 72-75.
4. Noell, W. K. (1979) *Photochem. and Photobio.* 29, 717-723.
5. Organisciak, D. T. and Noell, W. K. (1977) *Invest. Ophthal. and Visual Sci.* 16, 188-190.
6. Lanum, I. C. J. (1978) *Survey Ophthalm.* 22, 221-249.
7. Faber, D. S. (1969) Ph.D. Thesis, Buffalo.
8. Delmelle, M., Noell, W. K. and Organisciak, D. T. (1975) *Exp. Eye Res.* 21, 369-380.
9. Themann and Noell, unpublished.
10. Noell, W. K. (1965) *in* Biochemistry of the Retina, Suppl. of Exp. Eye Res., Graymore, C. L., ed., Academic Press, London.
11. Kuwabara, T. (1970) *Am. J. Ophthal.* 70, 187-198.
12. Noell, W. K. (1958) *Annal. N. Y. Acad. Sci.* 74, 337-361.

HORMONAL INFLUENCES ON RETINAL PHOTODAMAGE

W. Keith O'Steen

Bowman Gray School of Medicine
Wake Forest University
Winston-Salem, North Carolina 27103

Photic stimuli influence the function and behavior of animals in a number of ways. As analyzed by Wurtman (1), photic information serves at least three functions in mammals: 1) it acts as a stimulus for the complex phenomenon of vision, 2) it regulates optic autonomic reflexes, and 3) it controls certain "neurovegetative functions", such as gonadal maturation and some endocrine rhythms. Environmental photoperiodicity participates in the regulation of sexual development of the rat (2). Diurnal and seasonal photic changes can synchronize, or even generate, rhythms in endocrine function, such as reproductive hormone cycles. The origin and continuation of the rhythms are dependent in many instances on feedback mechanisms involving gonadal hormones and hypothalamic and pituitary vascular connections. Whether pituitary glands and gonadal hormones have feedback functions directly with the photoreceptor organ, the eye, is unknown at this time.

While studying the effects of continuous photoperiod and intraocular serotonin metabolism on reproductive cyclicity in the laboratory rat, a severe and statistically significant reduction in the number of retinal photoreceptors was observed in the eyes of animals exposed to low-intensity visible illumination (3). Animals reared in and exposed to cyclic photoperiods of the same intensity did not have this retinal defect. A search of the literature related to this study indicated that Noell et al. (4) previously had described this phenomenon after exposing several kinds of animals to more intense visible light, but its association with continuous exposure to a relatively low intensity level of illumination had not been observed previously. Several questions arose as the result of these reports and were related to the influence of photically-damaged retinal receptor cells on visual activity, photically regulated endocrine responses, visually and optically mediated and guided behavioral responses, and behavior of the animals, in general. Other considerations have been concerned with the causative mechanism of receptor damage and subsequent degeneration and whether the damage could be experimentally reversed or prevented by manipulating pharmacologically the light-exposed animals.

LOW-INTENSITY VISIBLE LIGHT AND RETINAL DAMAGE

Noell et al. (4) reported a functional deficit in the visual system of rats examined one week after exposure to high levels of illuminance (800 Ft-C) for 24 hours. Examination of the retina of these animals revealed a destruction of portions of the photoreceptor and pigment epithelial layers. The damaging effect of light exposure was shown to be related to the body temperature of the rats during the exposure period, the length of the exposure time and the intensity of the fluorescent light source. Similar changes were reported by Gorn and Kuwabara (5) in rats exposed to high levels of white (2000 Ft-C) and colored (200 Ft-C) light for periods ranging from two hours to more than seven days. A later electron microscopic study (6) of retinas from similarly treated rats showed that some of the earliest changes caused by intense light exposure include a separation and vacuolation of the lamellar structure of the outer segments of the photoreceptors. The damaged outer segments increased in diameter two-fold, and they were irregularly distorted in shape. When an extremely bright light was applied, or the environmental temperature was elevated, the destruction of the receptor cell body preceded the damage to the outer segment, an observation made earlier by Noell et al. (4) with the light microscope. After five days of exposure, pigment epithelial cells showed a marked increase in height and a lengthening of their microvilli. Kuwabara and Gorn (6) found that the retinal tissue showed a great capacity for structural and functional recovery after five days of exposure, if the animals were placed in dim light. These investigators did not describe the events related to the regenerative process.

Albino rats kept in constant illumination at an intensity comparable to that found in many animals rooms, that is, 18 Ft-C as reflected from the cage floor or 75 Ft-C direct lighting, develop a retinal degeneration, which in its early stages apparently is limited to the photoreceptor cells (3). Although receptor damage can be detected with the light microscope during the first few days of exposure, rapid destruction of these cells occurs during the first few weeks; after 30 days of exposure to this low-intensity illuminance, no intact photoreceptors were observed in the retina. Remnants of photoreceptors, particularly cone nuclei, have been described in the retinas of rats exposed for long periods of time (7). Contrary to the effects of high levels of illuminance and elevated body temperature (4), the pigment epithelium and innermost layers of the retina apparently are not irreversibly affected by these lower levels of light. The degenerative changes are reproducible, regardless of whether the light source is fluorescent or incandescent (8). In a study to determine the general range of spectral wavelengths that contributes to photoreceptor degeneration, Anderson, Coyle and O'Steen (9) found the blue spectrum to be the most effective in paralleling the effects of exposure to white light (10, 11). Green light was approximately one-half as effective as blue, and red light was the least effective of the spectral hues tested. These results are in agreement with the earlier report by Noell et al. (4).

In a later ultrastructural study, Shear, O'Steen and Anderson (12) observed reversible, "degenerative" changes in photoreceptors, including outer segment vacuolization and detachment, after less than 12 hours exposure to low level illuminance.

Although the studies which are described above have provided a partial description of the course and extent of the structural and functional degeneration of the retina, the mechanism by which light causes these deleterious effects is unknown. One of the mechanisms suggested was thermocoagulation of the photoreceptors because of the relatively high level of illumination used by some of the early investigators (4). However, Gorn and Kuwabara (5) showed that exposure to 2000 Ft-C illumination was insufficient to cause thermocoagulation. Other mechanisms that have been proposed (4, 13) include: 1) photosensitized oxidation of cell constituents, such as natural pigment; 2) adverse effects on a metabolic pathway essential for the maintenance of cell structure, especially at high body temperature; 3) damage from the accumulation of toxic photoproducts; and 4) the suggestion that vitamin-A might be involved. Noell et al. (13) showed that retinol, one of the metabolites of the phototransduction process, is transferred to the pigment epithelium after being released by receptor cells following the bleaching reaction. This observation, together with the finding that the pigment epithelium undergoes rapid degeneration when exposed to intense, continuous light (4), led Noell et al. (13) to evaluate the effect of low levels of retinol on light-induced retinal degeneration. They found that vitamin-A deficiency (retinol depletion) actually protects against damage to visible light and came to the conclusion that rhodopsin, either as the source of retinol or as the mediator of damage *via* a mechanism that does not involve retinol directly, is probably central to the degenerative mechanism.

The normal retina continuously synthesizes the lamellated discs of the receptor outer segments, which are then displaced outwardly toward the pigment epithelium (14). The discs become detached and are incorporated as phagosomes into the cytoplasm of the pigment cells (15, 16). Phagosomes containing lamellated outer segment discs are not found in the pigment cells of rats exposed to low intensity illumination for extended periods of time (4 to 6 months), because the damaged photoreceptors are removed from the retina by the end of 30 days of exposure (3, 11). During the earlier stages of retinal degeneration, phagocytes, which are not commonly seen in normal retinas, invade the retina and engulf cellular debris into cytoplasmic vacuoles. The action of these cells probably accounts for the rapid removal of disintegrating photoreceptor cells (4, 10, 17). A light and electron microscopic study of phagocytosis in the degenerating retina indicates that two populations of cells possess phagocytic properties: mononuclear cells of vascular origin and the pigment epithelial cells (18). The hematogenous origin of retinal phagocytes has been further demonstrated using tritiated thymidine autoradiography (19).

The question as to whether albino rats that have their photoreceptors damaged or destroyed by radiant energy exposure still have visual function has been the subject of several studies and will be discussed in detail by Dr. K. V. Anderson in this symposium. The existence of visual behavioral responses in the absence of intact photoreceptors was first suggested over 20 years ago by Karli and Bonaventure (20, 21, 22), who found that mice with inherited retinal degeneration could make visual discriminations; however, in these animals the light level had to be increased for discriminations to be made. In mice with hereditary retinal degeneration, the retina has been shown to develop normally from birth to 10-14 days of age and then begin to degenerate until, by 21 days of age, there is virtually no receptor layer remaining (23, 24, 25). Behavioral experiments with rats exposed to 30 days of low-level illumination have not revealed impairments in visually guided behavior, despite the fact that the retinas of such animals are extensively light-damaged. Animals with retinal damage are able to perform certain visual tasks prior to retinal degeneration as well as normal animals and can learn new visual discriminations at rates indistinguishable from control animals (26). In a later study, when response latency was used to assess performance, performance decrements were observed on the pattern tasks (27). Behavioral experiments in which rats are repeatedly tested during many months of exposure to continuous light have indicated that the level of discrimination performance depends in part upon the length of time the animal is exposed to light (28, 29). While performance in retinally degenerated mice does not appear to be as good as in light-degenerated rats, such mice have been shown to be capable of performing at least some kinds of visually guided behavior, such as responding efficiently on the elevated maze (30, 31), and learning a 6-unit T-maze (32).

CONTINUOUS PHOTOPERIOD AND ENDOCRINE RESPONSES

Hormonal Rhythms and Constant Photoperiod

Adult, female rats acclimated to controlled lighting and temperature following 90 days exposure to continuous illumination show the expected retinal degeneration and normal pituitary-adrenal function (33). Fiske and Leeman (34) have shown an abolition of plasma corticosteroid circadian periodicity in female rats exposed to constant light for 9 to 10 weeks. Cheifetz et al. (35) studying adult female rats in constant light for 7 weeks, found abnormal patterns with several daily peaks in plasma corticosteroids, but not decreased amplitude, in adult male rats raised in constant light for 21 days. Krieger (37) found that normal circadian periodicity of plasma corticosteroid levels, once established, does not persist if the animal is subsequently enucleated or exposed to either constant light (CL) or constant darkness (CD); animals reared in CL or CD from birth until autopsy do not manifest normal circadian periodicity of plasma corticosteroids, however, normal periodicity will appear in such animals after a period of exposure to normal light-dark photoperiod.

Reproductive Cyclicity and Continuous Photoperiod

A variety of conditions, including exposure to continuous light, will induce prolonged or constant estrus in rats. Frequent or incorrect vaginal examination can induce prolonged vaginal cornification. Extracts of the anterior pituitary gland, parabiosis, injection of oestrin, hypothyroidism, and vitamin-A deficiency cause constant estrus or prolonged vaginal cornification (38, 39, 40). Browman (41) in his classic study, produced prolonged vaginal cornification by exposing rats to constant light. Animals exposed to constant darkness, reversed day-night, and ordinary daylight conditions exhibited normal estrous cycles. Animals that were blinded by enucleation or by severing the optic nerves during day-night light conditions maintained normal estrous histories regardless of the light conditions. Hemmingsen and Krarup (42) also found that ovarian cycles were suppressed in rats by exposure to constant light. Dempsey and Searles (43) reported that some rats with persistent estrus as a result of constant light exposure would mate and ovulate.

Brown-Grant, Davidson and Grieg (44) exposed adult albino rats to constant illumination for 55 to 65 days and found no visible luteal tissue in their ovaries; more than 90% of the rats were sexually receptive. Ovulation could be induced by mating or injections of progesterone. Retinas of rats from similar experiments had extensive damage of the photoreceptor layer (45). When these rats were returned to a light-dark photoperiod, spontaneous ovulation was resumed; a similar observation was also reported by O'Steen and Anderson (8). When the rats were returned to constant light a second time, persistent estrus (vaginal cornification) developed at the same rate as in rats exposed for the first time, and ovulation was again suppressed. Pigmented rats also show persistent estrus after exposure to constant light (46), and such animals do not show retinal damage under conditions of lighting that severely damage the retinae of albino rats (11).

Several human degenerative retinal diseases have been associated with abnormal endocrine function and have been ameliorated by surgical or pharmacologic therapy. According to Cogan (47), retinitis pigmentosa, a retinal degenerative disease affecting the photoreceptors and pigment epithelium, may occur as an isolated disease in an otherwise normal individual, or it may occur with other abnormalities, or with systemic, metabolic disease associated with infundibulopituitary deficiencies. Lawrence-Moon-Biedle syndrome (or Bardet-Biedl syndrome) is an example of a pituitary deficiency in which retinitis pigmentosa occurs with mental retardation, hypogenitalism, and sometimes polydactyly.

Diabetic retinopathy, one of the most common causes of blindness, is associated with retinal vascular abnormalities, such as microaneurysms and increased permeability of the retinal vessels. Lyall and Innes (48) reported the improvement of the diabetic syndrome in a patient who developed pituitary disease. Paulsen (49) found a significant improvement in diabetes and a recovery of the associated retinopathy in a woman who developed postpartum pituitary necrosis. The most impressive changes after hypophysectomy occurred in the retinal blood vessels with the regression of neovascularization and resorption of retinal hemorrhages. Removal of the growth hormone source with hypophysectomy was thought to be related to the amelioration of the diabetic syndrome (50, 51). Whether hypophysectomy affects the vasculature of the degenerating retina of rats exposed to prolonged periods of fluorescent radiant energy is unknown; in fact, a direct influence of continuous photoperiod on retinal vasculature remains to be elucidated, although the endothelium of the choriocapillaries was apparently structurally unmodified by exposure of intact animals to low intensity illumination (52).

SEXUAL MATURATION AND RETINAL DEGENERATION

As described above, when adult rats are exposed to incandescent or fluorescent illuminance, the photoreceptors fragment and are engulfed by active phagocytosis; by 7 days after exposure, the fragmenting photoreceptors have been removed from the central retina, and the cells of the inner nuclear layer lie adjacent to the pigment epithelium or to Bruch's membrane. Changes such as these do not occur in retinas of rats younger than 5 weeks of age; exposed retinas of 6 to 7 week old rats have areas of localized damage in the central retina (53). A significant reduction in the thickness of the outer nuclear layer (photoreceptors) occurs in the 8-week and subsequent age groups (up to 24 months of age). A portion of the results presented by these authors substantiate the earlier report by Ballowitz and Dämmrich (54) that retinas of newborn rats were less damaged than those of adult animals after more prolonged periods of exposure. The utilization of a more intense radiant energy source (incandescent bulbs) than previously used resulted in retinal destruction after only 24 hours of exposure and allowed for a more precise determination of the development of the period of retinal susceptibility to damage.

Using this technique, retinal photoreceptor damage was detected early after eye opening, but the abnormal receptor cells were restricted to a specific focus on an area of the central retina adjacent to the optic papilla. The more peripheral retina appeared to be undamaged as judged from histologic examination and a morphometric analysis of outer nuclear layer thickness (ONL thickness) and entire retinal thickness measurements. By applying a statistical analysis of variance (multiple regression analyses) to these two parameters for the time sequence series of exposed animals, the age at which retinal susceptibility to damage occurred under these photoperiod conditions was determined to be between 6 to 8 weeks which is concurrent with the age of sexual maturation for the rat as indicated by the vaginal opening time at 41 to 42 days of age. This event signals that the ovaries have begun to elaborate elevated levels of estrogen in response to hypothalamic-hypophyseal stimulation of the gonads. The degree of photically-induced retinal damage in male animals was not statistically different from that in females. The simultaneous occurrence of severe susceptibility to light-induced retinal damage and the period of sexual maturation may be coincidental, or both events may be related to the maturation of the pituitary gland and target organ responsiveness to trophic hormones. Relatively little information exists in the literature on the subject of reproductive hormonal or sexual maturational influences on the neurons of the visual system, and none was available on retinal photoreceptors. Therefore, several studies were designed and are being conducted to investigate the influence of reproductive hormones on retinal photoreceptor damage. The remainder of this presentation will be devoted to a discussion of the relations of the ovaries and pituitary glands and their hormonal secretions to retinal susceptibility to light-induced damage.

OVARIAN HORMONES AND RETINAL DYSTROPHY

The early observation that ovariectomy, which removes the source of endogenous estrogen, and hypophysectomy, which eliminates the source of all pituitary trophic hormones, afford a degree of protection of light-induced retinal damage (55, 56), led to studies to determine if hormonal replacement could reverse the protection (56, 57). Each of the operations, if performed prior to puberty, delays or prevents sexual maturation, and results of experiments involving these procedures possibly contribute to an explanation of differences between retinal damage in young and old animals and of the role of hormones in photoreceptor destruction. Several of these experiments have been discussed recently in a review (58) and will only be summarized here.

Treatment of OVEX animals with low, near physiologic doses of estradiol benzoate (0.05 μg/day) effectively reversed the protection afforded by OVEX, but daily administration of larger, pharmacologic dosages (100 μg/day) significantly reduced the retinal damage, even more than by OVEX alone. Progesterone, another ovarian hormone was ineffective in modifying photoreceptor damage, either alone or in combination with estradiol. Therefore, estradiol may be considered to affect the susceptibility of retinal photoreceptors to damage by light exposure. The results led to the speculation that estrogen might directly regulate protein synthesis in the retina through activation of specific steroid receptor sites or indirectly control the induction or regulation of a retinal enzyme system during sexual maturation, as has been demonstrated by others in areas of the central nervous sytem (59). The biphasic response to low (0.05 μg) and high dosages (100 μg) of estradiol benzoate was contrary to results frequently observed on dose-response curves for drugs, but is not unprecidented for some hormones, such as estradiol. For example, estrogen has been shown, under certain circumstances, to have both an inhibitory and facilitatory action on the secretion of pituitary gonadotropic hormones (60, 61). To study further the biphasic response to estrogen in the retina, photoreceptor damage has been assessed in OVEX albino rats treated with 100 μg daily estradiol before, during, or after high intensity (500 Ft-C) photoperiod exposure. Retinas from rats given 100 μg per day 17 β-estradiol for 4 days prior to and during the 24 hours of intense light exposure, as described elsewhere (53), were significantly protected as compared to groups receiving estradiol for the same period after light exposure or only the oil vehicle for the same periods. The ONL thicknesses of retinas from each group were respectively 9.30 ± 0.87 μm, 3.38 ± 0.03 μm, and 1.65 ± 0.01 μm. Therefore, treatment of rats with pharmacologic dosages of estradiol before light exposure appears to protect the retina from damage even more substantially than when given after exposure. Since a degree of protection was apparent in the post-exposure treatment group, even after the damage had occurred, the steroid hormone may be influencing the rate of phagocytosis of the damaged photoreceptors. That is, circulating estrogen possibly is not only lessening the cellular destruction in the retina, but is slowing the rate of phagocytic removal of the debris.

TABLE 1

Influence of norgestrel and estradiol (OVRAL) on the retinas of female African Green monkeys

Groups	(n=)	ONL Thickness (μm)	Retinal Thickness (μm)
____, Butter	(12)	28.33 ± 0.90	103.73 ± 3.34
OVRAL, Butter	(9)	26.92 ± 1.05	87.63 ± 2.61*
____, Safflower Oil	(8)	27.78 ± 1.19	97.27 ± 4.75
OVRAL, Safflower Oil	(6)	27.11 ± 1.07	94.09 ± 3.52

*$P < 0.005$

Two recent studies of the effects of ovariectomy and estradiol on the retina present evidence which indicates that earlier results dealing with the protection of photoreceptor from light-induced damage by ovariectomy (55, 56) and its reversal with estradiol treatment may have broad, general applications to retinal photoreceptor metabolism. These studies, one with primates on a regulated diet regimen and the other with RCS rats with hereditary retinal dystrophy, demonstrate that estradiol and ovariectomy may influence retinal cells in animals kept in cyclic photoperiod. The primate experiments were designed to study the interaction of a cholesterol-fat diet with or without the estrogen-containing oral contraceptive, OVRAL (0.5 mg norgestrel and 0.05 mg ethinyl estradiol per tablet, Wyeth) on the structure and biochemistry of several organs and tissues, including the eye. A 2 x 2 factorial experimental design was selected: 21 monkeys were administered a diet containing butter, and 9 of these animals received OVRAL also; 14 monkeys were given a diet supplemented with safflower oil, and 6 of them were administered OVRAL also (Table 1). OVRAL was added to the diets for 3 weeks and then withheld for one week, as administered to human females, during the entire 22 month experimental period. At autopsy, the eyes were fixed in Bouin's solution and processed for histology as described for rat eyes in the above experiments. Average thicknesses of the entire retina and the outer nuclear layer were calculated from 8 measurements per section taken at 4.5 mm intervals around the retina from the periphery centralward (Table 1). Retinal thickness was reduced by 15.5% ($p < 0.005$) in monkeys on the OVRAL-butter fat diet as compared to the butter only group. A portion of the difference appeared to be the result of a reduction in ONL thickness (5%), although the reduction was not statistically significant as calculated from the average thicknesses of outer nuclear layers. Retinas of monkeys fed safflower oil or safflower oil-OVRAL diets did not differ in ONL or entire retinal thicknesses. Therefore, the retina of the African green monkey was influenced adversely by a diet including butter supplemented by an oral contraceptive containing estradiol ingested over a period of 22 months. The estradiol influence on the retina appeared to be related to the type of lipid in the diet.

The other experiment involving the influence of estradiol on the retina of animals in a cyclic photoperiod was performed on RCS rats with retinal dystrophy. According to previous microscopic studies, extensive retinal photoreceptor degeneration has occurred in this strain by 30 days of age (62, 63, 64), and continues until only a few, if any, intact photoreceptors remain in the retina. Groups of female RCS rats were either ovariectomized (n = 13), or sham-operated (n = 16) at 14 days of age and returned to a 14:10 hour photoperiod (light:darkness) until autopsy at 30 days of age. Thickness measurements of the entire retina and of the ONL indicate a significant reduction in the number of degenerated photoreceptors, or a protection from the expression of the genetic defect, in the

TABLE 2

Ovariectomy and retinal dystrophy in RCS rats. Ovariectomy at 14 days of age, autopsy at 30 days of age

Group (n=)	ONL Thickness (μm)	Retinal Thickness (μm)
Sham-operated (16)	23.97 ± 1.49	88.62 ± 2.43
Ovariectomized (13)	28.91 ± 0.87*	98.58 ± 2.16**

*P < 0.02 **P < 0.01

ovariectomized group. These results, along with those from ovariectomized animals exposed to continuous photoperiod, suggest that removal of the ovaries, and thus, the source of endogenous estrogen, beneficially influences retinal photoreceptors by moderating the severity of the photically-induced and genetically expressed retinal defect. The design of the experiment does not permit the determination of the mechanism involved, that is, whether estrogen directly affects the photoreceptor cells, indirectly influences the receptor by feedback to target tissues, or acts synergistically with other hormones available in intact animals. It must also be emphasized that endocrine manipulations, such as ovariectomy, hypophysectomy, or hormonal administration, have not yet prevented the development of photically-induced or genetically expressed retinal dystrophies, but only provides a degree of protection from or ameliorates the effect of, the causative agents.

Competitive uptake of tritiated estradiol by the retina. As the result of the evidence from these experiments that the ovary and estrogen or estradiol influence responses of photoreceptors to light exposure, it was felt that ovarian steriod receptors might occur on photoreceptor cells and be demonstrated by two methods: steroid autoradiography and competitive steroid binding with liquid scintillation spectroscopy. The demonstration of steroid receptors on photoreceptors would provide a direct correlation between estrogen and the cell responding to light stimulus and lead to speculation that estrogen might influence directly the metabolism of photoreceptors or be acting with other hormones, either synergistically or antagonistically, which act on photoreceptor cell membranes. Firstly, experiments on competitive binding of unlabeled versus labeled estradiol in the retina were begun with liquid scintillation methods, as utilized by McEwen and co-workers (65, 66) in hypothalmamic and pituitary gland studies. Albino female rats at either 60 to 70 days of age (postpubertal) or 25 to 30 days of age (prepubertal) were ovariectomized and returned to a cyclic light environment for 10 days. At that time, animals were divided into a first group which received 1000 μg unlabeled estradiol benzoate i/p 30 minutes prior to an i/p injection of tritiated 6,7-^{3}H estradiol and a second group which was administered the alcohol vehicle 30 minutes prior to the isotope. Two hours later at autopsy following decapitation, the hypothalamus, pituitary gland and a small sample of parietal cerebral cortex were removed from each frozen brain. After removal of the cornea and lens from each eye, the retina were gently extruded by ocular pressure. The radioactive isotope was extracted by agitation of the tissue sample in dichloromethane for 18 hours, and, after adding scintillation fluid, competing estradiol uptake was determined and calculated as counts per minute per milligram of sample (Table 3). Steroid hormone uptake in all tissues is expressed ratios of concentration (cpm/mg wet tissue weight) to blood concentration (cpm/mg wet weight). Table 3 also gives the blood plasma concentration in cpm/wet weight for each group. The results from both pre- and postpubertally ovariectomized rats indicate a statistically significant competitive uptake of tritiated estradiol in the retina, pituitary gland and hypothalamus, since pretreatment of the animals with unlabeled estradiol inhibited the uptake of the isotope. The percent change in isotope uptake (Table 3) in these experiments compares closely with that previously reported for the pituitary gland and hypothalamus by McEwen and co-workers (65). The cerebral cortex and blood plasma showed only insignificant changes in tissue radioactivity, and the correction of the isotope uptake in each of the other tissues with that of blood provides a more accurate index of steroid competition in the tissue. A significantly greater uptake occurred in tissues of the prepubertally ovariectomized rats as compared with that of the postpubertally operated animals.

Therefore, these experimental results from tissues of ovariectomized rats maintained in cyclic photoperiod indicate that exogenously administered estradiol is competitively bound to retinal cells in a manner similar to that previously described for the pituitary gland and hypothalamus, two tissues containing cells known to be influenced metabolically by estradiol (67) and associated functionally in hormonal feedback mechanisms. The exact localization of the labeled steroid to specific cells of the retina must be determined with autoradiographic methods.

PROLACTIN AND RETINAL DYSTROPHY

Prepubertal removal of the pituitary gland (hypophysectomy, HYPEX), which delays sexual maturation, significantly reduces retinal photoreceptor destruction induced by exposing the animals to continuous illumination (57). When crude pituitary gland homogenate is administered to adult HYPEX rats, or when a pituitary gland is implanted to the kidney capsule of HYPEX animals, the

TABLE 3

Competition by unlabeled estradiol (E_2) for radioactive estradiol ($^3H\text{-}E_2$) in adult and prepubertally ovariectomized (OVEX) rat retinas, pituitary glands, and brains. Minus signs indicate reduction in $^3H\text{-}E_2$ uptake due to the competing steroid. 1000 µg unlabeled E_2 given i/p 30 min prior to 100 µCi 6,7 -$^3H\text{-}E_2$.

	n=	Retina	Pituitary	Hypothalamus	Cortex	Blood
Post-Pubertal OVEX**						
$^3H\text{-}E_2$ + vehicle	5	8.92 ± 0.74*	53.72 ± 3.68	2.40 ± 0.72	0.94 ± 0.39	0.47 ± 0.03
$^3H\text{-}E_2$ + E_2	5	5.48 ± 0.37	8.20 ± 0.58	1.36 ± 0.02	0.76 ± 0.32	0.45 ± 0.02
% Change		-38.6	-84.7	-43.3	-19.1	-4.3
P value (t-test)		<0.001	<0.001	<0.001	n.s.	n.s.
Pre-Pubertal OVEX**						
$^3H\text{-}E_2$ + vehicle	5	20.60 ± 3.70	229.81 ± 38.82	5.22 ± 0.88	3.02 ± 0.28	1.09 ± 0.19
$^3H\text{-}E_2$ + E_2	5	11.34 ± 2.96	37.92 ± 15.36	2.45 ± 0.18	2.90 ± 0.38	1.01 ± 0.49
% Change		- 44.9	-83.5	-53.1	-4.0	-7.4
P value (t-test)		<0.05	<0.001	<0.001	n.s.	n.s.

* Estradiol uptake in all tissues expressed as ratios of concentration (cpm/mg wet weight) in tissue to blood concentration. Blood concentrations are given as cpm/mg wet weight.

** Ovariectomized at 21 days of age (prepubertal) or at 50-56 days of age (postpubertal); all groups autopsied at 70-80 days of age.

protective effect of pituitary gland removal is significantly reduced, and photoreceptor damage is more severe than that in HYPEX rats. Everett (68, 69) demonstrated that separation of the pituitary gland from the hypothalamus, and therefore, from the control of hypothalamic hormones, by implanting it to the kidney capsule would initiate an elevated level and prolonged period of prolactin secretion. However, implanted pituitary glands were unable to secrete other trophic hormones in levels sufficient to maintain a functional state in their target organs.

Prolactin administration to HYPEX animals was the most effective method of reversing the protection of photoreceptor cells from light damage (57). Prolactin, 2 mg per day, a pharmacologic dose administered to HYPEX rats during the continuous light exposure period, caused increased retinal damage. In an attempt to determine the minimal effective dose of prolactin needed to reverse the influence of HYPEX, a dose response series of animals was examined for retinal damage after exposure to continuous illumination for three weeks. Results indicate that 125 μg per day of ovine prolactin (NIH-P-S12, intraperitoneally), a more physiologic dose that 2 mg/day, reverse the protection afforded the retina during light exposure by removing the pituitary gland. Other dose levels (250 μg to 2000 μg/day) were equally effective as compared to the HYPEX control retinas. Therefore, the administration of exogenous prolactin to HYPEX rats in a broad range of dosages effectively increases the susceptibility of the retina to light damage.

In addition to the examination of the influence of endogenously secreted prolactin by pituitary implants on photically-induced retinal destruction as described above, two other experiments demonstrate the importance of high endogenous prolactin levels on the photoreceptor response to illumination. Firstly, endogenous prolactin levels were increased by injecting ovariectomized rats with pimozide (0.63 mg/kg body weight in 0.1M tartaric acid). Pimozide, a neuroleptic drug, blocks only dopamine receptors in the brain (70), and dopaminergic systems exert an inhibitory control on prolactin release (71); therefore, pimozide elevates endogenous prolactin produced by the pituitary gland by inhibiting the prolactin inhibitory factor (PIF) of the hypothalamus or by acting directly on the pituitary cells secreting the hormone (71). Injections of pimozide were given twice weekly during the three-week light exposure period to maintain elevated levels of plasma prolactin. Retinal destruction was significantly more severe in pimozide groups (57.60 ± 2.05 μm) than in vehicle injected groups (70.07 ± 1.56 μm; $p < 0.001$).

Secondly, endogenous prolactin levels have been elevated by maintaining rats in 48 hours of darkness prior to exposing them to an intense, short photoperiod. Since the elevated levels of prolactin are pineal gland-mediated responses (72, 73), the effect of pineal gland removal (Px) at 21 days of age on retinal damage was studied in immature, adolescent, and sexually mature rats (40, 50, and 60 days of age) (74). There was no significant difference in retinal damage between Px and sham-pinealectomized rats when they were subjected to light treatment at 40 and 50 days of age. However, 60-day old Px rats showed less photoreceptor damage than did sham-operated animals as the result of photoperiod exposure. Presumably, the sham-operated rats had high levels of plasma prolactin (72, 73), although titers were not assayed in this study. However, the results support the hypothesis that increased levels of prolactin are associated with augmented photoreceptor destruction (57). Also, direct effects of pineal gland hormones, such as melatonin and arginine vasotocin, on retinal photoreceptor cells cannot be discounted, particularly as they may be related to light damage mechanisms. However, the pineal gland does not appear to control outer segment disc shedding in rats on a cyclic photoperiod (75, 76, 77). Rod receptor cell outer disc shedding follows a cyclic pattern, which apparently is also independent of the pituitary gland, since HYPEX has no effect on the shedding cycle (78, 79).

Competitive uptake of ^{125}I-prolactin by the retina: autoradiographic study. The positive results discussed above which indicated a relationship between exogenous and endogenous prolactin levels and susceptibility to photically-induced photoreceptor damage raised the issues as to whether prolactin is bound to cells of the retina and is associated directly with the metabolism of retinal neurons. Preliminary results from autoradiographs of rat retinas labeled with iodinated ovine prolactin show a competitive localization of the hormone within the photoreceptor cells and cells within the choroid layer of the eye. Intact and HYPEX female albino rats were anesthetized with chloral hydrate, the abdomen was opened, and intracardiac injections of ^{125}I-labeled ovine prolactin (6 x 10^8 dis/min) were made transdiaphragmally; one-half of the animals also received an injection of 500 μg of unlabeled ovine prolactin (NIH-P-S12) at the same time. Five minutes later, the rats were perfused through the left ventricle with physiologic saline (200 ml) and then with Bouin's fixative (200 ml). Tissues, including the brain and both eyes, were embedded in paraffin and prepared for autoradiography (80). Walsh and co-workers (80) demonstrated specific prolactin binding sites localized on the ependymal cells of the rat choroid plexus and over specific areas of the brain using this autoradiographic competitive-binding assay. Autoradiographs in the present study confirmed the localization of prolactin binding sites over cells of the ependyma and the reduction of labeling, as indicated by reduced silver grains, in animals receiving both the isotope and unlabeled prolactin. Specific binding sites for prolactin were observed over the cells of and among the chorio-capillaries of the choroid coat of the eye and within the retina over the inner segments of the photoreceptors. A reduction in the number of silver grains was obvious at these sites in rats receiving the

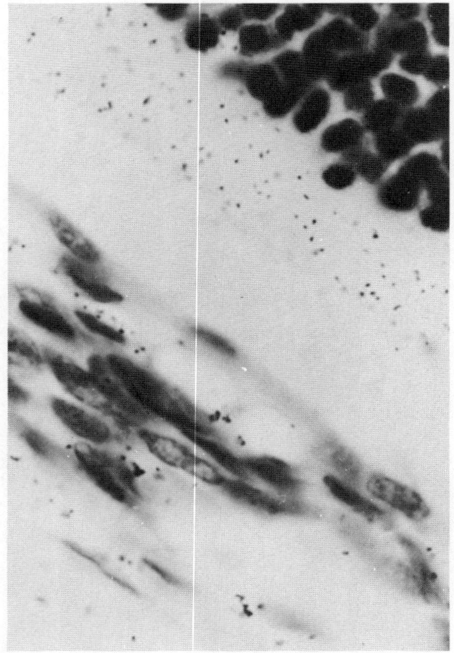

Fig. 1. Autoradiograph of retina from hypophysectomized female rat given an intracardial injection of ^{125}I - ovine prolactin. Reduced silver grains are located over the inner segments of photoreceptors (upper area) and over cells of the choroid coat (lower area), but only background labeling overlies the outer segments between the two areas. H & E. X1000. (Reduced 10% for purposes of reproduction.)

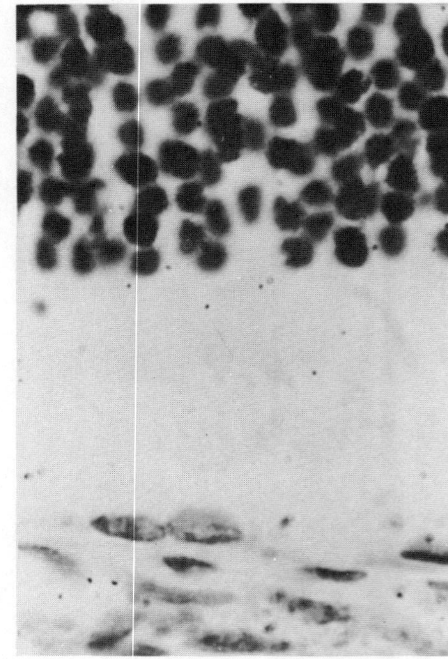

Fig. 2. Autoradiograph of retina from hypophysectomized female rat given intracardial injections of unlabeled ovine prolactin (500 μg) and ^{125}I - prolactin. Note the reduction in number of silver grains over both the inner segments of the photoreceptors (upper) and choroid coat (lower), as compared with that in Figure 1, indicating a competitive binding of the hormone. H & E. X1000. (Reduced 10% for purposes of reproduction.)

unlabeled prolactin (Figures 1, 2). The other layers of the retina were not conspicuously labeled and appeared to have only background levels of radioactivity. The central retinal artery and its branches were negative for radioactivity, an observation which possibly emphasizes the significance of specific binding sites in the choroid coat and photoreceptors.

These results apparently are the first demonstration of specific prolactin binding sites in the eye and the retina and provide an anatomic basis for an association of prolactin with photically-induced retinal damage.

The demonstration of estradiol and prolactin binding sites in the rat retina, using liquid scintillation spectroscopy and autoradiography, respectively, as described in this presentation, suggests that a relationship exists between the two hormones and cells of the retina. An increasing amount of evidence is implicating estradiol in the regulation of prolactin synthesis and secretion. Gorski, Maurer, and co-workers (81, 82) have shown that treatment of rats with estradiol specifically increases the incorporation of precursors into prolactin and results in accumulation of preprolactin mRNA in the pituitary gland. They have recently reported that primary pituitary cells respond to estradiol by increased incorporation of radiolabeled precursors into prolactin (83). Also, ovariectomy decreases, and estrogen replacement increases, prolactin binding sites in the liver of rats and mice (84, 85, 86), and anti-estrogen compounds reduce prolactin receptors in female rat liver (87). Interestingly, the response of hepatic prolactin receptors to estrogen may be sex dependent in some species, such as the mouse (88). Whether circulating or bound estrogen in the female rat retina is associated with the prolactin influence on retinal degeneration or with specific prolactin binding sites in the retina remains to be clarified.

SUMMARY

1. Retinal photoreceptor destruction by light exposure is significantly more severe in intact, adult rats than in prepubertal, young rats. The period of increasing susceptibility to damage coincides with that of pubertal and post-pubertal development, an age associated with maturation of the hypothalamohypophyseal-gonadal axis and readjustment of endocrine function.

2. Two hormones, prolactin and estrogen (estradiol), are implicated in the susceptibility of photoreceptors to light damage in the present studies.

a. Removal of the ovary, a source of endogenous estrogen, and of the pituitary gland protects the retina. Injections of crude pituitary extracts into light-exposed, HYPEX rats reverses the protection.

b. Pituitary gland implants to the kidney capsule, which produce high levels of prolactin, reverse the protection and cause more severe retinal damage than in HYPEX animals.

c. Prolactin injected during the light exposure period increases the susceptibility of receptors to damage as compared to uninjected HYPEX animals.

d. Pinealectomy and injection of pimozide, which influence plasma prolactin levels (pinex decreases; pimozide increases), also affect the degree of photically-induced destruction in the retina.

e. Competitive binding sites for estradiol using liquid scintillation spectrosocpy and for prolactin using autoradiography have been demonstrated in the retina and choroid coat of the rat eye.

3. Prolactin and estrogen, therefore, appear to have a regulatory influence on the severity of light-induced photoreceptor damage and are possibly associated also with the normal metabolic activity of photoreceptors and other CNS neurons.

REFERENCES

1. Wurtman, R. J. (1967) in Neuroendocrinology, vol. 2 (Martini, L. and Ganong, W. F., eds.) pp. 20-59, Academic Press, New York.
2. Fiske, V. (1941) Endocrinol. 29, 187-196.
3. O'Steen, W. K. (1970) Exp. Neurol. 27, 194-205.
4. Noell, W. K., Walker, V. S., Kang, B. S. and Berman, S. (1966) Invest. Ophthalmol. 5, 450-473.
5. Gorn, R. A. and Kuwabara, T. (1967) Arch. Ophthalmol. 77, 115-118.
6. Kuwabara, T. and Gorn, R. A. (1968) Arch. Ophthalmol. 79, 69-78.
7. LaVail, M. M. (1976) Invest. Ophthalmol. 15, 64-70.
8. O'Steen, W. K. and Anderson, K. V. (1972) Z. Zellforsch. 127, 306-313.
9. Anderson, K. V., Coyle, F. P. and O'Steen, W. K. (1972) Exp. Neurol. 35, 233-238.
10. O'Steen, W. K. and Lytle, R. B. (1971) Am. J. Anat. 130, 227-234.
11. O'Steen, W. K. and Anderson, K. V. (1971) Exp. Neurol. 30, 525-534.
12. Shear, C. R., O'Steen, W. K. and Anderson, K. V. (1974) Am. J. Anat. 138, 127-132.
13. Noell, W. K. Delmelle, M. C. and Albrecht, R. (1971) Science 172, 72-75.
14. Young, R. W. (1971) J. Cell Biol. 49, 303-318.
15. Young, R. W. and Bok, D. (1969) J. Cell Biol. 42, 392-403.
16. Young, R. W. and Bok, D. (1970) Invest. Ophthalmol. 9, 524-536.

17. Hansson, H. A. (1970) Virch. Arch. Abt. B. Zellpath. 6, 247-262.
18. O'Steen, W. K. and Karcioglu, Z. A. (1974) Am. J. Anat. 139, 503-518.
19. Gloor, B. P. (1974) Arch. Klin. Exp. Ophthalmol. 190, 183-194.
20. Karli, P. (1954) Arch. des Sciences Physiol. 8, 305-328.
21. Karli, P. and Bonaventure, N. (1970) Editions du Centre National de la Recherche Scientifique, 15, Quai Anatole-France, Paris VII[e], France.
22. Bonaventure, N. and Karli, P. (1961) Comptes rendus des seances de la Soc. de Biol. 155, 2015-2018.
23. Keeler, C. E. (1924) Proc. Nat. Acad. Sci. 10, 329-333.
24. Tansley, K. (1954) J. Hered. 45, 123-127.
25. Sidman, R. L. and Green, M. C. (1965) J. Hered. 56, 23-29.
26. Anderson, K. V. and O'Steen, W. K. (1972) Exp. Neurol. 34, 446-454.
27. Anderson, K. V. and O'Steen, W. K. (1974) Physiol. Behav. 12, 633-637.
28. Bennett, M. H., Dyer, R. F. and Dunn, J. D. (1972) Exp. Neurol. 34, 434-445.
29. Bennett, M. H., Dyer, R. F. and Dunn, J. D. (1973) Exp. Neurol. 38, 80-89.
30. McClearn, G. E. (1958) Am. J. Psychol. 13, 405.
31. McClearn, G. E. (1965) In Proceed. XI Internat. Cong. Genetics, vol. 3 (Geearts, S. J., ed.), Pergamon Press, New York.
32. Lindzey, G. and Winston, H. (1962) J. Comp. Physiol. Psychol. 55, 748-752.
33. Dunn, J., Dyer, R. and Bennett, M. H. (1972) Endocrinol. 90, 1660-1663.
34. Fiske, V. and Leeman, S. E. (1964) Ann. N. Y. Acad. Sci. 117, 231-243.
35. Cheifetz, P., Gaffud, N. and Dingman, J. (1968) Endocrinol. 82, 1117-1124.
36. Scheving, L. E. and Pauly, J. E. (1966) Am. J. Physiol. 210, 1112-1117.
37. Krieger, D. T. (1973) Endocrinol. 93, 1077-1091.
38. Witschi, E. and Pfeiffer, C. A. (1935) Anat. Rec. 64, 85-106.
39. Van Horn, W. M. (1933) Endocrinol. 17, 152-162.
40. Mason, K. E. and Ellison, E. T. (1935) J. Nutrit. 9, 735-755.
41. Browman, L. G. (1937) J. Exp. Zool. 75, 375-388.
42. Hemmingsen, A. M. and Krarup, N. B. (1937) Kgl. Danske Videlskab. Selskab. Biol. Medd. 13, 7.
43. Dempsey, E. W. and Searles, H. F. (1943) Endocrinol. 32, 119-128.
44. Brown-Grant, K., Davidson, J. M. and Greig, F. (1973) J. Endocrinol. 57, 7-22.
45. Glickstein, M., Brown-Grant, K. and Raisman, C. (1972) J. Anat. 111, 515.
46. Critchlow, V. (1963) In Advances in Neuroendocrinology, pp. 377-402 (Nalbandov, A. V., ed.) U. of Ill. Press, Urbana.

47. Cogan, D. G. (1975) In Neuro-ophthalmology, vol. 2 (Smith, J. O., ed.) p. 50. C. V. Mosby, St. Louis.
48. Lyall, A. and Innes, J. A. (1935) Lancet 1, 318-321.
49. Paulsen, J. E. (1953) Diabetes 2, 7-12.
50. Pearson, O. H., Dominguez, J. M., Greenberg, E., Pazianos, A. and Ray, B. S. (1960) Tran. Assoc. Am. Physicians 73, 217-226.
51. Luft, R. (1965) Sandoz J. Med. 7, 2-6.
52. O'Steen, W. K., Shear, C. R. and Anderson, K. V. (1972) Am. J. Anat. 134, 5-22.
53. O'Steen, W. K., Anderson, K. V. and Shear, C. R. (1974) Invest. Ophthalmol. 13, 334-339.
54. Ballowitz, L. and Dämmrich, K. (1972) Z. Kinderheilk. 113, 42-52.
55. Olafson, R. P. and O'Steen, W. K. (1976) Invest. Ophthalmol. 15, 869-872.
56. O'Steen, W. K. (1977) Exp. Eye Res. 25, 361-369.
57. O'Steen, W. K. and Kraeer, S. L. (1977) Invest. Ophthalmol. Vis. Sci. 16, 940-946.
58. O'Steen, W. K. (1979) Photochem. Photobiol. 29, 745-753.
59. Luine, V. N., Khylchevskaya, R. I. and McEwen, B. S. (1975) Brain Res. 86, 293-306.
60. Caligaris, L., Astrada, J. J. and Taleisnik, S. (1971) Endocrinol. 88, 810-815.
61. Swerdloff, R. S., Jacobs, H. S. and O'Dell, W. D. (1972) Endocrinol. 90, 1529-1536.
62. Bourne, M. C., Campbell, D. A. and Tansley, K. (1938) Brit. J. Ophthalmol. 22, 613-623.
63. Dowling, J. E. and Sidman, R. L. (1962) J. Cell Biol. 14, 73-109.
64. LaVail, M. M. and Battelle, B. A. (1975) Exp. Eye Res. 21, 167-192.
65. McEwen, B. S. and Pfaff, D. W. (1970) Brain Res. 21, 1-16.
66. McEwen, B. S., Weiss, J. M. and Schwartz, L. S. (1969) Brain Res. 16, 227-241.
67. Gorski, R. A. (1966) J. Reprod. Fertil., Suppl 1, 67-88.
68. Everett, J. W. (1954) Endocrinol. 54, 685-690.
69. Everett, J. W. (1956) Endocrinol. 58, 786-796.
70. Andén, N. E., Butcher, H., Corradi, K., Fuxe, K. and Ungerstedt, U. (1970) Eur. J. Pharmacol. 11, 303-314.
71. Ojeda, S. R., Harms, P. G. and McCann, S. M. (1974) Endocrinol. 94, 1650-1657.
72. Relkin, R. (1972a) J. Endocrinol. 53, 179-180.
73. Relkin, R. (1972b) Neuroendocrinol. 9, 278-284.
74. Rudeen, P. K. and O'Steen, W. K. (1979) Exp. Eye Res. 28, 37-44.
75. LaVail, M. M. (1976a) Exp. Eye Res. 23, 277-282.
76. LaVail, M. M. (1976b) Science 194, 1071-1073.
77. LaVail, M. M. (1978) ARVO Program Extracts, Suppl. to Invest. Ophthalmol. Vis. Sci. 17, 134.

78. Tamai, M., Tierstein, P., Goldman, A., O'Brien, P. and Chader, G. (1978a) ARVO Program Extracts, Suppl. to Invest. Ophthalmol. Vis. Sci. 17, 134.
79. Tamai, M., Tierstein, P., Goldman, A., O'Brien, P. and Chader, G. (1978b) Invest. Ophthalmol. Vis. Sci. 6, 558-562.
80. Walsh, R. J., Posner, B. I., Kopriwa, B. M. and Brawler, J. R. (1978) Science 201, 1041-1043.
81. Maurer, R. A. and Gorski, J. (1977a) Endocrinol. 101, 76-84.
82. Stone, R. T., Maurer, R. A. and Gorski, J. (1977b) Biochem. 16, 4915-4920.
83. Lieberman, M. E., Maurer, R. A. and Gorski, J. (1978) Proc. Nat. Acad. Sci. 75, 5946-5949.
84. Gelato, M., Marshall, S., Boudreau, M., Bruni, J. F., Campbell, G. A. and Meites, J. (1975) Endocrinol. 96, 1292-1296.
85. Posner, B. I., Kelly, P. A., Shiu, R. and Friesen, H. (1974) Endocrinol. 95, 521-531.
86. Posner, B. I. (1976) Endocrinol. 98, 645-654.
87. Kelly, P. A., Ferland, L., Labrie, F. and Delean, H. (1976) In Hypothalamus and Endocrine Function (Labrie, F., Meites, J., and Pelletier, G., eds.) p. 321. Plenum Press, New York.
88. Marshall, S., Bruni, J. F. and Meites, J. (1978) Proc. Soc. Exp. Biol. Med. 159, 256-259.

ACKNOWLEDGEMENT

Supported by Grant No. EY-02350 from NEI, National Institutes of Health. Technical assistance from Arcellia Angulo, Jan Donnelly, and Douglas Johnson is gratefully acknowledged.

ALTERATIONS IN MUSCLE AND RETINAL CELL STRUCTURE AFTER EXPOSURE TO LIGHT

Charles R. Shear

Department of Anatomy
University of Maryland School of Medicine
Baltimore, Maryland 21201

Exposure of albino rats to both incandescent and fluorescent low-level radiant energy causes damage and degeneration to the photoreceptor cells and pigment epithelial cells of the retina and to the muscle fibers of the extraocular muscles (1-4). The type of cellular damage is the same for both incandescent and for fluorescent light exposed animals (5). Our previous studies indicate that elevated environmental and orbital temperatures are not responsible for the damage seen in the retina or in the myofibers of the extraocular muscles. In order to better understand the relation between light exposure and damage in such diverse cell types, the sequences of ultrastructural changes has been examined.

MATERIALS AND METHODS

Adult female albino rats (Charles River Strain) were kept in an environment of either cyclic light (14 hours of light; 10 hours of darkness) or low level constant light. Illumination in the constant incandescent and fluorescent light environments was maintained at 1) 70 Ft-C or 2) 150 Ft-C measured as direct light from the center of the cage. Details of the incandescent and fluorescent light source and their placement have been described elsewhere in these proceedings (6). The animals were maintained in constant light for periods of 3 hours to 1 year and then either sacrificed or returned to a low intensity, cyclic photoperiod for 3.5 hours to 2 months. For electron microscopy the rats were overanesthetized with ether, and the orbital contents were removed and fixed in cold (4°C) 2.5% glutaraldehyde solution. Prior to eye removal, 1.0 ml glutaraldehyde solution was injected into the orbit. After 15 minutes, the eyes were enucleated, removed from the orbit, and the extraocular muscles and retinas separated and placed into a larger volume of the same fixative at 4°C for an additional 2 hours. After fixation the muscles and retinas were postfixed in 1% OsO_4 and embedded in Araldite. Semithin sections (1.0 μm) were mounted on glass slides and stained with alkaline 1.0% toluidine blue. Ultrathin sections were stained with uranyl acetate and lead citrate solutions. The sections were examined with either a Philips 300 or 201 electron microscope.

RESULTS

Rats exposed to short term (3.5 to 24 hr.) low intensity incandescent (70 Ft-C) and fluorescent (150 Ft-C) light show structural changes within the distal regions of all of the extraocular muscle fibers (EOM's) and the junctions between the retinal pigment epithelial cells (RPE). Muscle fiber damage was seen in all EOM's and in the levator palpebrae muscles of the light-exposed animals. The severity of the damage varied in relation to 1) the duration of the exposure period and 2) the intensity of the illuminance. The first morphologic indication of myofiber damage occurs at the insertion of the fibers adjacent to the eye. A single, 3.5 hour exposure to incandescent light (70 Ft-C) or fluorescent light (150 Ft-C) is sufficient to damage the muscle fibers. The damage is more severe if the animals are returned to a darkened or cyclic light environment for an additional 17.5 to 24 hours than if examined immediately after the exposure period. Each of the light exposure periods resulted in a wide spectrum of specific structural damage within a single muscle fiber.

Muscle Degeneration-Regeneration

A single 3.5 hour exposure to light is sufficient to damage the muscle fibers at their insertion, adjacent to the eye. Such damage is not apparent at the light microscopic level of observation until 17.5 to 24 hours after the initial light exposure. However, once the damage begins, the degeneration spreads rapidly and longer continuous light-exposure damages increasingly longer lengths of the muscle fiber. By 24 hours exposure the myofibers show damage along 30% of their length (Fig. 1). Fibers examined immediately after a 24 hour exposure to light have already undergone considerable degeneration (Figs. 1, 3 and 4).

At the insertion of the myofibers on the eye, the sarcoplasm becomes fragmented and the contractile filaments lose their longitudinal orientation. Proceeding deeper into the orbit toward the origin of the fibers and away from the eye (i.e., from most to least damaged), the sequence of cellular damage is as follows:
a) The Z discs are absent and the sarcomeres shortened to the extent

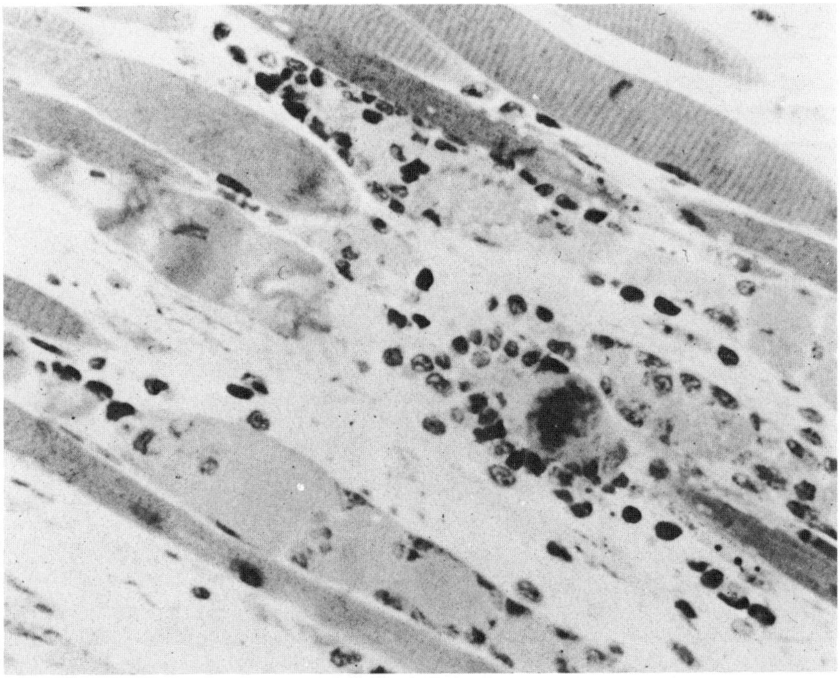

Figure 1. Damaged, degenerating lateral rectus muscle fibers prepared immediately after 24 h of continuous exposure to incandescent light. Leucocytes have infiltrated the area of degeneration. The fibers appear swollen and cross-striations are lost. X700.

that adjacent thin filaments overlap one another; the sarcoplasmic reticulum is dilated and fragmented and the transverse tubules are disoriented (Fig. 2). b) the Z discs are spread out; the sarcoplasmic reticulum and transverse tubules are dilated and there is less overlap between the contractile filaments than seen in sarcomeres closer to the surface of the eye (Fig. 3). c) The sarcoplasmic reticulum and transverse tubular system are dilated but remain intact (Fig. 4) and many of the mitochondria in this region contain flocculent densities and paracrystalline inclusions (Fig. 5-7). d) The least damaged region of the fiber has a slightly dilated sarcoplasmic reticulum, occasional shortened sarcomeres and regions of local Z disc spreading. The vascular supply appears undamaged along the entire length of the exposed muscles. Myofiber damage does not extend into the motor endplate region in muscles exposed to constant light for 17.5 hours or less.

Regeneration is rapid--the myofibers begin to regenerate 24 to 48 hr after injury. Between 24 and 48 hr after removal from

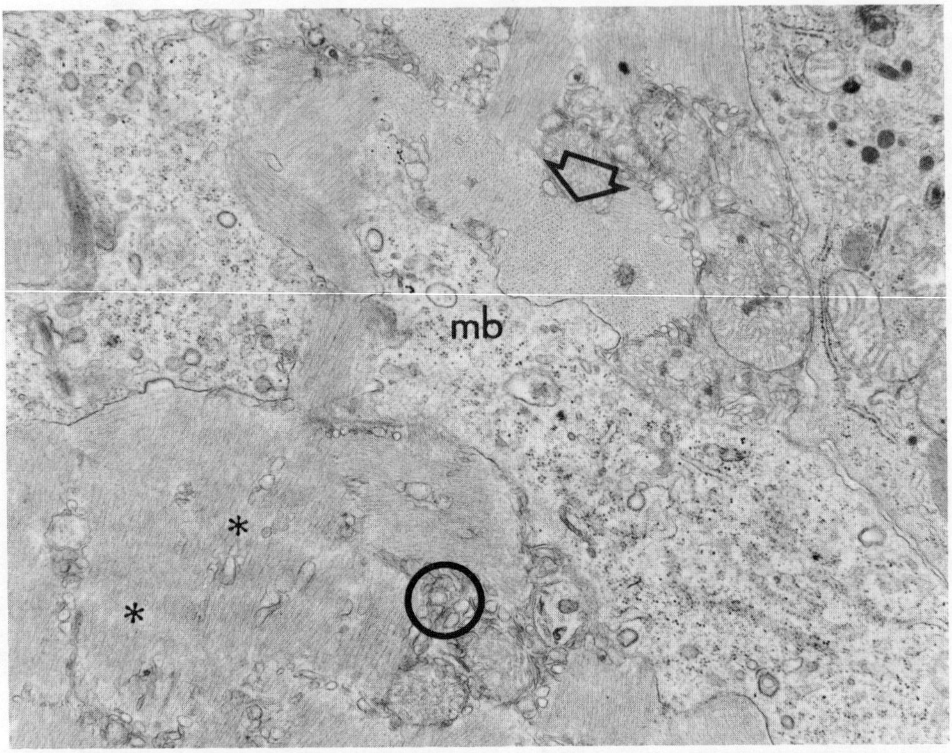

Figure 2. A longitudinal section of a single muscle fiber (mf) 72 h after a 24 h exposure to constant light. Within the fiber a myoblast (mb) is seen traversing the field. The myoblast cytoplasm only appears to contain myofibrillar material, when in fact the myofibrillar material bridges the myoblast process while remaining within the sarcolemma of the damaged fiber. Note the absence of Z discs (*) the perpendicular orientation of some of the contractile filaments (arrow), and the vesicular profiles of the fragmented sarcotubular system (circle). A portion of a macrophage cell is also seen (p). X21,850.

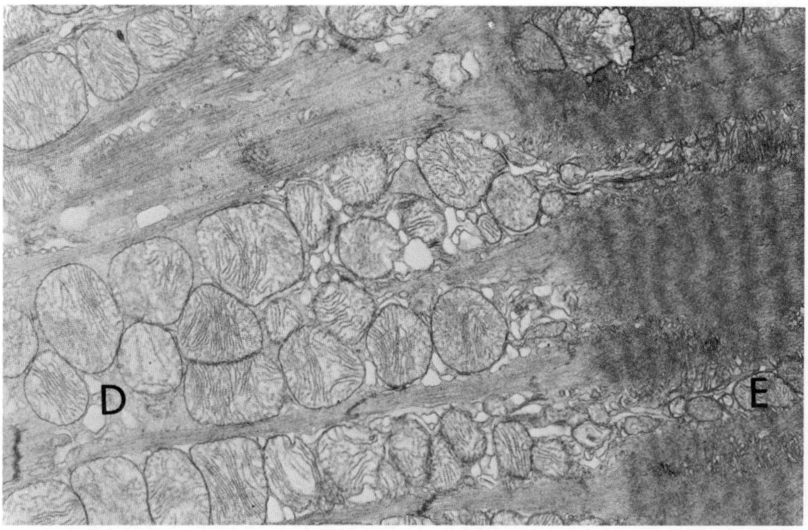

Figure 3. A longitudinal section of a medial rectus muscle fiber immediately after 24 h of continuous exposure to light, similar to the light photomicrograph of the swollen fibers seen in Fig. 1. The damage is more severe nearest the eye (E) than deeper in the orbit (D). Note the absence of Z discs and extensive thick and thin filament overlap adjacent to the eye. Further from the eye, Z discs are only occasionally seen, the sarcotubular complex (sarcoplasmic reticulum and transverse tubules) appears swollen and mitochondria are abundant. X11,900.

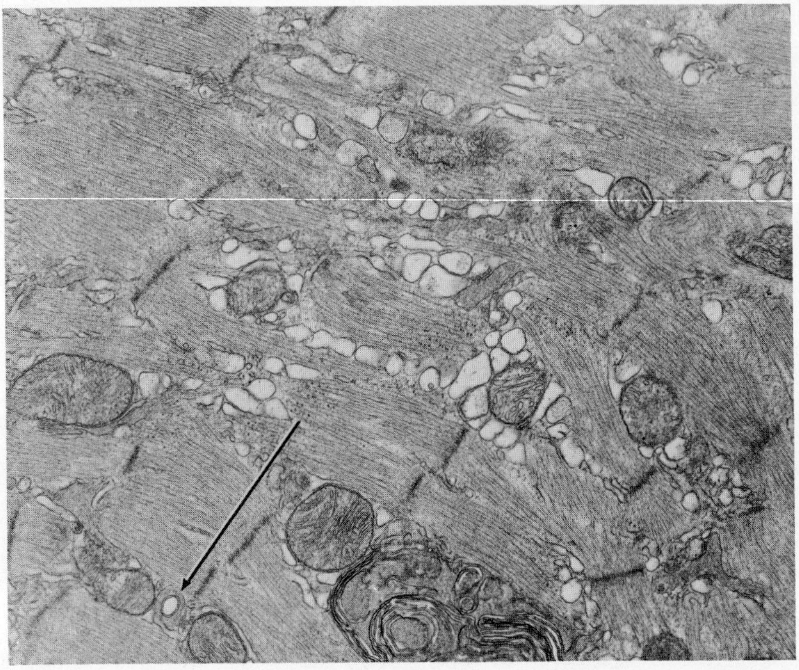

Figure 4. A longitudinal section of the same muscle fiber seen in Fig. 3, but showing an area deeper in the orbit than the preceding figure. Note the dilated sarcotubular complex. Occasional triads (one transverse tubule plus two terminal cisternae) remain intact (arrow) but most appear to have swollen and to have lost their normal organization. X20,900.

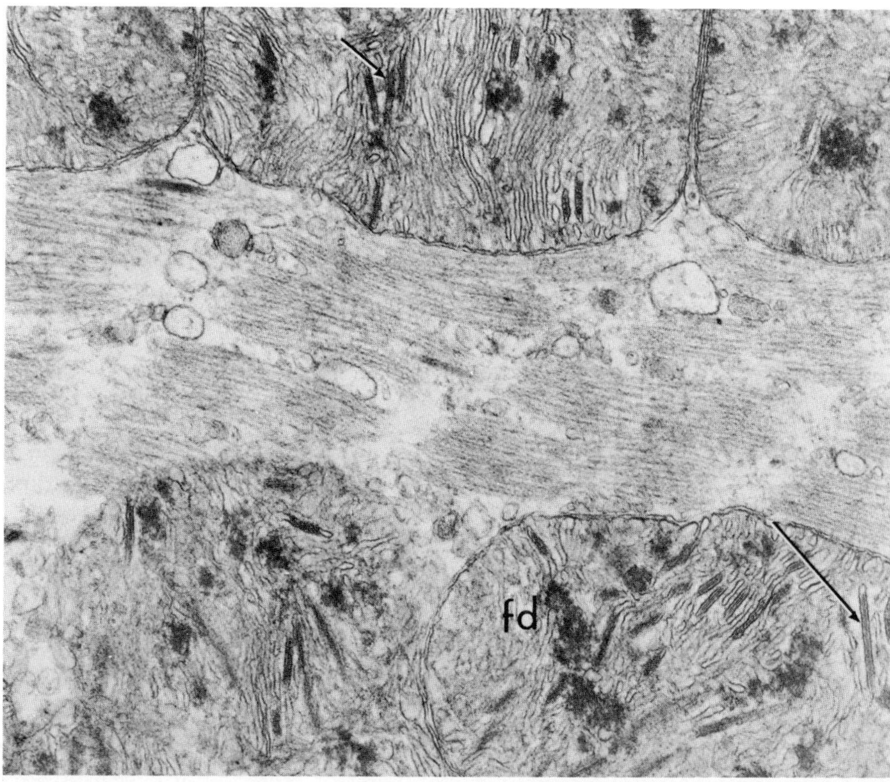

Figure 5. A longitudinal section of a medial rectus muscle fiber 24 h after a 24 h exposure to continuous light. Note the complete absence of Z discs. Also note the flocculent densities (fd), and the paracrystalline inclusions (arrows) of the intracrystal space within the mitochondria. X39,500.

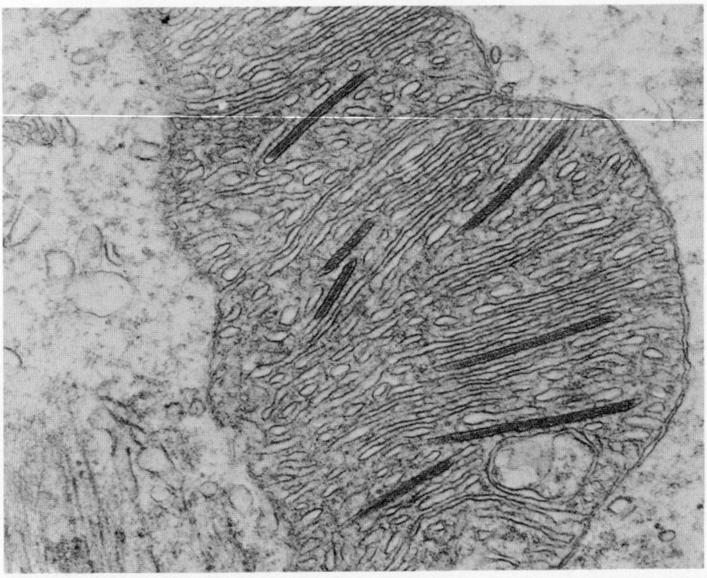

Figure 6. A mitochondrian from a lateral rectus fiber 24 h after a 24 h exposure to continuous light. Note the paracrystalline inclusions may vary in appearance depending upon the plane of sectioning. X42,000.

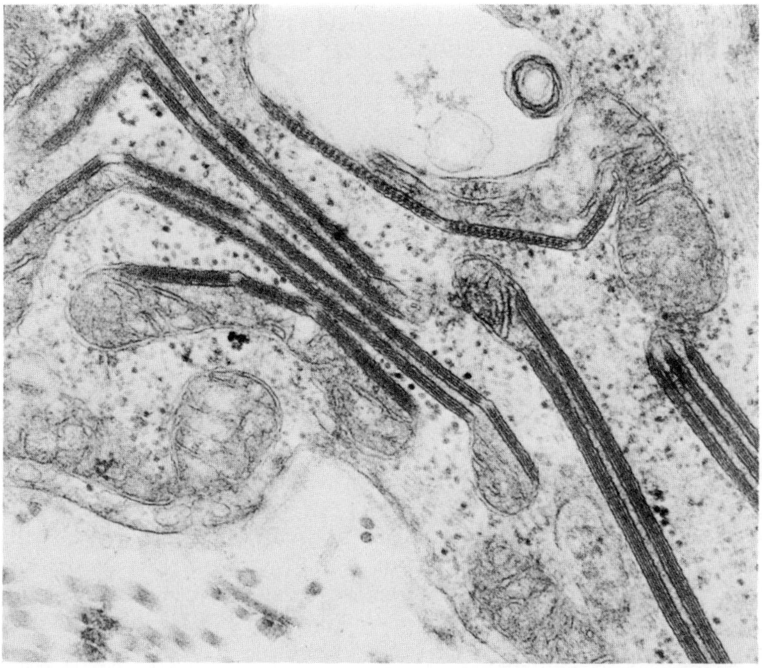

Figure 7. Mitochondria from the same muscle fiber as preceding (Fig. 6). Note the paracrystalline inclusions between the outer and inner mitochondrial membranes. The paracrystalline inclusions take on a different appearance depending upon the plane of sectioning. X50,400.

the incandescent energy source there is: a) a reduction in the number of mononuclear cells within the sarcolemmal tubes, b) less fragmented sarcoplasm, c) an increase in basophilic mononuclear cells, many undergoing mitotic division and resembling presumptive myoblasts, and d) the appearance of long, multinucleated myotubes containing myofilaments organized into discrete myofibrils and oriented along the original longitudinal axis of the muscle. By 7 days after injury, all trace of myofiber degeneration has disappeared and juvenile myofibers and occasional myotubes extend throughout the originally damaged area. By 14 days, regeneration is nearly complete, with most fibers having near normal cross-sectional area. The sarcoplasmic reticulum and transverse tubular system appears normal and crystalline inclusions are no longer seen in the mitochondria of the regenerate fibers. A double basement membrane (basal lamina) consisting of a new one lying within an old one is seen surrounding many fibers 7 to 14 days after injury (Figs. 8-10).

In our studies (3, 7) we find that the elevations in orbital and body temperatures following exposure to incandescent radiant energy and radiant energy plus increased environmental temperature were comparable to those seen after exposure to elevated environmental temperature alone (i.e., an increase in environmental temperature from 22° to 39°C, for 17.5 hr with or without concurrent exposure to radiant energy, causes rectal temperature to increase from 38° to 42.5°C and orbital temperature to rise from 37.2° to 40°C). Therefore, thermal energy by itself does not seem to be a critical factor in the induction of myofiber degeneration. Also, the extraocular muscles were undamaged when the eyes were shielded with black contact occluders that blocked transmission of the full spectrum of radiant energy but allowed for the conduction of thermal energy into the orbit. Furthermore, exposure to fluorescent illumination alone (i.e., without any increase in environmental, rectal or orbital temperature) results in an identical pattern of extraocular muscle damage and regeneration as is seen following exposure to incandescent illumination.

It should be noted that since our earlier studies did not reveal any degenerative changes in extraocular muscles of animals exposed to fluorescent light, and since exposure to incandescent light did cause extensive damage, it was concluded that the different spectral wavelengths emitted by the two light sources must be responsible for the observations. However, the more recent experiments in which rats were exposed to approximately twice the intensity of fluorescent illumination as used in earlier studies indicated that cool-white fluorescent light exposure could induce (extraocular muscle) damage. In these latter experiments, the appearance and time sequence of degeneration and regeneration were identical to those originally described for incandescent damage. In that study, rats were exposed for 3.5 to 24 hr and the muscles

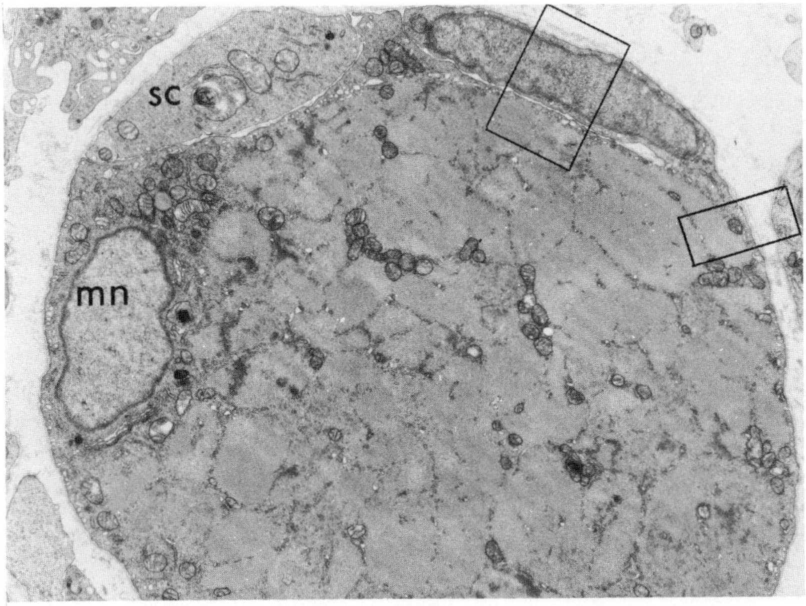

Figure 8. A cross-section of a medial rectus muscle fiber 14 days after a 24 h exposure to continuous light. Note the peripheral myonucleus (mn) has not yet become flattened and that a portion of a satellite cell (sc) lies under the basal lamina of the regenerated muscle fiber. Also note the occurence of a partially fused, presumably myogenic cell (arrow), lying beneath the basal lamina of the myofiber. Most of the regenerated muscle fibers are surrounded by a double basal lamina (see insets, Fig. 9, 10). X9,100.

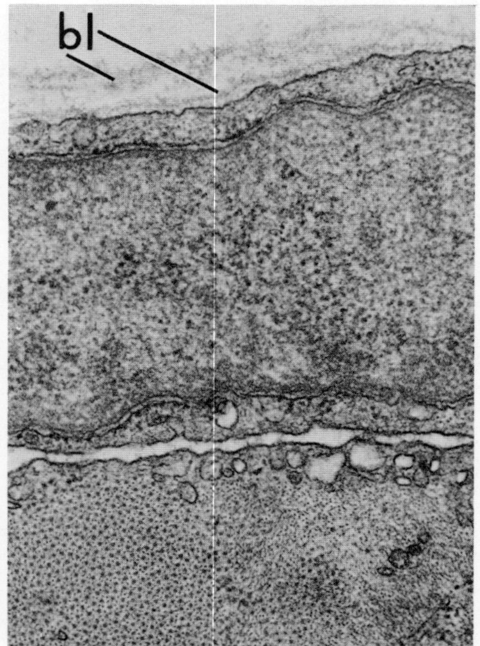

Figure 9. A cross-section at higher magnification of the double basal lamina (bl) of the area indicated in Fig. 8. X39,000.

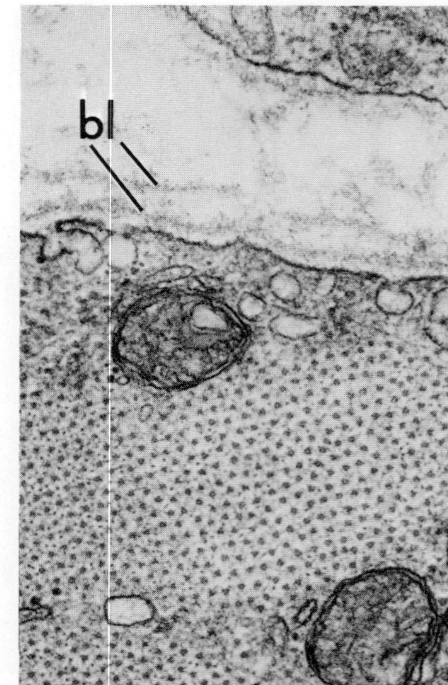

Figure 10. A cross-section at higher magnification of the double basal lamina (bl) of the area indicated in Fig. 8. X58,500.

were examined 0 to 14 days later while in the more recent fluorescent light studies, the animals were exposed continuously from 17.5 hr to 7 days. The long-term fluorescent studies revealed an interesting sequence of events. Exposure to cool-white fluorescent illumination for 1 day was sufficient to damage skeletal muscle, and signs of regeneration were observed during the second day of exposure. Even though the animals remained exposed to continuous illumination during the next few days, the degenerative changes apparently ceased, and regeneration progressed rapidly. The developmental myoblastic and myotube stages of regeneration appeared to be resistant to the damaging effects of fluorescent light. However, in tissues examined after 7 days of continuous light exposure, loci that resembled the pattern of primary muscle fiber destruction were observed again (Fig. 11). Only juvenile myofibers were susceptible to the effects of continuous exposure while less mature myotubes were undamaged (Fig. 11). Therefore, it seemed reasonable to speculate that in the extraocular muscles of adult rats, fully differentiated myofibers were susceptible to damage by continuous fluorescent illumination but that myoblasts and myotubes were more resistant to destruction. However, as the myotubes mature, lose their sarcoplasmic basophilia, and gain prominent cross-striations, but while muscle nuclei were still centrally rowed, the muscle fibers are again susceptible to damage by incandescent and fluorescent illumination.

Retinal Degeneration-Regeneration

Albino rats placed in continuous, low intensity, fluorescent light for 6, 12, and 18 hours showed structural changes both in retinal pigment epithelial cells and in the outer segments of the photoreceptor cells. Pigment epithelial cells of retinas exposed for 6 hours showed disorientation and retraction of the villous processes along their apical border, and separations (600 to 1000 Å in diameter) between adjacent epithelial cells, as compared to the dark-adapted retinas. The basal portion of the epithelial cells and Bruch's membrane appeared unchanged. The flattened apical discs of the dark-adapted photoreceptor outer segments become tubular after 6 hours of constant light (Fig. 12). After 12 hours of continuous light-exposure, the villous processes of the epithelial cells were more compacted against the cell's apical surface. The space between adjacent epithelial cells increased (5,000 to 27,000 Å) and were often filled with membranous bodies of indeterminate origin (Fig. 13). Those structural changes already mentioned were even more pronounced after 18 hours of constant light. The space between adjacent epithelial cells had increased (5,000 to 36,000 Å) and the lamellated discs of all photoreceptors appeared tubular along the entire length of the outer segments (Fig. 14). Photoreceptor cell inner segments, cell bodies, and connecting cilia showed no degenerative change after 6, 12 or 18 hours of light

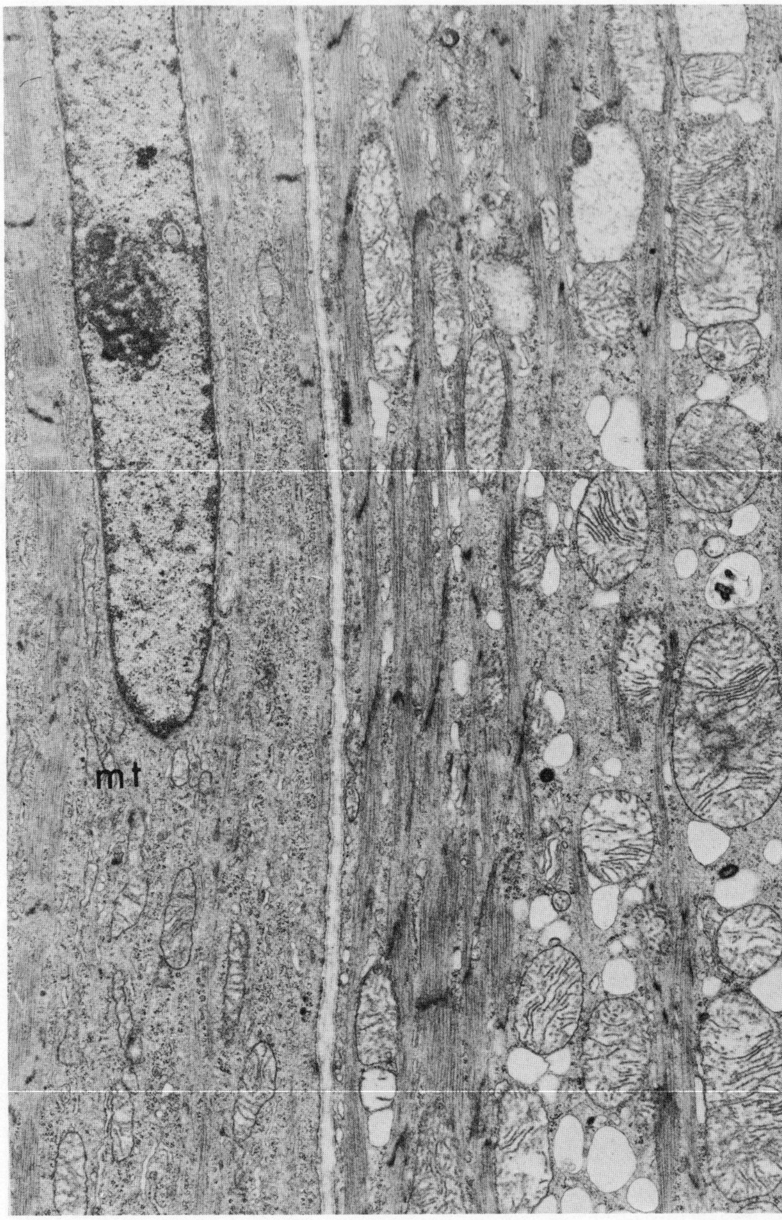

Figure 11. A longitudinal section of a superior oblique muscle fiber examined immediately after a continuous seven day exposure to continuous light. The regenerating myotube on the left (mt) is not visibly damaged by further exposure to light. The more mature muscle fiber on the right has undergone a secondary degeneration. X27,900.

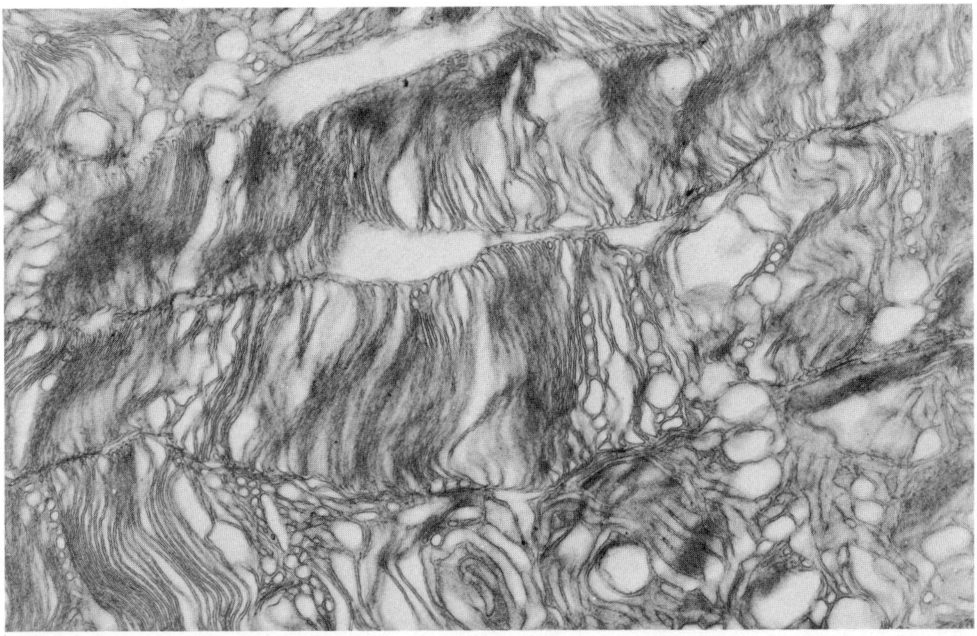

Figure 12. Photoreceptor outer segments from a rat exposed to constant light for 6 h. Note, the normally flattened apical discs of the dark adapted outer segments become tubular; by 18 h of constant light exposure the outer segments are tubular along their entire length. X23,000.

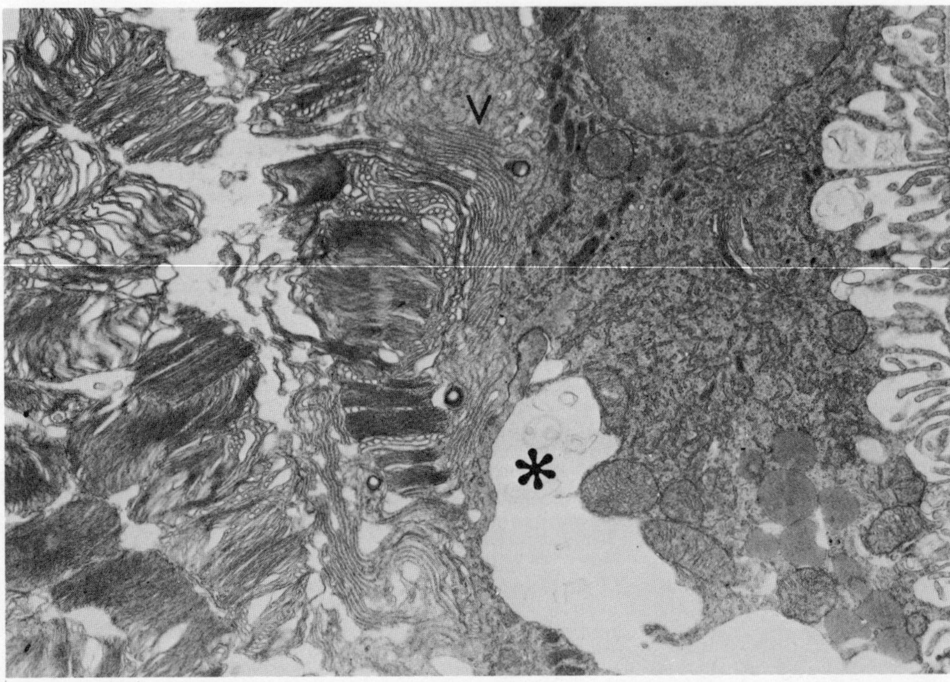

Figure 13. Retina from a rat exposed to continuous light for 12 h. The apical portion of many outer segments have become detached. Large intercellular spaces, often containing electron-dense materials occur between adjacent epithelial cells (*). Note, the villous processes (V) of the pigment epithelial cells, are compacted against the cell's apical surface. X14,000.

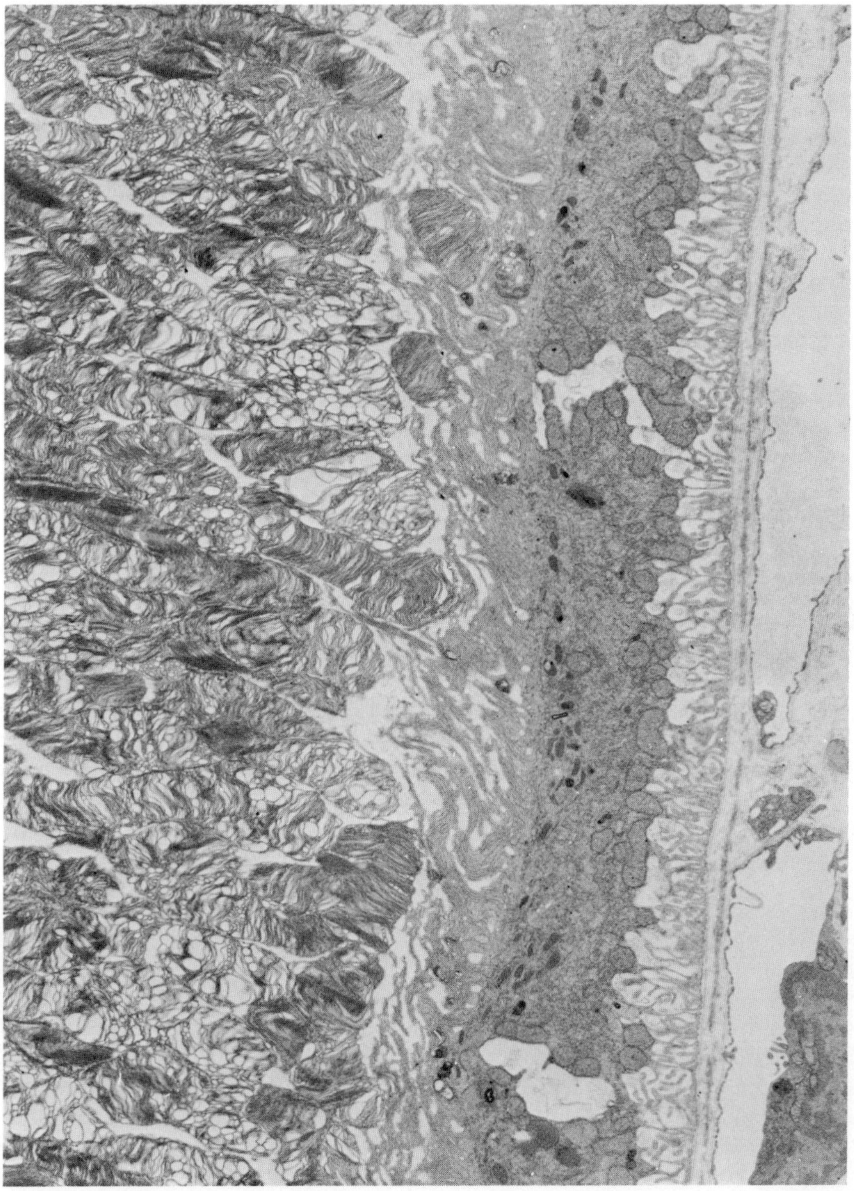

Figure 14. Retina from a rat exposed to continuous light for 18 h. The photoreceptor cell outer segments appear tubular along their entire length. The space between adjacent epithelial cells is greater than in retinas exposed for 6 or 12 h. Note, the mitochondria are located along the basal surface of the epithelial cells and along the lateral borders, between cells. X5,400. (Reduced 10% for purposes of reproduction.)

exposure. The inner layers of the retina were apparently structurally unaltered by short-term exposure to low levels of illumination. If the animals are placed in a darkened environment, a more typical epithelial morphology returns within 3 hours. No structural changes were seen in the choroidal layer after 6, 12, or 18 hours constant light exposure.

Rats exposed to long term continuous, low intensity (70 Ft-C), fluorescent light for periods of 30 days to one year developed an irreversible retinal degeneration. After 30 days of continuous exposure the photoreceptor cells are entirely lost. The villous processes of the retinal pigment epithelial cells were more compacted than in normal retinas and were restricted in position by the outer limiting membrane, which was now contiguous to them. Otherwise, the pigment epithelial cells appeared to be undamaged by the extended period of light-exposure and the large intercellular spaces that occur between adjacent epithelial cells during the first stages of light-exposure are absent (Figs. 15 and 16). The bipolar neurons, amacrine cells, and Müller cells in the inner nuclear layer apparently were unaffected by exposure to continuous light even after one year. The inner plexiform layer of the retina, consisting mainly of axons of bipolar neurons extending to and synapsing with dendrites of ganglion cells, was not structurally altered by continuous exposure. Also, the ganglion cells and ganglion cell layer in retinas exposed to continuous low level illumination were structurally identical to those of cyclic light-exposed animals. Some experiments were done with pigmented rats, also. The extraocular muscle fibers, retinal pigment epithelium, and photoreceptors of such animals were not affected by short term exposure to constant light at 70 Ft-C or 150 Ft-C.

DISCUSSION

Exposure for only a few hours to high intensity light (250-750 Ft-C) causes damage to the retinal pigment epithelium and photoreceptor cells of the vertebrate retina (8-11). Retinal damage and degeneration after continuous exposure to lower levels of illumination (70-150 Ft-C) follows a different pattern over a longer period of time (1, 9-10) and damages the extraocular muscle fibers as well (3, 4). The purpose of this paper is to highlight some of the more notable features of cellular damage and regeneration that occur in the albino rat retina and in its extraocular muscle tissue following exposure to low levels of light. Many of these findings are the result of collaborative studies between myself, Drs. O'Steen, Anderson, and recently Dr. Strum.

Typically in the relaxed vertebrate muscle fiber, each thick filament is surrounded by six thin filaments. However, in many of the radiant energy-damaged myofibers, each thick filament is

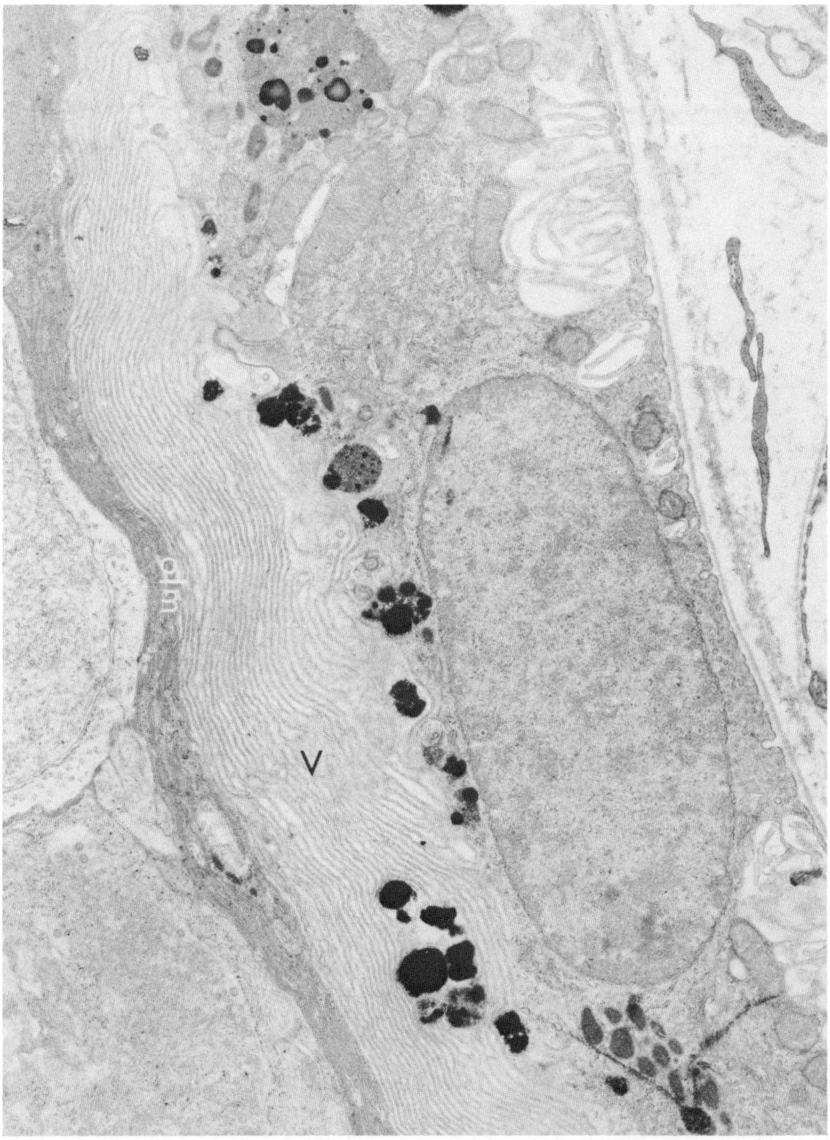

Figure 15. Retina from a rat exposed to constant light for five months. The photoreceptor cells are lost. The villous processes (V) of the pigment epithelium are more closely compacted than in retinas of rats exposed for 6 to 18 h of constant light. Note, the intercellular spacing between adjacent epithelial cells is more normal than in retinas exposed for 6 to 18 h. The outer limiting membrane (olm) consisting of Müller cell end-feet lies in close apposition to the villous processes of the pigment epithelium. X17,300. (Reduced 10% for purposes of reproduction.)

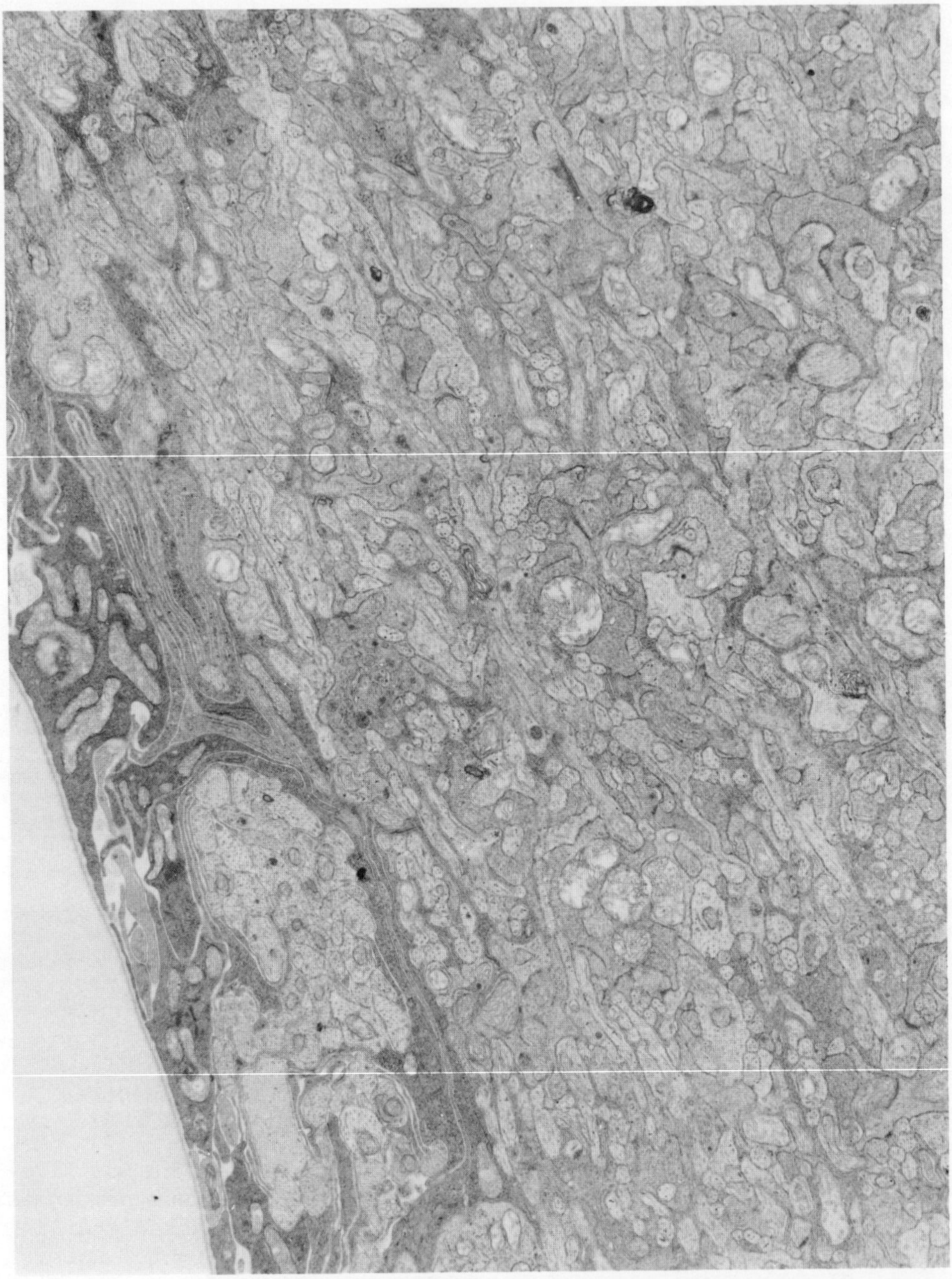

Figure 16. Retina from a rat exposed to constant light for 1 year. Note, the inner plexiform layer, nerve fiber layer and inner limiting membrane are similar to control animals. X10,300.

surrounded by 10 to 12 thin filaments. The increased thin-to-thick filament ratio indicates sarcomere shortening with extensive double, thin contractile filament overlap. Such overlap would be expected if the dilated and fragmented sarcoplasmic reticular system released normally bound Ca^{2+} into the sarcoplasm (3). At elevated sarcoplasmic Ca^{2+} levels, troponin inhibition of myosin ATPase would be repressed, allowing actinomyosin crossbridge interaction and extensive shortening to occur (11-15).

It is possible that elevated Ca^{2+} levels are responsible for certain other structural changes seen in radiant energy-damaged myofibers. In a study by Busch et al. (16), a protein fraction was isolated that caused Z-disc removal from rabbit skeletal muscle myofibrils in the presence of Ca^{2+}. Evidently, Z-disc removal by this protein fraction requires Ca^{2+} levels higher than 0.1 mM. Damage or degradation of other myofibrillar elements was not observed. Busch and colleagues suggest in their original report that the level of unbound (Ca^{2+} in the sarcoplasm may regulate the activity of this protein fraction *in vivo*. In a more recent study by Dayton et al. (17), evidence is presented indicating that a Ca^{2+} activated protease exists in normal, living mammalian muscle fibers but is not activated at the low Ca^{2+} levels (10^{-5}M or less) typical of healthy myofiber sarcoplasm. The published ultrastructural findings of Busch et al. (16) and Dayton et al. (17) show myofibril changes at increased Ca^{2+} levels, nearly identical to those found in radiant energy-damaged extraocular myofibers. A native myofibrillar protease has also been identified in dystrophic mice and in tumor-bearing rats (18). Unfortunately, it is not possible to compare the identity of the protease reported by Mayer et al. (18) with the Ca^{2+} activated protease described by Busch et al. (16), and Dayton et al. (17).

In light of the findings described above, it is tempting to speculate that the loss of Z-discs observed in radiant energy damaged myofibers is the result of selective myofibrillar degradation, catalyzed by an endogenous muscle protein, whose activity is triggered or regulated by the high Ca^{2+} levels resulting from the fragmentation of the sarcoplasmic reticulum or from an influx of external Ca^{2+}. At regions along the fiber showing less degeneration, i.e., normal Z-discs remaining, the sarcoplasmic reticulum is dilated but not fragmented and intact terminal and lateral cisternae remain associated with adjacent transverse tubules (3). Although the mechanism of intracellular protein degradation and its control in muscle is only poorly understood, it is possible that myofiber degeneration follows a common pattern, based on cellular economy, regardless of the cause.

Our most recent findings of fluorescent radiant energy-induced myofiber damage reveal that myofiber regeneration begins even while the muscles are exposed to continued levels of damaging radiant energy. Furthermore, the participating myoblastic cells and regenerating myotubes do not appear to be susceptible to further damage until 7 days after the initial damage, i.e., until all originally damaged myofibers have regenerated to the juvenile myofiber stage. The reason for the apparent immunity of the newly developing, regenerate myofibers is not understood. However, in a previous study (19) on developing tonic and phasic muscle, a significantly greater ratio of sarcoplasmic reticulum to myofibrillar protein was found in the juvenile fibers compared with mature fibers of the same muscle. Since the SR is known to store intracellular Ca^{2+}, it is tempting to speculate that in regenerating extraocular myofibers, an extensive SR is able to maintain sufficiently low sarcoplasmic Ca^{2+} levels and so prevent further damage. It is not surprising that in our original study (3), 3 to 4 week old rats were subjected to nearly the same extraocular myofiber damage as were the mature animals since postnatal differentiation of 80% of muscle fibers is already complete by 3 to 4 weeks (20).

In a study by Hess (21) the Ca^{2+} level in retinal pigment epithelial cells was reported as 15 mmol Ca^{2+}/kg wet weight. As this author points out, this value is higher than for most cells and tissues that are not engaged in calcification. Indeed it is higher than values reported in maximum superprecipitated rabbit myofibrils (4.6 mmol/mg wt) equilibrated in 10^{-5} M Ca^{2+} (22). Retinal pigment epithelial cells hyperpolarize on illumination, and this has been correlated with the c-wave of the ERG (23), the action spectrum of which coincides with that of rhodopsin in rod retinas. One current hypothesis of the ionic basis of the c-wave is that the potassium concentration is decreased between the photoreceptors and the apical border of the retinal epithelial cells during illumination (24). Based on these findings Hess (21) suggests that in both pigmented retinal epithelial cells and photoreceptors an internal release of Ca^{2+} may be related to the external fall in potassium, perhaps through the potassium leak mechanism. Thus Ca^{2+} appears to play a significant role in the electrophysiological relation between the pigment epithelium and the photoreceptor cells of the retina. It is possible that the pigment epithelium is unable to cope with periods of high cytoplasmic Ca^{2+} for more than 3 to 4 hours. Certainly the increase in retinal pigment epithelial intercellular spacing shown in our studies suggests these cells are engaged in selective permeability metabolism. If Ca^{2+} were begin removed from these cells, it might be expected to diffuse across the choroid and raise the level of external Ca^{2+} surrounding the extraocular muscles at their insertion. How this might raise the intraocular level of myofiber Ca^{2+} is currently being investigated.

REFERENCES

1. O'Steen, W. K., Shear, C. R. and Anderson, K. V. (1972) Am. J. Anat. 134, 5-22.

2. Shear, C. R., O'Steen, W. K. and Anderson, K. V. (1973) Am. J. Anat. 138, 127-132.
3. O'Steen, W. K., Shear, C. R. and Anderson, K. V. (1975) J. Cell Sci. 18, 157-177.
4. Shear, C. R. (1979) in Muscle Regeneration (Mauro, A., ed.) pp. 249-263, Raven Press, New York.
5. O'Steen, W. K., Kraeer, S. L. and Shear, C. R. (1978) Invest. Ophthalmol. Visual Sci. 17, 847-856.
6. O'Steen, W. K. (1979) These proceedings.
7. Shear, C. R. and O'Steen, W. K. (1976) J. Cell Biol. 70, 101a.
8. Noell, W. K., Walker, V. S., Kang, B. S. and Berman, S. (1966) Invest. Ophthal. 5, 450-473.
9. O'Steen, W. K. (1970) Expl. Neurol. 27, 194-205.
10. O'Steen, W. K., Anderson, K. V. and Shear, C. R. (1974) Invest. Ophthalmol. 13, 334-339.
11. Bailey, K. (1948) J. Biochem. 43, 271-279.
12. Ebashi, S. and Endo, M. (1968) Progr. Biophys. Molec. Biol. 18, 123-183.
13. Huxley, A. F. and Taylor, R. E. (1958) J. Physiol. (Lond.) 144, 193-200.
14. Peachey, L. D. (1970) in "Physiology and Biochemistry of Muscle as a Food" (Briskey, E. J., Cassens, R. G. and Marsh, B. B., eds.) Vol. 2, pp. 213-310, University of Wisconsin Press, Madison.
15. Peachey, L. D. and Porter, K. R. (1959) Science 129, 721-722.
16. Busch, W. A., Stromer, M. H., Goll, D. E. and Suzuki, A. (1972) J. Cell Biol. 52, 367-381.
17. Dayton, W. R., Reville, W. J., Goll, D. E. and Stromer, M. H. (1976) Biochemistry 15, 2159-2167.
18. Mayer, M., Amin, R., Milholland, R. J. and Rosen, F. (1976) Exp. Mol. Pathol. 25, 9-19.
19. Shear, C. R. and Goldspink, G. (1971) J. Morphol. 135, 351-372.
20. Chiakulas, J. J. and Pauly, J. E. (1965) Anat. Rec. 152, 55-62.
21. Hess, H. H. (1975) Exp. Eye Res. 21, 471-479.
22. Winegrad, S. (1969) in Mineral Metabolism (Comar, C. L. and Bronner, F., eds.) vol. 3, pp. 191-232, Academic Press, New York.
23. Steinberg, R. H., Schmidt, R. and Brown, K. T. (1970) Nature, (London) 227, 728-730.
24. Steinberg, R. H. and Miller, S. (1973) Exp. Eye Res. 16, 365-372.

THE EFFECTS OF CONSTANT LIGHT ON VISUALLY GUIDED BEHAVIOR

Kenneth V. Anderson and Vance Lemmon[†]

Department of Anatomy
Emory University
Atlanta, Georgia 30322

[†]Department of Physiology and Biophysics
Washington University
St. Louis, Missouri 63130

Adult, albino rats kept in continuous light have been shown to develop severe retinal degeneration. Both light and electron microscopic evidence has indicated that substantial destruction occurred in the receptor cell layer when low levels of light (less than 20 Ft-C) were used to produce the degeneration (1-10) and that both the receptor cell and pigment epithelial layers were seriously damaged when high levels of light (more than 100 Ft-C) were used to induce degeneration (11-13). Incandescent lighting was only slightly less effective than fluorescent lighting in producing light-induced, retinal degeneration (8). Spectral sensitivity studies have suggested that the shorter wavelengths of visible light produce the most damaging effects on the retina (14-16).

In addition to the anatomical studies cited above, which describe various histological changes in the retina that have been induced by exposure to light, neurophysiological studies have shown that retinal physiology was markedly changed in animals exposed to constant illumination. Electroretinograms were substantially altered in animals exposed to 2 or more days of constant, intense light (13, 17, 18, 19). Evoked potentials to flashes of light recorded in the central visual system were abnormal or absent in animals exposed to 30 or more days of constant, low-intensity light (7, 20). In addition to the alterations in evoked activity, it has been shown that spontaneous, rhythmic activity was abolished throughout the visual system after only 4 days of exposure to constant, low-intensity illumination (1, 2). In some instances, the anatomical and electrophysiological changes that can be induced by exposure to constant light can be at least partially reversed by maintaining an animal in total darkness (19).

While most previous neurophysiological studies have found little evidence of physiological activity in albino rats exposed to constant light for long periods of time, at least one recent investigation has reported the presence of visually responsive cells in animals whose retinas exhibited massive degeneration of the photoreceptor cells (21). In that study, Lemmon and Anderson found visually responsive cells in both the lateral geniculate nuclei and the visual cortex of albino rats that have been housed in constant light for periods of up to 24 weeks. With increasing exposure times, it was found that cells in the lateral geniculate nuclei and the visual cortex had longer response latencies and less vigorous responses and that it was necessary to illuminate larger areas of the visual field in order to produce responses. The spectral sensitivities of cells in both control and experimental rats were similar in shape and peak sensitivity to the spectral absorption properties of rhodopsin. The spectral sensitivity curves of some of the cells in control and light-exposed animals had small side peaks at 600 nm, which may have been due to some additional visual pigment molecule.

Taken together, the anatomical and neurophysiological studies cited above might lead one to assume that rats exposed to constant light would not be able to perform as well as normal animals on visual discrimination tasks or might even be blind. This would not appear to be the case, however, since behavioral studies performed in several different laboratories have shown that rats with substantial retinal damage can still perform well on a variety of visually guided tasks. The goals of the present chapter are both to describe some of the behavioral experiments we have performed over the past eight years, each of which shows that rats with retinal degeneration retain a substantial amount of sophisticated visual competency, and to place these experiments in the context of the results from similar experiments from other laboratories.

METHODS

The animals in all of our behavioral investigations have been adult, albino rats weighing about 250 g. The animals were tested in a T-maze apparatus or in a fully automated apparatus that permitted one to test a wide variety of visually guided behaviors. Both the stem and T-portion of the maze were gray and measured 102 x 12 x 16 cm. The visual discriminanda were attached to 12 x 25 cm panels that could be slipped into place at the entrance to each goal box. Each panel had a small door in the middle, through which the rats could pass to the goal box after correct responses. Incorrect responses did not provide access to the goal box. Visual cues covered the entire panel in all cases, so as to avoid cue-response contingulty problems. Animals in the T-maze experiments were tested on either a black-white (BW) discrimination or on a pattern (P) discrimination.

In the BW discrimination, responses to the white panel were rewarded by permitting the rat access to the goal box where it could consume ground, moistened rat pellets for 5 sec. The P discrimination consisted of panels with alternating black-and white stripes oriented horizontally or vertically. Two pattern tasks were used, one of which was relatively easy for normal rats to learn and one which was somewhat more difficult for them to learn. In the easier pattern task the stripes were 0.8 cm wide and in the more difficult task they were 0.4 cm wide. For choices made at the junction of the stem and T-portion of the maze, a 0.8 cm wide stripe would subtend about 4 degrees of retinal angle in the rat, and a 0.4 cm wide stripe about 2 degress of retinal angle. Since rats cannot discriminate lines closer together than 1 degree (22), the narrower stripes should require discrimination performance near the rat's limit of acuity, while the wider stripes should be readily discriminated. All rats were rewarded for making responses to the horizontally striped panel. The discriminanda were changed from side to side on successive trials according to a Gellerman series (23). Each animal received 20 trials a day and was trained until it reached a criterion of 90% correct responding per day for 3 consecutive days, a rather strict level of performance for rats. To eliminate olfactory and taste cues food was placed behind both the correct and incorrect panels and each panel was wiped clean after each trial. In some experiments, the olfactory bulbs were removed before animals were trained and tested in the T-maze. To control for auditory cues the panels were removed and replaced after each trial, even if they were designated to remain on the same side on the next trial. Two types of response measures were used in the maze experiments, a time-independent measure and a time-dependent measure. The time-independent measure consisted of recording an animal's percent correct responses and, thus, is basically an index of the accuracy of an animal's choice. The time-dependent measure was a latency measure, which we defined as the time taken to select one of the two visual discriminanda after leaving the start-box.

In all, more than 130 rats were used in various T-maze experiments. Each rat was tested before and after exposure to 30 or more days of constant, fluorescent light at 18 Ft-C (184 lux) intensity. The details of the lighting system used to produce retinal degeneration, the techniques used to measure light intensity, and the details of the methods for training and testing rats in the maze have been described previously (1, 7). The cyclic photoperiod for the control animals in the maze experiments, as well as in all other experiments, was 14 hours of light and 10 hours of darkness, with the light-period beginning at 5 a.m. All experimental animals in the maze studies were exposed for at least 30 days, because that time period produced a retina that exhibited extremely severe degeneration in the receptor cell population.

The automatic testing apparatus consisted of a gray experimental chamber that measured 35 cm on a side and was housed in a dark room. During training and testing, white noise was used to mask extraneous sounds. A clear plexiglass nose panel, called the start panel, was mounted on one wall of the experimental chamber. Behind the start panel was a rear projection system containing a variety of visual stimuli. On the opposite wall of the chamber were two nose panels, called choice panels, with rear projectors behind them. Between the two choice panels was a pellet delivery tube. All animals were pretrained to press the panels using traditional shaping methods. At the beginning of an experimental trial, a visual stimulus was projected onto the start panel. When the animal pressed the start panel the start stimulus was turned off and two stimuli were simultaneously projected onto the choice panels. Pressing either choice panel turned off the visual stimuli and initiated a 10-second waiting period before the next trial. The visual stimulus projected onto the start panel was always the same as the stimulus that was rewarded when projected onto the choice panels. As was the case in the maze experiments, the stimuli projected onto the choice panels were alternated from side to side, on successive trials, according to a Gellerman series. In fact, 18 different Gellerman series were used and the series was changed each day. During training each animal received 30 trials per day and was trained, as in the maze experiments, until it had reached a criterion of 90% correct responses on three consecutive days.

Each animal in these latter experiments was trained on one or more of the following tasks: (1) an intensity discrimination; (2) a pattern discrimination; (3) a color discrimination. If an animal were tested on more than one task, it was tested on each task on alternate weeks. Once an animal reached the 90% criterion level of performance on all tasks to which it had been assigned, it was placed in a constant-light environment. The animals were then tested 5-7 days a week for up to 350 days.

In the intensity task the animals were required to identify the brighter of two stimuli when the brightest of the stimuli was changed in a series of graded steps. The pattern discriminanda consisted of horizontal or vertical stripes. The stripes were alternating black and white bars 3 mm wide and the total flux of the two targets was equal. The horizontally striped target was designated as the correct target. In the color task, animals were required to discriminate between blue and red targets which consisted of circles that were 3.5 cm in diameter. The blue stimulus was rewarded. The relative brightness of the blue and red stimuli were changed each day in order to prevent the rats from using brightness cues to make the discrimination. The technique of varying the brightness and the detailed procedures used in training the animals in the automatic behavioral apparatus have been described elsewhere (24).

RESULTS

T-Maze Experiments

Time-Independent Measures. The basic results for several groups of animals used in this series of experiments are shown in Tables 1-4. The general finding was that exposure to constant light for 30 days did not impair performance on any of the discrimination tests used in these investigations, even though histological analyses showed that receptors were nearly lacking in all light-exposed animals. All animals trained on the black-white discrimination before exposure to constant light (group 1, Table 1) retained the discrimination following exposure. No animal in this group required more than 20 post-exposure trials to reestablish criterion

TABLE 1

Trials to Criterion for Rats Trained on the Black-White Discrimination

Group 1			Group 2	
Animal No.	Pre-Exposure Learning	Post-Exposure Testing	Animal No.	Post-Exposure Learning
1	20	0	11	80
2	60	20	12	60
3	80	20	13	20
4	40	0	14	20
5	20	0	15	60
6	60	20	16	40
7	40	0		
8	20	20		
9	40	20		
10	20	0		

TABLE 2

Trials to Criterion for Rats Trained on the Pattern Discrimination (.8 cm stripes)

Group 3			Group 4	
Animal No.	Pre-Exposure Learning	Post-Exposure Testing	Animal No.	Post-Exposure Learning
17	40	0	27	60
18	80	20	28	40
19	120	0	29	100
20	20	0	30	80
21	80	20	31	20
22	80	20	32	100
23	120	20		
24	100	0		
25	80	0		
26	80	20		

TABLE 3

Trials to Criterion for Rats Trained on the Pattern Discrimination (.4 stripes)

Group 5			Group 6	
Animal No.	Pre-Exposure Learning	Post-Exposure Testing	Animal No.	Post-Exposure Learning
33	80	0	43	100
34	80	20	44	140
35	120	0	45	160
36	100	20	46	80
37	140	20	47	100
38	100	0	48	100
39	120	0		
40	80	0		
41	180	20		
42	140	20		

TABLE 4

Trials to Criterion for Control Rats Trained on the Black-White (BW) or Pattern (P) Discrimination and Tested After 30 Days in A Cyclic-Light Environment

Group 7(BW) Task			Group 8(P) Task (.8 cm stripes)		
Animal No.	Original Learning	Post-Cyclic-Light Testing	Animal No.	Original Learning	Post-Cyclic-Light Testing
49	40	0	55	60	0
50	20	0	56	100	20
51	60	20	57	80	20
52	20	20	58	40	0
53	20	0	59	40	0
54	40	20	60	60	20

performance. Comparison of the post-exposure, testing scores for animals in group 1 (Table 1) with the post-cyclic-light, testing scores for animals in group 7 (Table 4) showed that rats exposed to constant light reestablished criterion performance as quickly as did control animals kept in cyclic-light environment for 30 days prior to testing ($p > 0.1$). In addition, rats trained on the black-white discrimination for the first time after exposure to constant light (group 2, Table 1) were found to reach criterion performance at the same rate as control animals (group 7, Table 4). Statistical analyses indicated that there were no reliable differences ($p > 0.1$) between the post-exposure learning scores for animals in group 2 (Table 1) and the original learning scores for animals in group 7 (or pre-exposure learning scores for animals in group 1).

Results similar to those for the black-white discrimination were obtained with both of the pattern discrimination tasks. All animals trained on a pattern discrimination before exposure to constant light (groups 3 and 5, Tables 2 and 3) reestablished criterion performance after exposure in 20 or fewer trials, even though it originally took longer to learn the pattern task with the narrow stripes than to learn it with the wider stripes. Statistical analyses showed that the post-exposure testing scores for animals in groups 3 and 5 (Tables 2 and 3) were not reliably different ($P > 0.1$) from the post-cyclic-light testing scores for control animals in group 8 (Table 4). As was the case for animals trained on the black-white discrimination, rats trained on the pattern discriminations for the first time after exposure to constant light (groups 4 and 6, Tables 2 and 3) reached criterion performance at the same rate as did control animals (group 8, Table 4; see, also, pre-exposure scores for groups 3 and 5, Tables 2 and 3). Statistical comparisons showed that there were no reliable differences ($P > 0.1$) between the post-exposure learning scores for animals in groups 4 and 6 (Tables 2 and 3) and the original learning scores for groups 3 and 5 (Tables 2 and 3).

In these and subsequent experiments, many animals were enucleated and/or had their olfactory bulbs surgically removed to control for nonvisual sensory cues that might be expected to play a role in rodent discriminative behavior. Animals enucleated before training were never able to learn the discrimination tasks and those enucleated after learning various tasks exhibited performance that fell to chance levels immediately after enucleation. Removal of the olfactory bulbs was never observed to influence discriminative performance in our experiments, regardless of whether the removals occurred before training began or after criterion performance levels had been achieved.

Time Dependent Measures. The results for all groups of animals for which both accuracy and latency measures were taken are shown in Tables 5-8. While the exposure to 30 days of constant light did

TABLE 5

Trials to Criterion and Latency Scores for Rats Trained on the Black-White Discrimination Before and After Exposure to Constant Light

Animal No.	Pre-exposure Learning Score	Pre-exposure Latency (sec)	Post-exposure Testing Score	Post-exposure Latency (sec)
1	40	2.0	0	2.1
2	20	3.1	0	2.8
3	20	2.1	0	1.9
4	40	2.1	20	2.3
5	60	2.8	20	3.0
6	20	3.7	0	3.1
7	20	4.3	20	3.8
8	40	2.3	0	2.2
9	40	2.7	0	2.4
10	20	3.0	0	2.7

TABLE 6

Trials to Criterion and Latency Scores for Rats Exposed to Constant Light and Then Trained on the Black-White Discrimination

Animal No.	Post-exposure Learning Score	Post-exposure Latency Score (sec)
11	40	3.2
12	20	2.6
13	20	2.3
14	60	4.1
15	40	3.7

not impair the accuracy of visually guided behavior on any of the discrimination tests used in these experiments, it did <u>increase</u> the time required to make correct choices on the pattern problem. Table 5 shows that all animals trained on the black-white discrimination before exposure to constant illumination retained the discrimination following exposure and no animal in this group required more than 20 post-exposure trials to re-establish criterion performance. Further, the post-exposure latency scores for animals in Table 5 were not reliably different from the pre-exposure latency scores ($p > .05$). In addition, rats trained on the black-white discrimination for the first time after exposure to constant light reached criterion performance levels at the same rate and with similar latency scores as control animals ($p > .05$; compare post-exposure learning and latency scores for animals in Table 6 with pre-exposure learning and latency scores for animals in Table 5).

The findings on the pattern discrimination task were similar to those on the black-white problem when the percentage of correct responses to the visual targets was used as the performance measure. That is, as shown in Table 7, all animals trained on the pattern discrimination before exposure to continuous light retained the discrimination following exposure. Further, rats trained on the pattern problem for the first time after exposure to constant light reached criterion performance levels at rates similar to those exhibited by control animals ($p > .05$; compare post-exposure learning scores for animals in Table 8 with pre-exposure learning scores for animals in Table 5).

The principal finding in this series of studies related to the effects of retinal damage on the latency to make both correct and

TABLE 7

Trials to Criterion and Latency Scores for Rats Trained on

The Pattern Discrimination Before and After Exposure to Constant Light

Animal No.	Pre-exposure Learning Score	Pre-exposure Latency Score (sec)	Post-exposure Testing Score	Post-exposure Latency Score (sec)
16	120	8.6	20	18.7
17	80	8.3	0	23.8
18	220	6.8	20	22.4
19	240	9.3	20	17.1
20	180	9.5	0	19.6
21	100	11.8	0	27.3
22	80	6.0	0	18.0
23	140	7.7	20	16.4
24	180	8.4	20	20.6
25	160	10.2	0	23.4

TABLE 8

Trials to Criterion and Latency Scores for Rats Exposed to Constant Light and Then Trained on the Pattern Discrimination

Animal No.	Post-exposure Learning Score	Post-exposure Latency Score (sec)
26	100	29.6
27	180	32.4
28	160	26.3
29	80	23.8
30	120	19.7

incorrect choices on the pattern discrimination task. As shown in Table 7, post-exposure latency scores were significantly longer than pre-exposure latency scores for all animals that were first trained on the task, then exposed to continuous illumination for 30 days, and finally re-tested on the pattern problem ($p < .01$). Whereas the average latency was only 8.6 seconds prior to exposure, it was elevated to 20.7 seconds after animals sustained retinal damage as a result of exposure to the continuous light. When comparisons were made _between_ groups of animals the influence of retinal damage on the latency of responses was even more evident. For example, animals trained on the pattern task for the first time _after_ exposure to constant light (as in Table 8) not only took significantly longer to make correct choices than did control animals that learned the task prior to exposure ($p < .01$; Table 7), they also had longer average latencies than animals with retinal damage that were trained on the task _prior_ to exposure ($p < .05$; compare post-exposure latency scores for animals in Table 8 with post-exposure latency scores for animals in Table 7).

Matching-to-Sample Experiments

In the studies with the automated behavioral apparatus, experimental animals were tested on visual tasks for up to 350 days following exposure to constant light. In general, an animal's performance on the visual tasks declined in proportion to the length of time they were exposed to light. The task that was most resistant to light-induced, retinal degeneration was the brightness discrimination. Most animals performed at or near criterion levels until they were exposed for at least 200 days. For these animals, the critical period seemed to be somewhere between 300-350 days, when their performance fell to near chance levels in all instances.

Figure 1 shows the results from 6 animals that were trained only on a pattern discrimination task consisting of horizontal stripes vs. vertical stripes, with horizontal stripes being the positive stimulus. The animals were trained for periods ranging from 180-230 days, following exposure to constant light. Animals 16 and 17, at the top of the figure, are control animals that were housed continuously in cyclic light, and animals 8, 21, 13, and 23 are experimental animals that were first trained to criterion and then placed in constant light. Day zero on the abcissa represents the time at which the experimental animals were placed in constant illumination. As can be seen, the control animals maintained high levels of discriminative performance throughout the rather lengthy period of testing, which was about 230 days in the case of rat 16. By contrast with the control rats, the performance of the experimental animals began to decline, somewhere between 100 and 150 days of exposure and by 200 days, all animals were discriminating at chance levels. Thus, after 150 days of exposure to constant light, the scores of the experimental animals indicated that they could no longer discriminate effectively between horizontal and vertical stripes.

Figures 2 and 3 show the results for 6 rats that were trained and tested on both the pattern discrimination task and on the color task. Animals 35, 40 and 42, shown on the left of the figures, are control animals, and rats 31, 37 and 41, shown on the right of the figures, are experimental animals. The testing of these animals showed that while the performance of the control animals remained at a high level, the performance of the light-exposed animals decreased. As can be seen in Figure 2, there was a decrease in the performance of the experimental animals that was first noticeable in the first week of exposure. In fact, the difference between the experimental and control animals, during this period, was significant at the 0.001 level. Following 5 weeks in constant light, however, the scores of the experimental animals were not significantly different from controls. By contrast, seven weeks of exposure to constant light again resulted in the scores of light exposed animals being significantly lower ($p < .05$) than those of the control animals. The scores of the light-exposed animals continued to decrease, so that following 23 weeks of exposure to continuous light the scores were only slightly above chance. These results are similar to those for animals that were trained only on the pattern discrimination task and illustrate that the performance of animals that are rigorously trained on at least two, demanding visual tasks is not unlike the performance of animals trained on only one task.

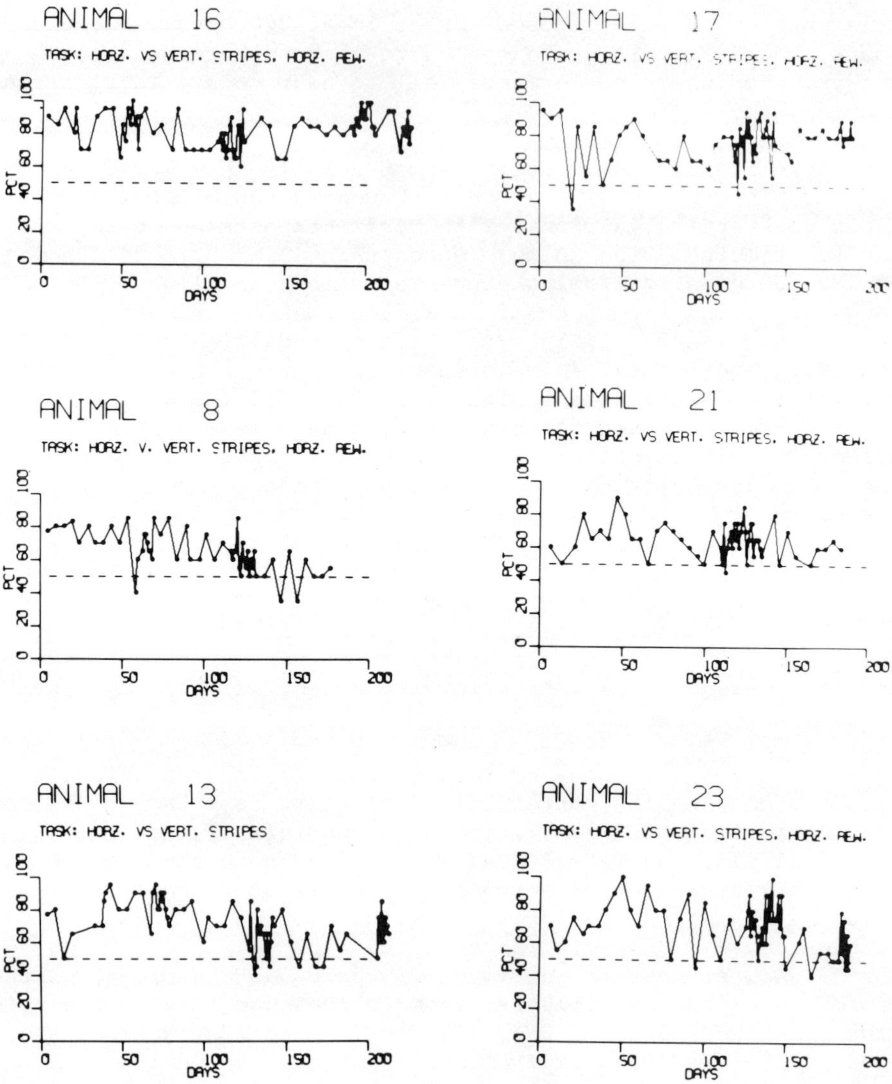

Figure 1: Scores of experimental and control animals tested for ability to discriminate between horizontal and vertical stripes in the matching-to-sample studies. Experimental animals were exposed to constant light beginning on day zero. Animal 16 served as the control for animals 8 and 13. Animal 17 served as the control for animals 21 and 23. The abscissa represents time in days. The ordinate indicates the percentage of correct responses in 30 trials per day. The dashed line indicates the 50% (chance) level of performance. After approximately 150 days of exposure to constant light, the scores of the experimental animals had decreased to near chance levels. See text for details.

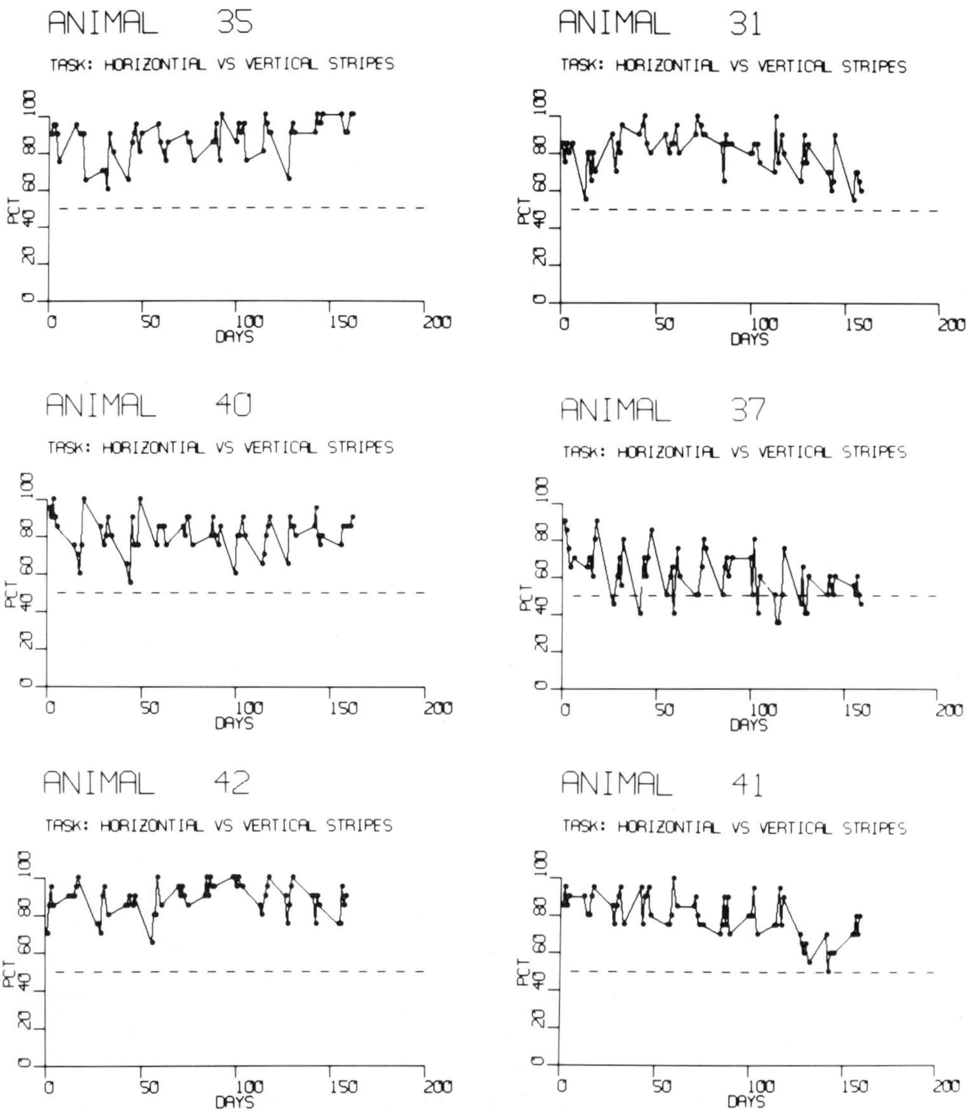

Figure 2: Performance of animals trained and tested on both the pattern and color tasks in the matching-to-sample experiments. The scores shown here are for animals tested on discriminating between horizontal and vertical stripes. Animals 35, 40 and 42 are controls that were housed in cyclic light, and animals 31, 37 and 41 are experimentals that were housed in constant light.

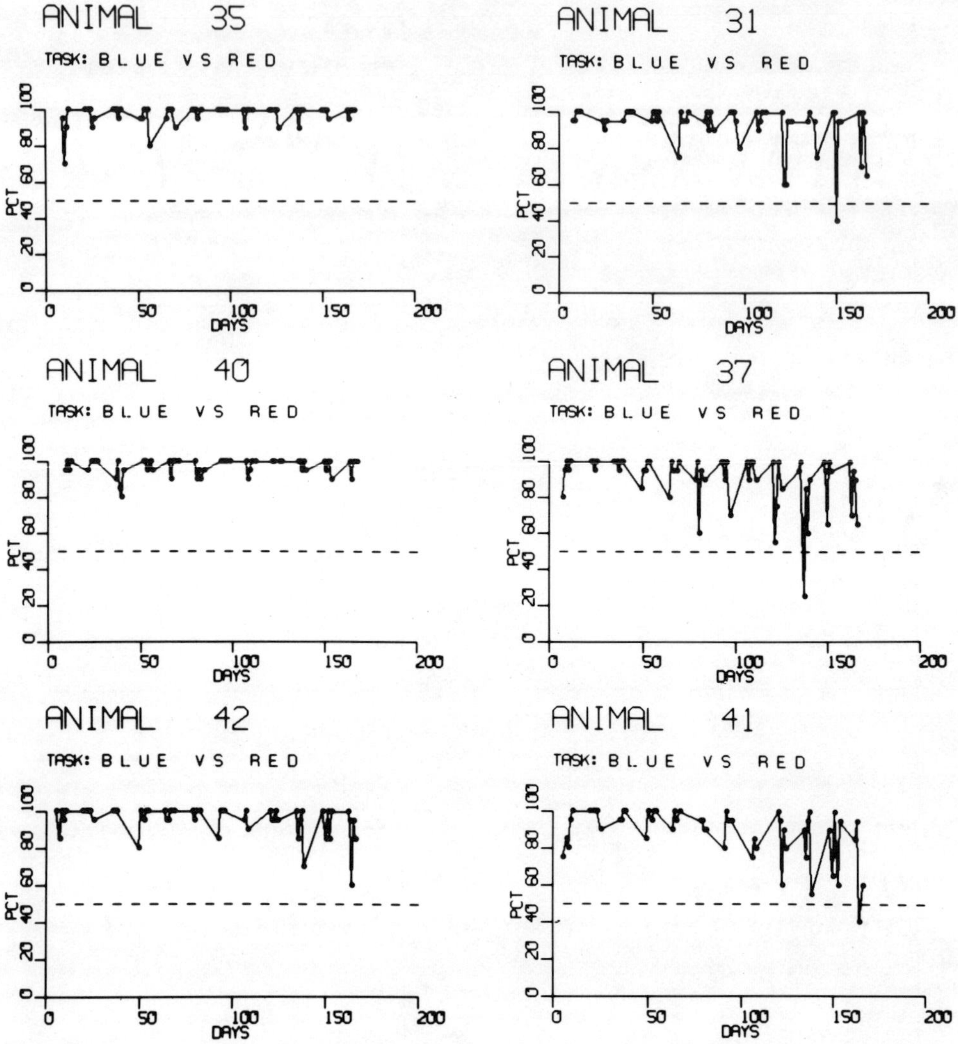

Figure 3: Performance of animals trained and tested on both the pattern and color tasks in the matching-to-sample experiments. The scores shown here are for animals tested for discrimination on between red and blue stimuli. Animals 35, 40 and 42 are controls that were housed in cyclic light, and animals 31, 37 and 41 are experimentals that were housed in constant light.

The performance of the 6 rats on the color task can be seen in Figure 3. As was the case on the pattern discrimination, it is apparent that an animal's level of performance on the blue vs. red color task is related to the length of time each animal was exposed to continuous illumination in the home environment. It can be seen that with increased exposure durations there was an increase in the variability of the scores of the light exposed animals, while the control animal's scores remained stable. Rats 31 and 41 began to show their first serious decrements in performance between 100 and 150 days of exposure, while rat 37 exhibited an inability to perform at criterion levels within the first three weeks of exposure. The performance of rat 37 had begun to approach chance levels by 100 days of exposure, while the performance of rats 31 and 41 did not seriously decline until around 125 to 150 days of exposure.

Histology

At least one eye from each animal used in the behavioral studies was prepared for histological analysis. In all animals, an increasing exposure to constant light produced a progressive decrease in the number of intact photoreceptor cells in the retina. There were also small, but significant, changes in the number of cells in the inner nuclear layer (INL) and in the ganglion cell layer (GCL) that resulted from exposure to constant light.

The retinas of control animals had all of the traditionally described layers of the retina, including layers of well defined inner and outer segments and outer plexiform layers (OPL). In the posterior retina the outer nuclear layer (ONL) was normally about 12 nuclei thick. The ONL decreased in thickness peripherally so that in the extreme periphery it was only 5-6 nuclei thick. The nuclei of cells classified as rods were round and about 5 microns in diameter. The nuclei of cells classified as cones were oval and normally at least 7 microns on the long axis. They also had 3 or 4 small clumps of heterochromatin. Figure 4 shows examples of nuclei that were classified as rod and cone nuclei and also illustrates the general type and extent of light-induced retinal damage that results from exposure to 1, 4, 10, 16, or 24 weeks of continuous illumination at approximately 18 Ft-C intensity.

After one week of exposure to constant light (Fig. 4B), the inner and outer segments of the photoreceptors were disrupted and the distance between the ONL and the pigment epithelium (PE) was decreased from the normal condition. The OPL was still present. The thickness of the ONL had decreased, being only 2-3 nuclei thick in the posterior retina. The retinas from animals exposed to 4 weeks of constant light (Fig. 4C) were qualitatively similar to the 1-week animals. A small bacillary layer of inner segments was

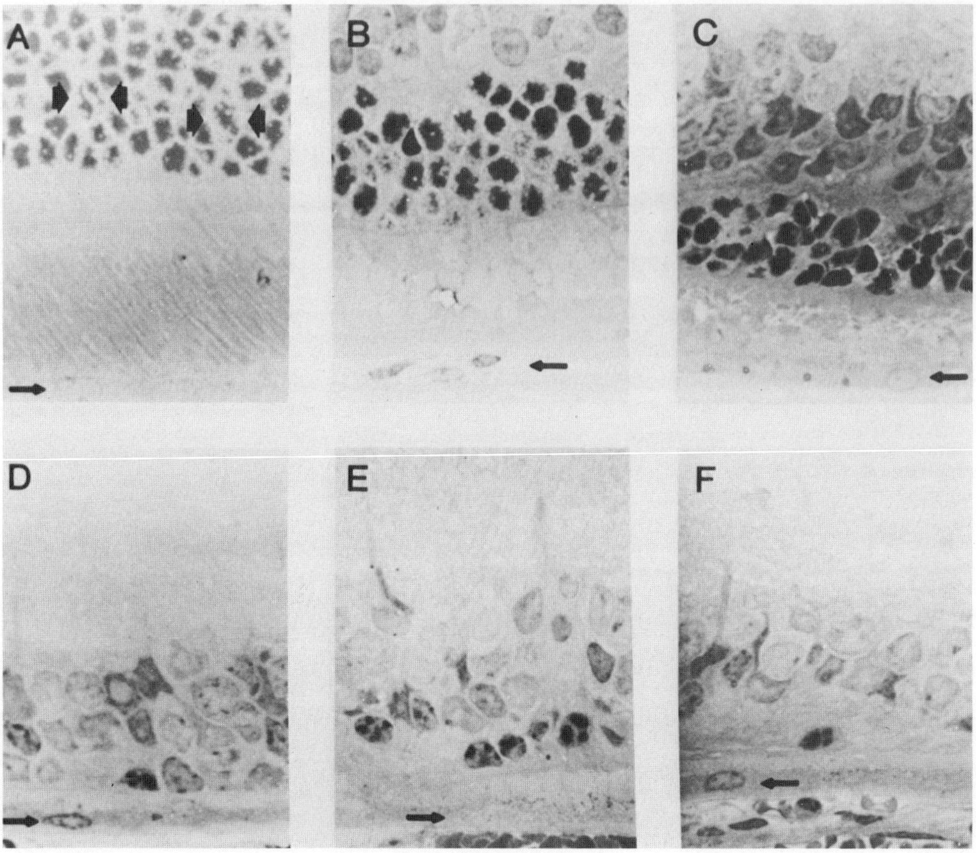

Figure 4: Sections from plastic-embedded retinas. Frames A-F show selected regions of retinas from a control rat and from rats exposed to 1, 4, 10, 16 and 24 weeks of constant light, respectively. The pictures were taken at a magnification of 250X and show sections of retina approximately 1 mm from the ciliary body. The frames were selected to show examples of photoreceptor nuclei at each exposure duration. The long, narrow, black arrows indicate the pigmented epithelium (PE). A: the bold, short arrows indicate two cone nuclei. Note the oval shape of these two nuclei and that they have more than one clump of densely staining heterochromatin. These two nuclei are also oriented perpendicular to the PE. The inner and outer segments of the photoreceptors are well defined and separate the PE from the photoreceptor nuclei. B: the disruption of the inner and outer segments and the decrease in the distance between the PE and the photoreceptor nuclei. C: after 4 weeks of constant light there is extensive disruption of the inner and outer segments. D: one photoreceptor nucleus is seen immediately adjacent to the PE. This nucleus had three small clumps of heterochromatin. E and F

still present, in some animals, and an OPL still separated the remaining photoreceptor nuclei from the INL. With 10 or more weeks of exposure (Fig. 4D-F), there was a marked difference in the condition of the retina. The INL was immediately adjacent to the PE and the few remaining photoreceptor nuclei were scattered along the retina, intimately touching both the PE and the INL. Rarely, when 2 or 3 photoreceptor nuclei were clumped together, there appeared to be a very small plexiform layer between them and the INL. The photoreceptor nuclei were pyknotic, being about 4 microns in diameter and were frequently elongated along a line parallel to the PE. Many of the photoreceptor nuclei had two clumps of heterochromatin, rather than one, as observed in control cone nuclei.

A number of eyes were embedded in paraffin and were examined to determine the number of cells present in the ONL, INL and the GLC. Table 9 shows the number of photoreceptor nuclei in the posterior, intermediate and peripheral retina at each exposure period. The data suggest that the photoreceptor degeneration occurs at a faster rate in the posterior retina than in the periphery, during the first four weeks. However, from 10 to 24 weeks the degeneration appears to proceed at the same rate throughout the retina. The number of cells present at all experimental times were significantly different from the controls at the 0.01 level.

Following long term exposure to constant light, there was a small but significant change in the number of cell bodies in the INL and the GCL. Table 10 shows the mean number of cells in these layers in the intermediate region of the retina at various exposure periods. A one-sided t-test was used to compare the control retinas to those of light-exposed animals. After 24 weeks there were differences from controls that were significant at the 0.05 level in 2 of the 3 regions of the INL and the GCL.

The cell counts of the ONL in the plastic-embedded retinas produced some extremely interesting results. Unfortunately, only one or two retinas at each time point provided usable data for cell numbers in the posterior, intermediate and peripheral retina. The primary reason for this was the extreme difficulty of obtaining the 2 micron sections that included the entire retina from the

illustrate retinas from animals exposed to 16 and 24 weeks of constant light, respectively. No intact photoreceptors can be seen. In fact, few presumed receptor cell remnants can be visualized. Nuclei with more than one clump of heterochromatin can occasionally be observed, but it should be noted that, when observed, these nuclei are not oval, nor are they oriented perpendicular to the PE, as are the cone nuclei in control animals. Thus, it is possible that the nuclei in retinas exposed to prolonged periods of light that contain multiple clumps of heterochromatin are from degenerating rods, rather than cones.

TABLE 9

NUMBER OF NUCLEI IN 200 MICRON LENGTHS OF THE OUTER NUCLEAR LAYER

EXPOSURE DURATION IN WEEKS		0 (CONTROL)	1	4	10	16	24
NUMBER OF EYES EXAMINED		7	6	5	5	5	4
POSTERIOR RETINA							
	MEAN	410	106	73.6	1.34	1.92	1.09
	S.E.M.	32.8	14.04	20.08	.50	.73	.88
	P	-	.001	.001	.001	.001	.003
INTERMEDIATE RETINA							
	MEAN	383	105	55	2.1	2.8	1.7
	S.E.M.	31.3	18.82	20.12	.94	.71	.84
	P	-	.001	.001	.001	.001	.003
PERIPHERIAL RETINA							
	MEAN	327	119	83.3	3.4	2.1	1.2
	S.E.M.	15.8	24.74	23.21	1.01	.63	.70
	P	-	.001	.001	.001	.001	.003

TABLE 10

NUMBER OF CELLS IN 200 MICRON LENGTHS OF THE INNER NUCLEAR LAYER AND THE GANGLION CELL LAYER IN THE INTERMEDIATE RETINA

EXPOSURE DURATION IN WEEKS		0 (CONTROL)	1	4	10	16	24
NUMBER OF EYES EXAMINED		7	6	5	5	5	4
INL	MEAN	128	129	133	115	109	107
	S.E.M.	3.93	9.06	6.26	2.73	8.32	6.05
	P	-	>.1	>.1	.012	.022	.006
GCL	MEAN	15	13	17	15	13	11
	S.E.M.	.75	1.91	.44	1.28	1.68	1.10
	P	-	.042	>.1	>.1	.016	.002

TABLE 11

MEAN NUMBER OF ROD NUCLEI AND CONE NUCLEI IN 100 MICRON LENGTHS OF THE INTERMEDIATE RETINA

EXPOSURE DURATION IN WEEKS	0 (CONTROL)	1	4	10	16	24
NUMBER OF EYES EXAMINED	1	2	1	1	2	2
ROD NUCLEI	145	33	33	0	0	0
CONE NUCLEI	7	10	12	2	1	1

ciliary body to the optic disk. As a result, there were not enough data collected to perform a meaningful statistical analysis. However, when similar regions of the retina were compared, the number of nuclei in the ONL at any particular time point was within the range found in the paraffin embedded eyes. By using the criterion of one clump of densely staining heterochromatin as characterizing a rod nucleus, it was found that, after 10 weeks of constant light, there were almost no rod nuclei remaining. A careful examination of an entire section would usually reveal no more than one or two pyknotic rod nuclei. On the other hand, the counts of presumptive cone nuclei were most remarkable, because in two of the three regions of the retina it was found that there was an apparent increase in the number of cone nuclei present after an exposure period of 4 weeks. Following longer exposure durations, there was a decrease in the number of presumptive cone nuclei. Table 11 lists the number of cone and rod nuclei counted in 100 micron lengths of the intermediate retina.

DISCUSSION

The findings in the maze and matching-to-sample experiments, in our laboratory, have clearly shown that rats can perform at relatively high levels on a variety of visual tasks, even when their retinas have undergone massive amounts of light-induced degeneration. In fact, when rats were exposed to only 30 days of constant light and were tested in a T-maze, they not only retained a visual habit learned prior to the degeneration, but they were able to learn new discriminations at rates that were indistinguishable from control animals. Similar findings resulted in the matching-to-sample studies with short-term exposure periods. In these latter experiments, animals exposed to less than 40 days of constant light performed as well as control animals on several visual tasks, including: intensity discriminations, pattern tasks, color discriminations and a test involving inverted and upright triangles.

In the T-maze experiments, comparisons between time-dependent and time-independent measures of performance were particularly interesting, since they revealed the special importance of experimental design variables in studies where one is rigorously assessing performance on sensory tasks. These studies clearly showed that whether or not one observes a performance decrement in animals with retinal damage depends at least in part upon the type of response measure one uses to assess performance and, equally importantly, upon the nature of the discrimination task itself. For example, in the maze experiments, performance decrements were observed on the pattern task when response latency was used to assess performance, but not when the percentage of correct responses to the visual targets was the primary response measure. In contrast to the findings on the pattern task, response latencies were not elevated in animals with retinal damage when visually guided behavior was tested with a black-white descrimination task. This latter finding suggests that the response-latency measure is sensitive not only to the functional state of the retina during visual performance, but it is also sensitive to the relationship between the state of the retina and the type of visually guided behavior being tested.

Our maze experiments have also shown that the type of experimental design utilized in testing the behavior of animals is a factor that should be carefully considered in evaluating the results of experiments designed to study the relationship between the nervous system and perception. For example, retinal damage led to longer latencies in animals that were trained for the first time after exposure to constant light than in animals that had received training prior to being exposed to continuous illumination. It would seem that experimental design variables are likely to have important consequences whenever one is attempting to study the function of portions of the nervous system. Our maze studies indicate that one should proceed with caution when using behavioral techniques to study the role of the nervous system and should obtain data from groups where each animal has been used as its own control and should also obtain data that would permit performance comparisons to be made between groups of animals.

The results from our matching-to-sample studies on intensity, pattern and color vision showed that exposure to constant light for extremely long periods was required to produce a significant change in the discriminative abilities of albino rats. Performance on the intensity task seemed to be the most resistant to disruption and that on the pattern task was most readily influenced. Surprisingly, the visual task that was the most resistant to influence was the color discrimination. In nearly every animal, at least 125 days of exposure to constant light was required to disrupt behavior on the color task. These results are consistent with those of previous investigators, whose work suggests that very long exposure durations are required to reveal behaviorally determined visual deficits in constant-light exposed rats. For example, Bennet, et al. (5, 25) found a decrease in the ability of rats to make a light-dark discrimination after about 90 days in constant light.

It was particularly interesting, to us, that rats that were able to perform color discriminations were incapable of performing above chance levels on a pattern task. At first glance, this observation might appear to give support to the suggestions of LaVail (26) and Cicerone (27) that residual cone fragments might be mediating vision in light-exposed rats. Neurophysiological and histological studies conducted in our laboratory, however, suggest that cone fragments are not the primary light-transducing element in the retinas of light exposed rats (21, 24). It is obviously of special importance to be able to identify the retinal elements that are capable of transducing light into a neural signal in the eyes of animals that exhibit massive amounts of light-induced degeneration. It must be said, however, that such an identification cannot be made, at this point in time. Future experiments in our laboratory will be designed to help answer two intriguing questions: (1) which anatomical elements play an important role in the transduction of light in retinas without intact photoreceptor cells; and (2) what is the mechanism by which retinal degeneration occurs in animals that have been exposed to periods of constant light?

REFERENCES

1. Anderson, K. V. and O'Steen W. K. (1971) Expl. Neurol. 30, 555-564.
2. Anderson, K. V. and O'Steen, W. K. (1971) Expl. Neurol. 32, 502-505.
3. Anderson, K. V. and O'Steen, (1972) Expl. Neurol. 34, 446-454.
4. Bennett, M. H., Dyer, R. F. and Dunn, J. D. (1972) Expl. Neurol. 34, 434-445.
5. Bennett, M. H., Dyer, R. F. and Dunn, J. D. (1973). Expl. Neurol. 38, 80-89.
6. O'Steen, W. K. (1970) Expl. Neurol. 27, 194-205.
7. O'Steen, W. K. and Anderson, K. V. (1971) Expl. Neurol. 30, 525-534.
8. O'Steen, W. K. and Anderson, K. V. (1971) Z. Zellforsch. 127, 306-313.
9. O'Steen, W. K. and Lytle, R. B. (1971) Am. J. Anat. 130, 227-234.
10. O'Steen, W. K., Shear, C. R. and Anderson, K. V. (1972) Am. J. Anat. 134, 5-22.
11. Grignolo, A., Orzalesi, N., Castelazzo, R. and Vittone, P. (1969) Ophthalmologica, 157, 43-59.

12. Kuwabara, T. and Gorn, A. (1968) Arch. Ophthal. 79, 69-78.
13. Noell, W. K., Walker, V. S., Kang, B. S. and Berman, S. (1966) Invest. Ophthal. 5, 450-476.
14. Anderson, K. V., Coyle, F. P., and O'Steen, W. K. (1972) Expl. Neurol. 35, 233-238.
15. Ham, W. T., Mueller, H. A. and Sliney, D. H. (1976) Nature, 260, 153-155.
16. Moon, M. E., Clarke, A. M., Ruffolo, J. J., Mueller, H. A. and Ham, W. T. (1978) Vis. Res. 18, 1573-1577.
17. Gorn, R. A. and Kuwabara, T. (1967) Archs. Ophthal. 77, 115-118.
18. Noell, W. K. and Albrecht, R. (1971) Science, 172, 76-80.
19. Kaitz, M. and Auerbach, E. (1977) Ophthal. Res. 9, 351-356.
20. Rose, G. H. and Gruenau, S. P. (1973) Dev. Psychobiol. 6, 69-77.
21. Lemmon, V. and Anderson, K. V. (1979) Expl. Neurol. 63, 50-75.
22. Walls, G. (1942) The Vertebrate Eye, Cranbrook Institute of Science, Bloomfield Hills, Michigan.
23. Gellerman, L. W. (1933) J. Genet. Psychol. 42, 206-208.
24. Lemmon, V. and Anderson, K. V. (1979) Expl. Neurol. 63, 35-49.
25. Bennet, M. H., Dyer, R. F. and Dunn, J. D. (1973) Expl. Neurol. 40, 652-660.
26. LaVail, M. M. (1976) Invest. Ophthal. 15, 64-70.
27. Cicerone, C. M. (1976) Science, 194, 1183-1185.

NEUROPHYSIOLOGICAL PROPERTIES OF VISUAL NEURONS IN RATS WITH LIGHT DAMAGED RETINAS

Kenneth V. Anderson, Vance Lemmon* and W. Keith O'Steen†

Department of Anatomy
Emory University
and
*Department of Physiology and Biophysics
Washington University
and
†Department of Anatomy
Bowman Gray School of Medicine

In recent years it has been shown by a number of investigators that the photoreceptors in albino rats undergo an irreversible degeneration as a result of long-term exposure to continuous illumination. In fact, by adjusting the intensity of the light source and varying the spectral properties of the light, it has been shown that the retinal degeneration can be experimentally graded in both amount and spatial extent (1-8). In our laboratory we have been interested in the perceptual and neurophysiological implications of light-induced retinal degeneration. Accordingly, we have pursued a broad program of research to help clarify the relationship between light-induced retinal degeneration and visually-guided behavior and to relate retinal and perceptual changes to alterations in the response properties of neurons in various portions of the visual system.

Our behavioral experiments have been described in another chapter in this book. In general, our behavioral studies have shown clearly that animals with extensive, light-induced retinal degeneration are not, as one might expect, blind. In fact, in both T-maze and matching-to-sample experiments, rats with massive retinal degeneration have been shown to be capable of performing at high levels on a variety of visual tasks, including: flux discriminations, pattern tests and color discriminations. It does appear, however, that rats become progressively more perceptually debilitated as they are exposed to longer and longer periods of continuous illumination. In the matching-to-sample studies, for example, animals exposed to constant light first began to show difficulties in performing pattern tasks. After exposure to 150 to 200 days of constant light, most animals were beginning to show lowered performance levels on flux and color discriminations.

Our electrophysiological studies have been designed to correlate changes in the response properties of visual neurons with progressive changes in retinal structure during exposure to various periods of constant illumination. Naturally, we have been most interested in relating observed neurophysiological changes to the decrements in visual performance that one observes in animals that have been exposed to prolonged periods of constant light. Our neurophysiological investigations have involved both macroelectrode and microelectrode recording methods and have been directed toward an analysis of three types of neuronal activity in the visual system of albino rats: (1) slow-wave potentials that occur spontaneously, (2) visually evoked slow-wave potentials, and (3) spontaneous and evoked single unit activity.

RESULTS

Experiments Measuring Spontaneous Slow-Wave Potentials

These investigations were among the first neurophysiological studies we performed on rats with retinal degeneration. We chose to study the rhythmic, slow-wave potentials, as a first approach to correlating visual function with retinal changes, because they are so pervasive in the visual system of several species of mammals, and because of their remarkable regularity and the ease with which the features of the potentials can be quantitatively characterized. As time went on, we realized that another, perhaps more important, reason related to the possibility that the study of spontaneous, rhythmic potentials may provide important clues regarding the neuronal mechanism by which slow-waves themselves are generated.

In general, rhythmic potentials have been demonstrated most readily in animals that have been injected with various kinds of anesthetics. The drugs that have been most effective in producing spontaneous rhythms are the hypnotic barbiturates, including hexobarbital, pentobarbital, phenobarbital and thiopental (9). The general features of spontaneous rhythmic potentials have depended both upon the species of mammal used in the investigation and the region of the brain from which potentials were recorded. For example, in the red nucleus of the rabbit (10) and cat (11) potentials have been found that range in frequency from 40 to 30 Hz. In contrast, rhythmic potentials with relatively narrow range of from 8 to 15 Hz have been observed in cats in the ventral tegmental region of Tsia and the lateral habenular, entopeduncular, and subthalamic nuclei (12, 13, 14).

Rhythmic potentials recorded in the visual system seem to be the most complicated of those reported to date, in that several kinds of rhythms have been clearly identified. In the visual systems of cats (15) and monkeys (16), rhythmic potentials with a frequency of 3-16 Hz have been recorded from the optic nerve, optic tract, and lateral geniculate nuclei. Rhythmic potentials recorded in the visual system of albino rats have a broader range of frequencies than those found in cats and monkeys and have been found in more widespread regions of the visual system (1, 2, 5, 17, 18). In our studies, the most predominant potential had a frequency of 9-40 Hz and could be recorded from the optic tract, lateral geniculate nuclei and visual cortex. A second kind of potential had a frequency of 90-220 Hz and was most commonly found in the optic tract and lateral geniculate nuclei. A third kind of potential had a frequency of 0.3-0.5 Hz and was restricted almost exclusively to the visual cortex.

In all of our studies of spontaneous, rhythmic potentials, the rats were anesthetized with pentobarbital. Animals with light-induced retinal degeneration were exposed to constant light for varying periods of time ranging from four days to more than 300 days, although 30 days of exposure was generally sufficient to disrupt or abolish all signs of spontaneous, rhythmic potentials in the rat visual system. In all of our studies of rhythmic potentials, recordings in the central nervous system were made with concentric, bipolar electrodes with a tip-to-barrel separation of 0.5 mm. In a few instances, cortical recordings were made with silver-ball electrodes. Recordings from the eye were made with silver-ball or wick electrodes.

Figure 1 shows examples of the relative kinds of rhythmic or oscillatory potentials that were recorded in control rats anesthetized with pentobarbital. Frame A shows an example of one of the faster of the three basic rhythmic potentials recorded in our experiments. The particular rhythm in A is approximately 120 Hz and, thus, falls into the 90-220 Hz category. Frame B shows a rhythm of 15 Hz and illustrates the 9-40 Hz range observed in our experiments. Frame C illustrates a 0.4 Hz rhythm and represents an example of the slowest basic rhythm found in our studies.

Interestingly, the presentation of a continuous light stimulus did not significantly influence either the frequency or amplitude of the rhythmic potentials, as can be seen in Fig. 2 by comparing frames A and B. Flashes of light, however, abolished the rhythms at all levels of the visual system, but only during the interval when an evoked potential to the flash was present, as can be seen in frame C. Following the completion of an evoked potential, the rhythms returned. The simultaneous presentation of continuous light and flashes disrupted all three kinds of rhythms and in some instances nearly abolished them, as can be seen in frame D of Fig. 2.

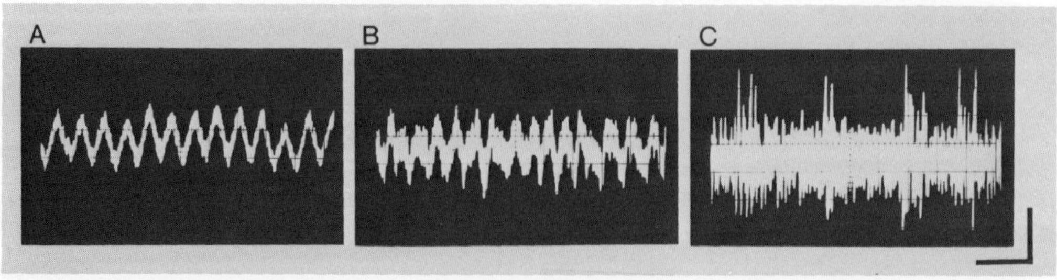

Figure 1. Three kinds of spontaneous, rhythmic potentials recorded from the visual system of control rats. Frame A illustrates one of the faster potentials we observed and measures about 120 Hz. Frame B shows the 9-40 Hz rhythm by illustrating a 15 Hz oscillatory potential. Frame C illustrates a 0.4 Hz rhythm and is an example of one of the slowest potentials observed in our studies. The time calibration represents 20 msec in A, 200 msec in B, and 2 sec in C. The amplitude calibration is 100 μV.

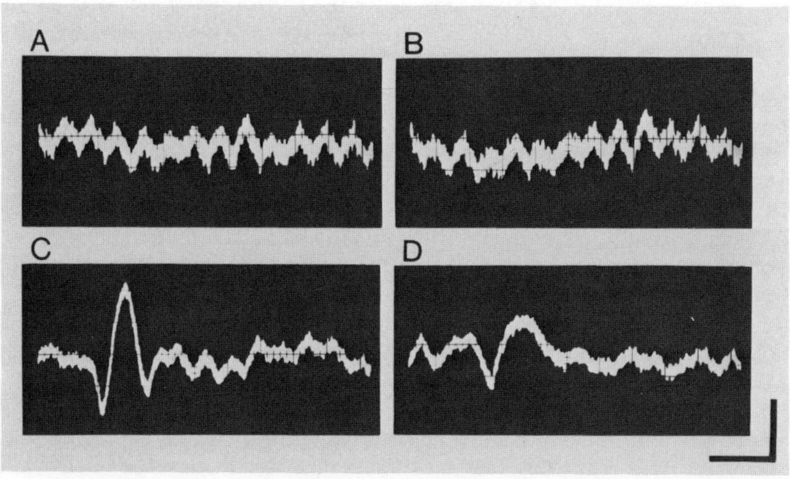

Figure 2. Spontaneous, rhythmic potentials in the 9-40 Hz range recorded from the LGN of two control rats. A and B show a 12 Hz potential when the ambient light level in the recording room was low and when continuous light was shone directly on the animal's eyes, respectively. Note that exposing the retina to brief episodes of continuous light did not alter the characteristics of the rhythmic potential. C and D show the effects of flashes on rhythmic potentials. In C flashes were presented when the ambient light level was low and in D when it was high. Note that the rhythmic potentials were abolished only during the interval when an evoked potential to the flash was present. The time calibration represents 200 msec in A and B, and 1000 msec in C and D. The amplitude calibration is 100 μV.

Somewhat surprisingly, removal of the retinas, via enucleation, abolished the rhythmic potentials throughout the visual system, indicating that the potentials somehow originated in the eye.

In order to evaluate the effects of various degrees of retinal degeneration on the integrity of rhythmic visual potentials, albino rats were exposed to continuous light for periods of 4 days, 14 days, or 30 days. As can be seen in Fig. 3, these exposure intervals produced a retina that had degenerative features that ranged from modest to significant. The constant light environment had an illumination, measured as reflected light from the cage floor, of 18 Ft-C (=194 lux) from two GE(F15T8.CW) 15-w fluorescent tubes (waves of 380-750 nm, peaking at 600 nm), located 40 cm from the top of the animal's cage.

Figure 4 shows the effects of light-induced retinal degeneration on rhythmic potentials recorded from the surface of the eye. Frame A shows a well-defined oscillatory potential with a frequency of about 190 Hz that was recorded from a control rat with an intact retina. B, C and D illustrate recordings made from the eyes of rats exposed to constant light for 4, 14 and 30 days, respectively. As can be seen, the characteristics of spontaneous activity recorded from the eyes of animals exposed to constant light depended upon the duration of exposure and upon the amount of retinal degeneration caused by the exposure. After 4 days of exposure, at a time when the outer segments of receptor cells were badly disrupted, all three kinds of rhythmic potentials found in control rats were abolished (Fig. 4B). Nonrhythmic activity, however, appeared to be relatively normal at this time period. After 14 or 30 days of exposure, when retinas had undergone massive degenerative changes, rhythmic potentials were completely absent and nonrhythmic activity was either absent or only occasionally observed (Fig. 4C and D).

The changes in rhythmic potentials recorded in the central nervous system are shown in Figs. 5, 6 and 7. Each figure has the same organizational features and shows a rhythmic potential recorded from a control rat in frame A, and potentials recorded from the optic tract (OT), lateral geniculate nucleus (LGN) and visual cortex, respectively. Figure 5 shows the results after 4 days of exposure, and Figs. 6 and 7 after 14 and 30 days of exposure, respectively. As can be seen in Fig. 5, the effects of 4 days of exposure on potentials in the OT, LGN and VC were similar to those observed in recordings taken from the eye. That is, animals in this group exhibited an absence of rhythmic potentials, although spontaneous, nonrhythmic activity was readily observed throughout the central visual system. The presence of nonrhythmic activity at this time period probably reflects the fact that degenerative changes in the receptor cell layer of the retina are just getting well under way after four days of exposure. For example, the retinal changes after 4 days of exposure consist of a developing destruction of the

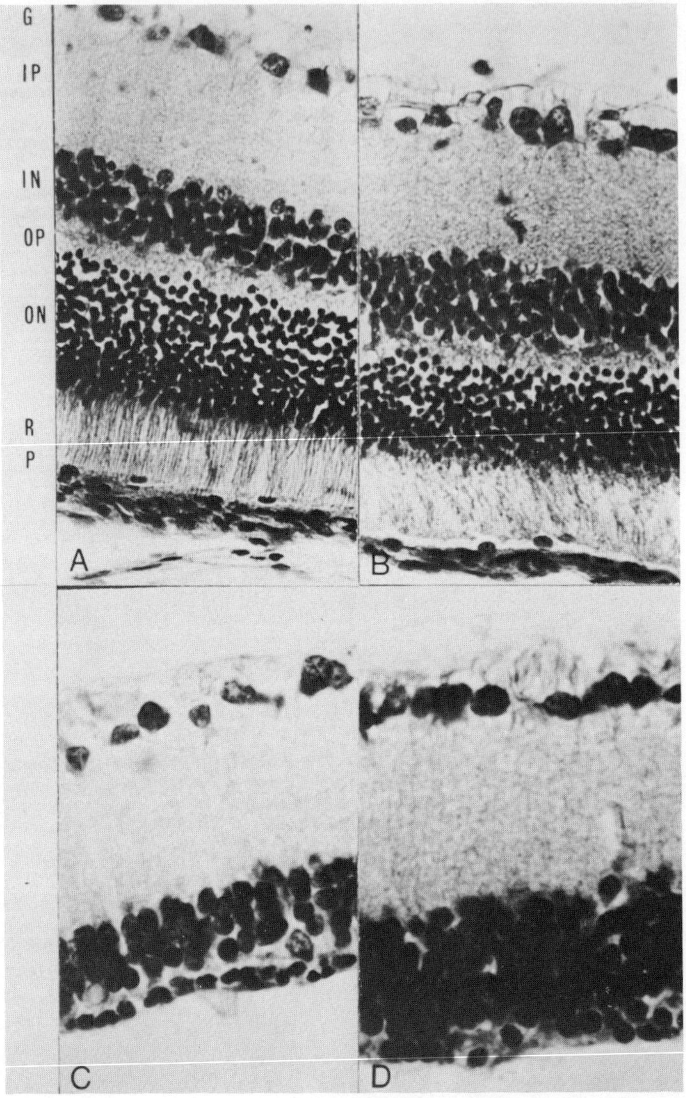

Figure 3. Frame A shows a control retina exposed to cyclic light. B. Retina from a rat exposed to constant light for 4 days. Note that light damage appears to be confined to fragmentation of the inner and outer segments in the receptor cell layer. C. Retina from a rat exposed to continuous light for 14 days. Pyknotic nuclei (arrow) are receptor cell nuclei. D. Retina from a rat exposed to 30 days of constant light. Layers of the retina are: G, ganglion cell; IP, inner plexiform, IN, inner nuclear; OP, outer plexiform; ON, outer nuclear; R, receptor cell; P, pigment cell layer. Magnification: A and B, 400X; C and D, 1,000X. (Reduced 10% for purposes of reproduction.)

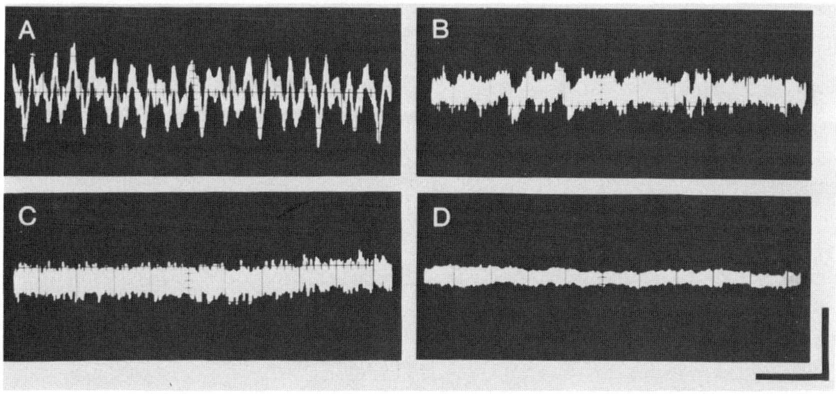

Figure 4. Frame A shows a well-defined oscillatory potential recorded from the eye of a rat with an intact retina. B, C and D are recordings made from the eyes of rats exposed to constant light for 4, 14 and 30 days, respectively. Note the absence of rhythmic potentials in B, C and D. Note also the presence of nonrhythmic spontaneous activity in B and, to some extent in C, but its absence in D after rats have been exposed to 30 days of continuous illumination. The time calibration represents 20 msec. The amplitude calibration represents 100 μV in A and 50 μV in B, C and D.

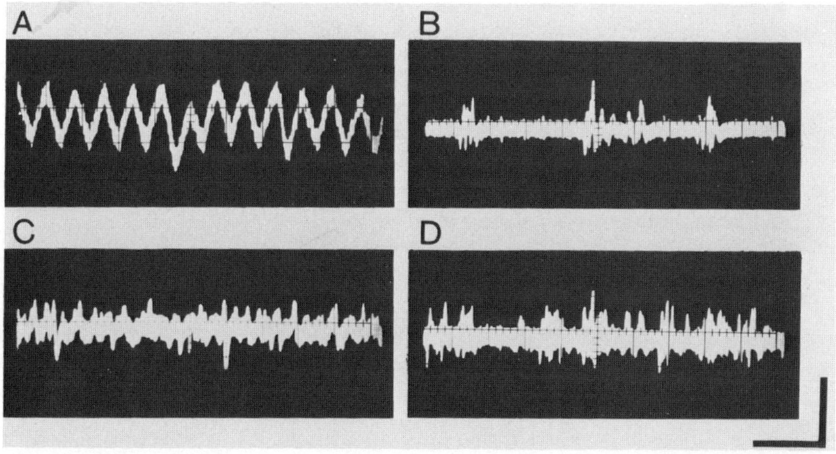

Figure 5. Frame A shows a well-defined oscillatory potential recorded from the visual cortex of a rat with an intact retina. B, C and D are recordings made from the OT, LGN and VC, respectively, of a rat with modest retinal degeneration following exposure to 4 days of continuous illumination. Note the absence of well-defined, rhythmic potentials in B, C and D, although spontaneous, nonrhythmic activity is prevalent. The time calibration represents 20 msec in A and 200 msec in B, C and D. The amplitude calibration is 100 μV.

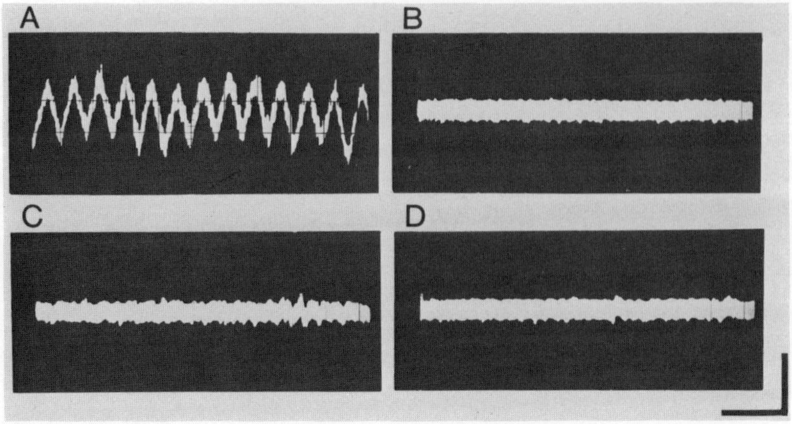

Figure 6. Frame A shows a well-defined, rhythmic potential recorded from the VC of a rat with an intact retina. B, C and D are recordings made from the OT, LGN and VC, respectively, of a rat with severe retinal degeneration following exposure to 14 days of continuous light. Note the absence of both rhythmic and nonrhythmic activity in B, C and D. The time calibration represents 20 msec in A and 200 msec in B, C and D. The amplitude calibration is 100 μV.

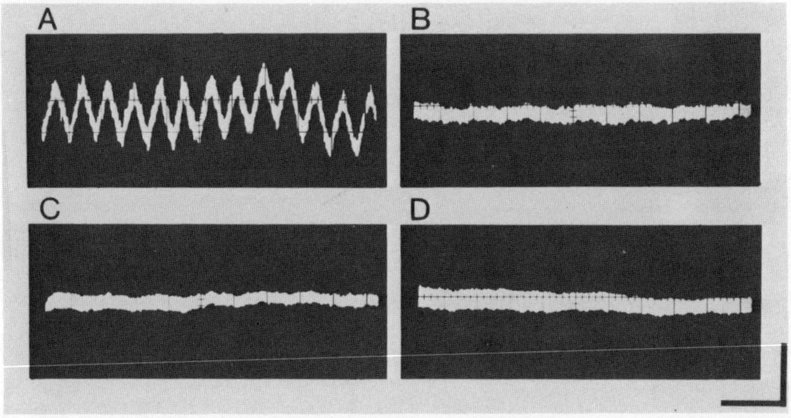

Figure 7. A. Well-defined rhythmic potential recorded from the VC of a rat with an intact retina. B, C and D are recordings made from the OT, LGN and VC, respectively, of a rat with extensive retinal degeneration following exposure to 30 days of continuous illumination. Note the absence of both rhythmic and nonrhythmic activity in B, C and D. Note also that the base lines in B, C and D are thinner than in Figs. 5 and 6, suggesting a generalized reduction in the amount of neural activity in the visual system. The time calibration represents 20 msec in A and 200 msec in B, C and D. The amplitude calibration is 100 μV.

outer segments of the receptor cells and pyknosis of some nuclei in the outer nuclear layer. While this incomplete destruction is sufficient to block the production of rhythmic potentials, it is apparently not extensive enough to abolish other types of retinally originated neural responses.

After 14 to 30 days of exposure to constant light, spontaneous, rhythmic potentials became undetectable in the OT, LGN and VC. In addition, there were few signs of any kinds of neural activity, at least as recorded with macroelectrodes. These substantial changes in spontaneous activity in central visual structures would seem to correlate fairly well with the developing destructive changes occuring in the retina between 14 and 30 days, as can be seen in Fig. 3. While the retinas of rats exposed for 4 days did not reveal significant destruction, the retinas exposed for 14 to 30 days did. For example, after 14 days of exposure, retinas exhibited severe damage in the receptor cell layers, particularly in the posterior region of the eye (Fig. 3C). Receptor cells were absent from this region, but were still occasionally present in the more rostral and peripheral areas of the retina. Following 30 days of exposure, however, all receptor cells in the posterior eye had disappeared and the inner nuclear layer was in close proximity to the choriocapillary coat throughout the retina (Fig. 3D). In general, the two layers were separated only by a thin zone of glial tissue.

In the course of our studies of rhythmic potentials in rats we came to wonder how well they might be synchronized in visual structures on both sides of the brain and what the relationship might exist between the slow-wave potentials and unitary activity occurring in individual neurons. To pursue these questions we chose to examine the characteristics of the oscillatory activity of the superior colliculi. We chose the superior colliculi for study for two reasons. First, our preliminary studies suggested that rhythmic potentials might be particularly well developed in these nuclei and it seemed that this fact might make it more likely that we could simultaneously record, from both sides of the brain, both slow-wave and unitary potentials. Second, the colliculi receive direct projections from the retina and we felt that it might be important to study central neurons driven by the retina, since the various rhythms apparently originated in the retina.

Figure 8 illustrates some of the general features of rhythmic slow-wave potentials recorded from the superior colliculi. In all animals, the rhythmic potentials were dramatically evident; they were of large amplitude (usually more than 100 μV) and remained present for many minutes, sometimes even hours, at a time. The frequency of the potentials ranged from 9-26 Hz. The recordings shown in Fig. 8 dramatically indicate the degree to which the spontaneous, rhythmic potentials in the left and right colliculi

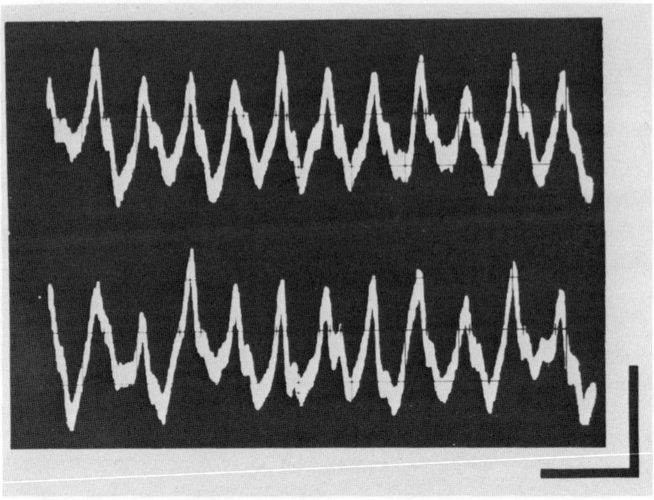

Figure 8. The upper and lower traces show well-defined, 22 Hz rhythmic potentials recorded simultaneously from the left and right superior colliculi, respectively, of a single rat. Note the phase-locked relationship between the two rhythmic potentials. The time calibration represents 50 msec and the amplitude calibration is 100 μV.

were synchronized with one another. In the experiment from which these tracings were derived, concentric, bipolar electrodes were placed into mirror-image positions in both the right and left superior colliculi, and simultaneous recordings were made of the slow-wave potentials generated in these regions. As can be seen clearly, the rhythmic potentials occurring in the two colliculi were in close, if not perfect, phase with each other.

Single cell recordings revealed that numerous collicular neurons were responding rhythmically, with discharge frequencies generally coinciding with those established for collicular slow-wave potentials (i.e., 9-26 Hz). Of 226 units studied in the colliculi, 174 showed rhythmic discharges. All units that showed spontaneous activity also exhibited a vigorous response to flashes of light and/or other types of visual stimuli, such as moving patterns. In most instances, collicular neurons responded in bursts of two or three discharges at at time. When this was the case, the interburst period was used to determine the frequency of rhythmicity of unit responding. Figure 9 shows the close relationship between the response patterns of single collicular neurons and the rhythmic, slow-wave potentials. In this experiment, slow-wave and single-cell discharges were recorded simultaneously through bipolar and pipette electrodes inserted into adjacent regions of the right and left colliculi. For the left and right colliculi, respectively, frames

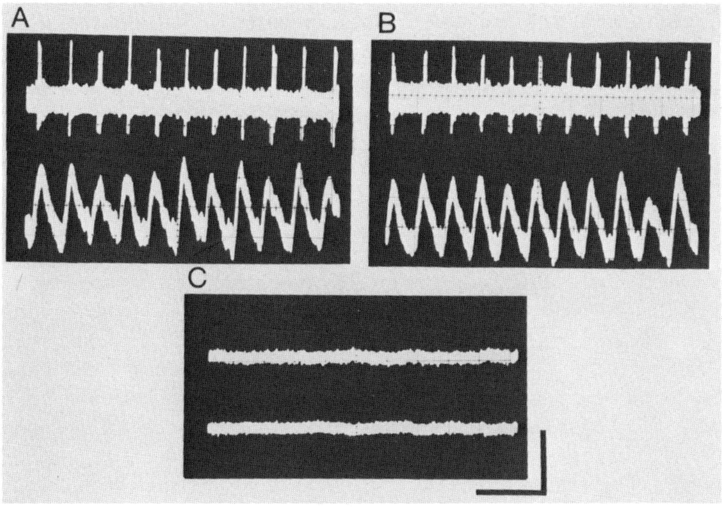

Figure 9. Shows the close relationship between the response patterns of single collicular neurons and the rhythmic slow-wave potentials. Frames A and B show collicular slow-wave rhythms (bottom traces, 22 Hz) and single cell bursts (top traces, 22 Hz) recorded from the left and right colliculi, respectively. C shows abolition of both slow-wave and unit rhythmicity in the left superior colliculus immediately after enucleation of the right eye. All records made from a single rat. The time calibration represents 50 msec, and the amplitude calibration is 200 μV for the top traces and 100 μV for the bottom traces.

A and B show collicular slow-wave rhythms (bottom traces) with a frequency of 22 Hz and associated single-cell bursts (top traces) with interburst periods also corresponding to 22 Hz. Frame C in Fig. 9 shows that removal of the right (or contralateral) eye immediately abolishes both slow-wave and unit rhythmicity in the left colliculus, indicating the retinal origin of the rhythmic potentials. While not shown, removal of the left eye abolished rhythmic activity in the right colliculus.

Studies of Evoked Slow-Wave Potentials

In our investigations involving evoked, slow-wave potentials in the rat visual system, we analysed electrophysiological activity recorded from a variety of loci, including: the retina, the optic tract, the superior colliculi, the lateral geniculate nuclei (both dorsal and ventral divisions) and the visual cortex. The experiments directed toward an analysis of retinal potentials had a twofold purpose. First, they were designed to enable us to determine the extent to which retinal potentials deteriorated as a function of exposure to constant light, and, thus, as a function of the extent of retinal degeneration. Second, they were designed to permit us to determine whether or not rats whose electroretinograms (ERGs) deteriorated with exposure to constant light could recover, if the animals were placed for various periods of time in constant darkness.

In order to study the deterioration rate of ERGs, retinal potentials were evaluated in animals that had been exposed to constant light for periods ranging from one to 14 days. All animals exposed to constant light had some deteriorations of the ERG, the extent of which appeared to be correlated with the length of time the animals were exposed to light. Rats maintained in continuous illumination for 24 hours had a 40-50% decline in the amplitude of the retinal a-wave and b-wave, while animals kept in constant light for longer periods had progressively larger declines in ERG amplitude. The a-waves were absent in animals exposed to constant light for 8 or more days, and b-waves were only barely detectable after 8 days of exposure. Animals kept in constant light for 10 or more days showed no ERG at all. The details of the temporal progression of ERG changes with increasing exposure times are shown graphically in Fig. 10 and quantitatively in Table 1. The typical features of retinal a- and b-waves in control animals can be seen in frame A of Fig. 10. Frames B, C, D, E and F illustrate the

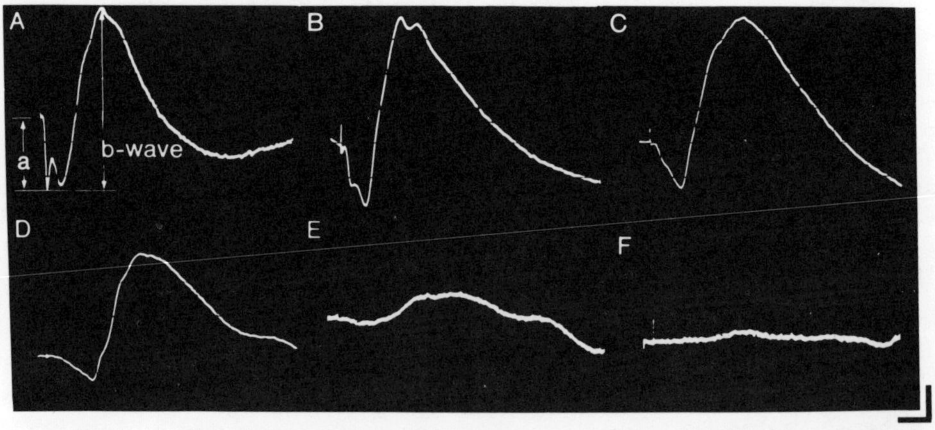

Figure 10. Illustrates the progressive deterioration of the ERG in rats exposed to continuous light. Frame A shows the ERG of a control animal maintained in a cyclic light environment. B-F are ERGs from animals exposed to constant light for 1, 2, 3, 4 and 6 days, respectively. The time calibration represents 20 msec and the amplitude calibration represents 200 μV in A and 100 μV in B-F.

TABLE 1

AVERAGE CHANGE IN THE AMPLITUDE OF THE a-WAVE AND b-WAVE COMPONENTS OF THE ERG AS A FUNCTION OF DURATION OF EXPOSURE TO LOW LEVEL CONTINUOUS ILLUMINATION (n = 43)

	Exposure Time (hours)						
	Control	24	48	72	96	144	196
"a-wave" (μV) ± SD	473 ± 29	200 ± 48	116 ± 47	80 ± 18	None	None	None
	n = 4	n = 4	n = 6	n = 7	n = 12	n = 5	n = 5
"b-wave" (μV) ± SD	1170 ± 42	700 ± 174	412 ± 153	426 ± 218	117 ± 49	60 ± 46	17 ± 5
	n = 4	n = 4	n = 6	n = 7	n = 12	n = 5	n = 5

changes in ERG potentials that occur after exposure to one, 2, 3, 4 and 6 days of constant light, respectively. As can be seen, both the latency and waveform of the various components of the ERG are radically altered after only 3 days of exposure and, by eight days of exposure the a-wave has disappeared and only a hint remains of a b-wave.

To determine the extent to which deteriorated ERGs could recover, we first exposed animals to 1-14 days of constant light, then placed the animals in constant darkness for 13 days, and then compared the ERGs taken after light exposure with those obtained after subsequent exposure to darkness. In general, animals that were maintained in darkness for 13 days, following prior exposure to constant illumination, exhibited partially recovered ERGs. As might be expected, the amount of recovery was related to the length of time the animals were kept in constant light. Animals kept in constant light for more than 4 days showed little or no recovery of the ERGs after being maintained in 13 days of darkness. Thus, it would appear that whatever retinal damage occurs after three to four days of light exposure incapacitates the retina, at least as measured *via* ERG methods, in a relatively permanent manner. The details of ERG recovery characteristics are shown graphically in Fig. 11 and quantitatively in Table 2. In Fig. 11, each column shows a sequence of 3 ERGs from the same subject. Frames A, D and G are recordings made from the 3 animals prior to their exposure to constant illumination. B, E and H are ERGs from the three animals following exposures of 1, 3 and 4 days, respectively. C, F and I are the ERGs after the 13 days of recovery in the darkened environment, which followed the previous light exposure of 1, 3 and 4 days. As one can see, even three days of exposure to constant light significantly alters the latency and waveform characteristics of the ERG, although recovery appears to be rather good after a period in darkness. The animal exposed to 4 days of light shows virtually no recovery, as can be seen by comparing frames G, H and I in Fig. 11. It would be interesting to know whether periods of time significantly longer than 13 days in darkness might help to promote additional recovery in animals exposed to 4 or more days of constant light.

In addition to the evoked, slow-wave potentials recorded from the eyes of animals exposed to constant light, we also studied the changes in electrophysiological activity in various central visual areas. Recordings were made with the same type of concentric, bipolar electrodes described earlier in animals that were exposed to cyclic light conditions (control animals) or to periods of constant light that ranged from 1-30 days. In general, the reduction in the amount of neurophysiological activity in light-exposed animals mirrored that seen in the studies of spontaneous, rhythmic activity and in the investigations of ERG potentials. That is,

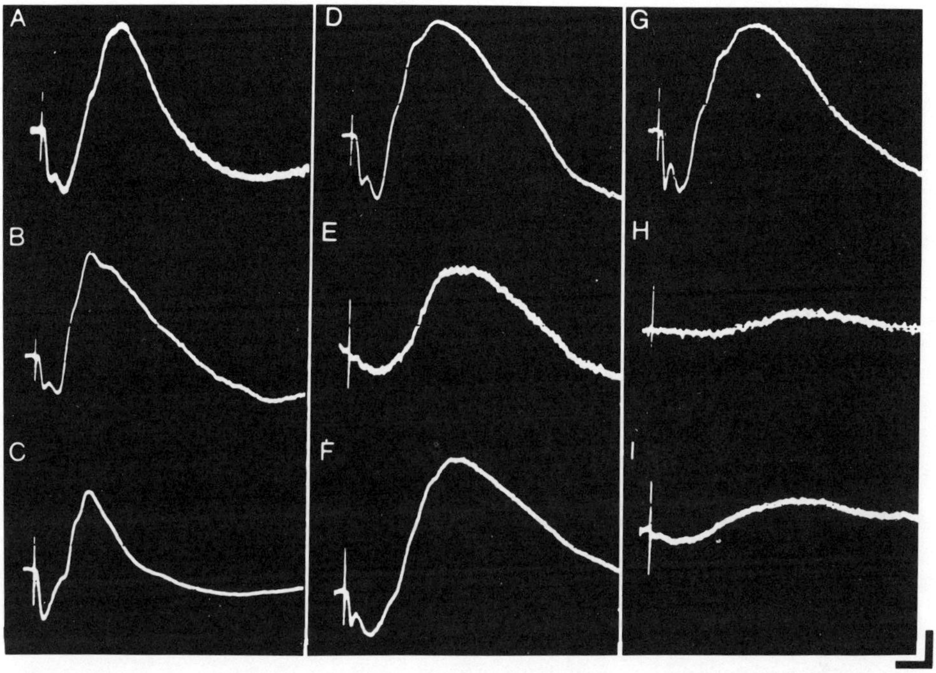

Figure 11. Examples of the recovery of the ERG in rats housed in a darkened environment following exposure to constant light. Each column shows a sequence of 3 ERGs from the same subject. A, D and G are recordings made from the 3 animals prior to their exposure to constant illumination. B, E and H are ERGs from the 3 animals following exposures of 1, 3 and 4 days, respectively. C, F and I are the ERGs after 13 days in a darkened environment, which followed light exposure of 1, 3 and 4 days. The time calibration is 20 msec. The amplitude calibrations are: 500 μV in A and C; 200 μV in B, D, F and G; and 100 μV in E, H and I.

TABLE 2

CHANGES IN COMPONENTS OF THE ERG OCCASIONED BY A RETURN TO A DARKENED ENVIRONMENT FOLLOWING EXPOSURE TO CONTINUOUS, LOW LEVEL, ILLUMINATION

Days of Exposure to Constant Light	Animal	Amplitude of ERG Components (in μV): After Exposure to Constant Light "a-wave"	"b-wave"	After 13 Days in Dark "a-wave"	"b-wave"
1	63	240	820	600	1500
2	66	95	350	340	900
2	71	100	380	110	440
2	73	95	400	200	620
3	48	100	320	280	1120
3	54	40	220	0	85
3	59	75	350	125	420
3	60	80	340	140	500
3	64	80	270	80	120
3	72	55	110	40	150
4	49	0	130	0	130
4	61	0	100	25	80
4	62	0	100	40	125
4	69	0	100	0	50
4	70	0	150	60	190

while the evoked responses in the visual system of rats that had been exposed to as little as 4 days of light varied somewhat according to the recording site in the brain, they were typically of low amplitude and exhibited long latencies. In addition, extensive changes in the waveform of the evoked potentials was observed. After 14 days of exposure to light, it was somewhat difficult to find clear signs of evoked visual activity, whether one was recording from the OT, the LGN or the VC. When activity was observed, however, it rarely exhibited the characteristics of control responses, usually having radically altered waveforms and showing latencies that were greater than control latencies by at least 100%. Figure 12 shows the effects on evoked, visual potentials of 30 days of exposure to constant light at 18 Ft-C intensity. Frames A, C and

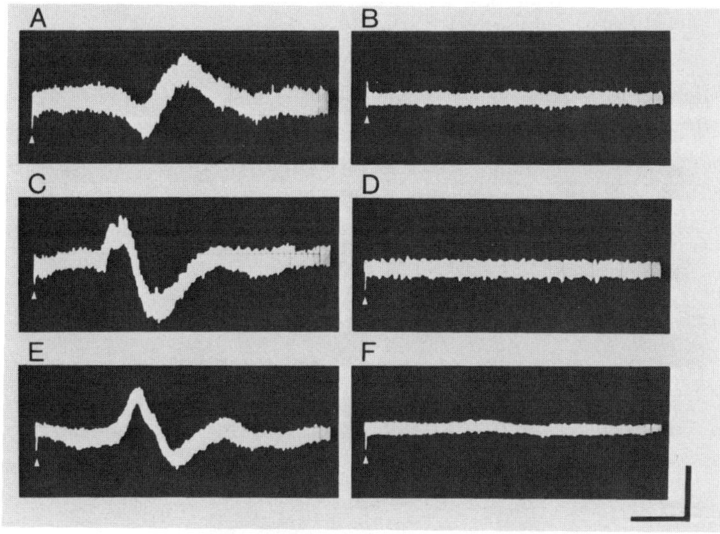

Figure 12. Evoked slow-wave potentials recorded from the OT, LGN and VC in control rats (A, C and E, respectively). The time calibration represents 10 msec in A and B and 20 msec in C-F. The amplitude calibration is 100 μV. Three successive traces are superimposed in each frame to show the response variability.

E show potentials recorded from the OT, LGN and VC in control rats, respectively, and frames B, D and F show analogous recordings made in animals exposed to 30 days of constant light. Three successive traces are superimposed in each frame to show the response variability.

Even though flashes of light failed to evoke significant activity in the visual system of animals exposed to 14-30 days of constant light, this failure was not due to neuronal incapacities within the brain itself. Rather, it seemed to simply reflect the fact that the extensive retinal degeneration in animals exposed to 14 or more days of light impaired the ability of the retina to generate visual signals in a way that could be detected with macroelectrode recording methods. The fact that central visual neurons still function well in animals exposed to 30 days of constant light was determined in a number of rats by comparing potentials recorded in the LGN and VC following stimulation of the OT with similar potentials obtained from control rats. Such an experiment is illustrated in Fig. 13. This latter figure shows evoked potentials recorded in the LGN and VC, respectively, after electrical stimulation in the OT in control rats, in frames A and C, and in rats exposed to 30 days of constant light, in frames B and D. As can be seen, the waveform and relative time course of both the LGN and VC potentials are unchanged in the

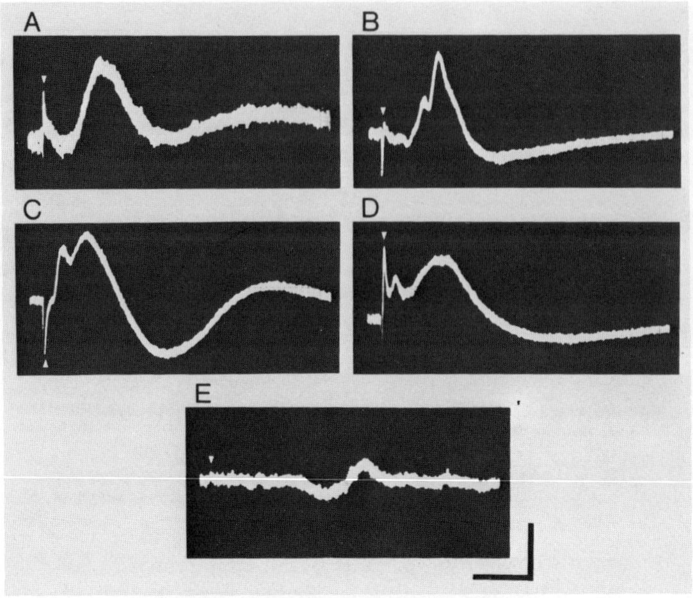

Figure 13. Evoked, slow-wave potentials recorded in the LGN and VC, respectively, after electrical stimulation in the OT in control rats (A and C) and in rats exposed to 30 days of continuous light (B and D). Frame E represents an evoked potential of longer latency and of smaller amplitude than that in control rats, recorded from the VC of a rat exposed to 30 days of constant light. The time calibration represents 10 msec in A and B; 4 msec in C and D; and 20 msec in E. The amplitude calibration is 100 μV for all frames.

animal with massive retinal degeneration. Frame E of Fig. 13 shows one of the few evoked responses, to a flash, observed in the brain of animals exposed to 30 days of continuous illumination. The potential in E was recorded in the VC and was of smaller amplitude and of longer latency than potentials observed in the VC of control animals. As was the case for Fig. 12, each frame in Fig. 13 is composed of the superimposition of three consecutive neurophysiological traces.

Studies of Single Neurons in the Rat Visual System

Taken together, the data derived from the various studies described in the previous sections of this chapter strongly indicated that animals exposed to 30 or more days of continuous illumination were unable to process visually evoked information. Nevertheless, it was clear from a variety of behavioral experiments from our laboratory (4, 19, 20) and from other labs (7, 8, 21) that animals with massive retinal degeneration, following exposure to 200 or more days of light, were not necessarily blind. Since they are not blind, it is apparent that some visual information must be processed by the retinas of such animals and that this information is received and translated by the appropriate regions of the central visual system. The failure to find reliable indicators of visual activity using ERG methods or macroelectrode recordings in the central visual pathway could be more a reflection of the inadequacy of the methods themselves than of the potential real absence of such activity. In our continuing efforts to clarify the neurological mechanisms involved in the residual vision in animals with massive retinal damage, we decided to pursue a series of microelectrode recording investigations in which we would examine the excitability of single neurons in the LGN and VC of rats exposed to various period of constant light. In most cases, the animals that were studied with microelectrodes had been previously trained and tested on a variety of visually-guided tasks, including: (1) a brightness discrimination, (2) a pattern task, consisting of alternating black and white stripes oriented either vertically or horizontally, and (3) a red _vs_ blue color discrimination (for details of the relationship between discrimination performance and exposure time, see chapter in this book by Anderson and Lemmon). Thus, we were often able to correlate changes in the firing properties of visual neurons in the LGN and VC, following light exposure, with changes in the performance of certain visual tasks.

All single-unit electrophysiological experiments were performed on albino rats housed in either cyclic light (control animals) or in constant light. The latter animals were exposed to continuous illumination for periods ranging from one to 24 weeks. The light intensity in these experiments was somewhat higher than that used in the previous studies, being 440 lux. All animals were placed in a specially designed head holder that provided a minimum of obstruction of the visual fields. Visual stimuli were projected onto a tangent screen that was oriented in such a way as to cover most of the visual field of the eye under study. The visual stimuli used in these experiments consisted of flashes of light, of various sizes, and a number of geometrical images, such as circles, annuli, bars and squares, to name a few. The images could be presented at any angle by the use of an image rotator built into the projector. In addition to analyzing the firing pattern of LGN and VC neurons to various visual targets, we also conducted receptive field studies of visually responsive neurons. In a separate series of studies we analyzed the spectral sensitivity of cells in the LGN of animals exposed to constant light.

The responses of LGN cells in control rats to flash stimuli were quite vigorous, with any given cell responding to almost every flash. LGN cells were classified as being principal cells (P cells) or interneurons (I cells) based on the number of spikes in the initial discharge following a flash. P cells had 1-3 spikes following a flash, while I cells had at least 5 spikes following each flash (22). The response patterns of P cells can be separated into three classes. In the first class (E cells), the excitatory period following a flash was followed by a return to the spontaneous level of firing. In the second class (E-S cells), the initial period of excitation was followed by a suppression of firing. The third class (S cells) was characterized by a suppression of firing as the first response following a flash. I cells all had similar response patterns, consisting of an early discharge followed by a period of no activity, which was followed by one or more late discharges. Figure 14 presents typical response patterns of the three classes of P cells and of I cells.

There was a fairly large degree of variability among the response patterns of VC neurons in control animals. Approximately one-half of the cells responded with a short latency, short duration period of excitation similar to the response patterns shown in the top of Fig. 15. The remaining cells had responses that could not be categorized simply. Some exhibited a generalized excitation, while others showed a period of suppression followed by a return to

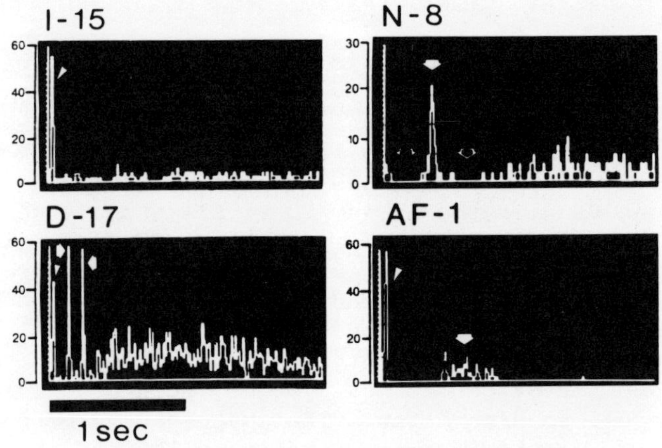

Figure 14. Poststimulus time histograms of LGN neurons in control rats in response to a flash. I-15: the response properties of an E cell; there was a short latency excitation (arrowhead), followed by a return to the spontaneous rate of firing. N-8: the response pattern of an S cell; after the flash there was a period of suppression (open arrow), followed by a late period of excitation (marked by solid arrow). D-17: the response pattern of an E-S cell; an early discharge (arrowhead), followed by a period of suppressed firing. AF-1: response pattern of an I cell to a flash; the early discharge (arrowhead) was followed by a late discharge having a variable latency.

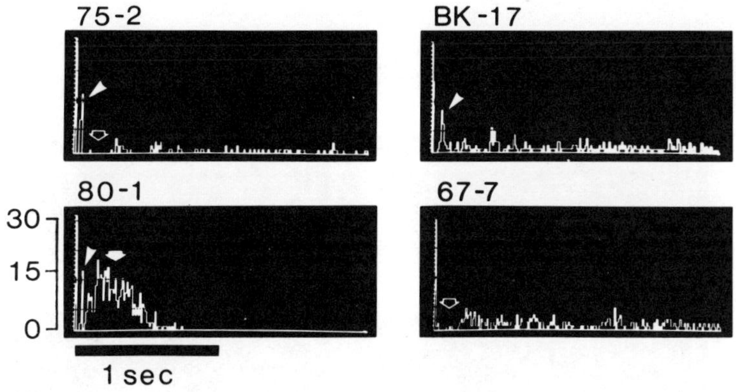

Figure 15. Response patterns of visual cortex neurons in control rats, showing the PSTHs after a flash stimulus. 75-2 and BK-17 responsed with a short-latency, short-duration period of excitation (arrowheads). 75-2 had a period of decreased activity (open arrows) after the period of excitation. 80-1 produced an extended burst of spikes (solid arrow) after a flash, the first spike (arrowhead) of the burst having a fixed latency. 67-7 responded to a flash with a short-duration reduction of ongoing activity.

the spontaneous rate of firing. The post-stimulus time histograms (PSTHs) shown in the bottom of Fig. 15 are examples of the types of response in this latter group.

The responses of visual cells in light exposed animals were less vigorous than those in normal animals and the response latencies were also more variable. This was particularly obvious for geniculate cells. The responses of many cortical cells are variable in normal animals, so the variability seen in experimental animals was not remarkable. Figures 16 and 17 show typical response patterns for geniculate and cortical cells of animals exposed to constant light for 10 and 16 weeks. With increasing exposure durations, it became more difficult to find cells in the cortex that had visually evoked responses. This observation was in contrast to the findings noted in the LGN, where it was fairly easy to find responsive cells, even after 24 weeks of exposure to constant light.

The response latencies to an intense flash stimulus were determined for cells in the LGN and VC. The response latency was defined as the time between the stimulus and the peak of the first excitatory period following the stimulus. These values were obtained by examining the PSTHs that were generated with data from 25 consecutive stimulus presentations.

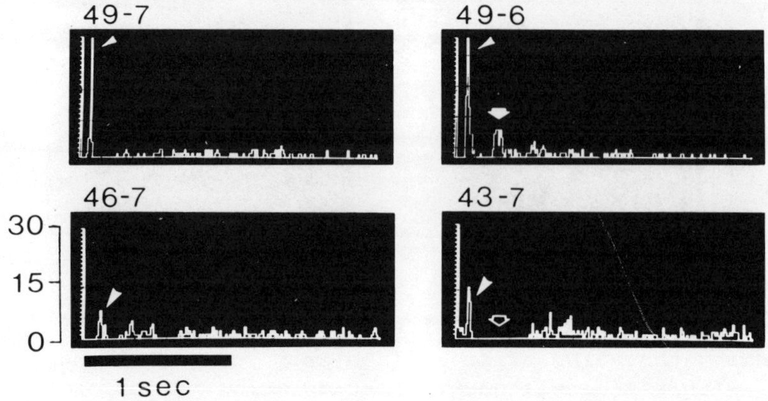

Figure 16. Response patterns of visual cells in rats exposed for 10 weeks to constant light. 49-7 and 49-6 are records from geniculate cells, and 46-7 and 43-7 are from cortical cells. 49-7 was a P cell, and 49-6 was an I cell. Note the vigorous responses (arrowheads) of these two cells.

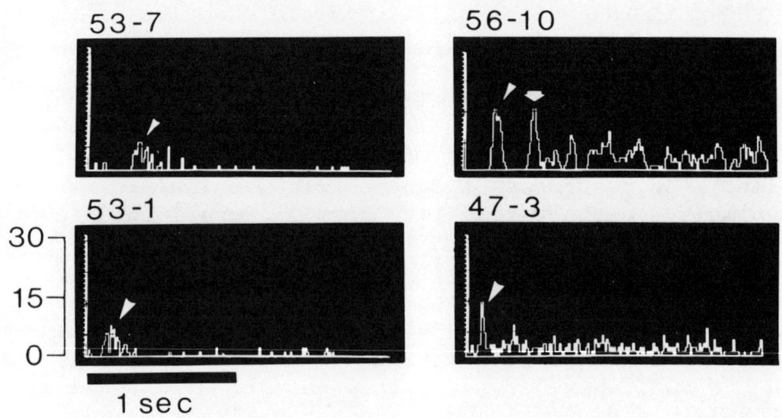

Figure 17. Response patterns of visual cells in rats exposed to 16 weeks of constant light. 53-7 (a P cell) and 56-10 (an I cell) are records from geniculate cells, and 53-1 and 47-3 are from cortical cells.

The mean response latency for LGN cells in control animals was 78 ± 31.27 msec (mean ± s.e.m.). If the response latencies of two S cells were excluded from the calculation, the values dropped to 32.70 ± 1.80 msec. The median and mean response latencies of the geniculate cells in control and light exposed animals are shown in Table 3, as are the standard error of the mean, and the minimum and maximum response latencies. The response latencies of cells studied in light exposed animals were compared to those from control animals with a one-sided Mann-Whitney U test. It was found that the latencies of all experimental cells were different from those of the control cells at the .01 level. The P values showing the levels of significance are also shown in Table 3.

Cells in the VC had longer latencies than cells in the LGN. The average response latencies of VC cells at the different exposure times are shown in Table 3. The response latencies of VC cells in control animals were shorter than those of VC cells in experimental animals. Statistical analysis revealed, however, that the response latencies of cortical cells in experimental animals were significantly different from those of cortical cells in the control animals at the 0.1 level for all time points, except 24 weeks, where only 2 cells were studied.

In addition to the study of response patterns, an attempt was made to examine the receptive field properties of the cells in the LGN and VC. The normal procedure consisted of studying a cell's response patterns before examining its receptive field properties. Because of the time involved in generating a PSTH, many cells were lost before a careful receptive field analysis could be completed.

Of the geniculate cells studied, it was found that 70% of the receptive fields consisted of a single, dominant region that produced either an "off" response or an "on" response. Thirty percent of the geniculate cells in control animals had peripheral regions of the receptive field that gave rise to an antagonistic response, when compared to the center of the receptive field. Of the cortical cells studied in normal animals, no receptive field could be found for 38% of the cells. Cortical cells that had identifiable receptive fields were classified as being either simple or complex. Simple cells had receptive fields that could be separated into areas that gave "on" or "off" responses to stationary, flashing stimuli (23). Complex cells responded to stationary, flashing stimuli, if at all, with an "on-off" discharge (23). In control animals, 27% of the cells were found to have simple receptive fields and 33% had complex receptive fields. These geniculate and cortical cells normally could be activated by stimuli that were only 2-4 visual degrees in diameter and had receptive fields that were from 20-40 degrees in diameter.

TABLE 3

LATENCY ANALYSIS OF CELLS IN CONTROL AND EXPERIMENTAL ANIMALS

EXPOSURE DURATION IN WEEKS	# OF CELLS	MEDIAN	MEAN	STANDARD ERROR OR THE MEAN	MIN	MAX	P
	RESPONSE LATENCIES OF LGN CELLS IN MSEC						
CONTROL	14	33	78	31.27	25	400	--
1 WEEK	22	105	190	32.19	40	500	.0001
4 WEEKS	12	80	135	39.55	55	450	.0005
10 WEEKS	17	90	119	22.56	60	200	.0003
16 WEEKS	10	120	175	41.74	70	500	.0011
24 WEEKS	8	190	192	42.07	150	400	.0031
	RESPONSE LATENCIES OF VC CELLS IN MSEC						
CONTROL	23	62.5	150	30.6	30	500	--
1 WEEK	15	200	232	35.3	60	500	.027
4 WEEKS	17	100	187	45.6	50	620	.061
10 WEEKS	15	210	230	36.4	70	630	.016
16 WEEKS	17	130	146	12.6	90	250	.079
24 WEEKS	2	230	230	29.7	200	260	.138

TABLE 4

RECEPTIVE FIELD TYPES

EXPOSURE DURATION IN WEEKS	CONTROL	1	4	10	16	24
LGN						
CENTER ONLY	7	7	5	10	4	4
CENTER + SURROUND	3	3	0	0	0	0
CORTEX						
SIMPLE	5	4	5	4	4	0
COMPLEX	6	2	0	2	0	0
NO RECEPTIVE FIELD	7	2	5	3	2	2

The results of receptive field analyses of cells in light-exposed animals are shown in Table 4. In the LGN, no cells having antagonistic surround were found after four or more weeks of exposure to constant light, while cells giving "on" or "off" responses were found at all exposure times. With increasing exposure durations it was necessary to use larger visual stimuli in order to activate cells. For example, following 10 weeks of constant light it was necessary to use stimuli that were 40-90 degrees in diameter. In the cortex there was a decrease in the number of cells having simple receptive fields and the decrease in simple cells appeared to have a longer time course than the decrease in complex cells. As in the LGN, extremely large stimuli were needed to activate cortical cells in animals exposed to long durations of constant light. An additional observation was that visually responsive cells were found in the cortex of only 2 of 7 animals exposed to one week of constant light. This observation was in spite of the fact that numerous electrode penetrations were made in what appeared to be healthy, undamaged cortices.

Once a geniculate cell's response pattern to a flash had been studied and the receptive field analyzed, a series of tests were performed to determine if the cell had opponent color properties. Of the seven cells studied in this way in control animals, one cell had opponent color properties. The responses of this cell are shown in Fig. 18. None of the 22 cells examined in the light exposed animals had opponent color properties. The responses to different colors of a cell in an animal exposed to 16 weeks of constant light are shown in Fig. 19. It should be noted that red stimuli were very ineffective in activating cells in animals exposed to constant light.

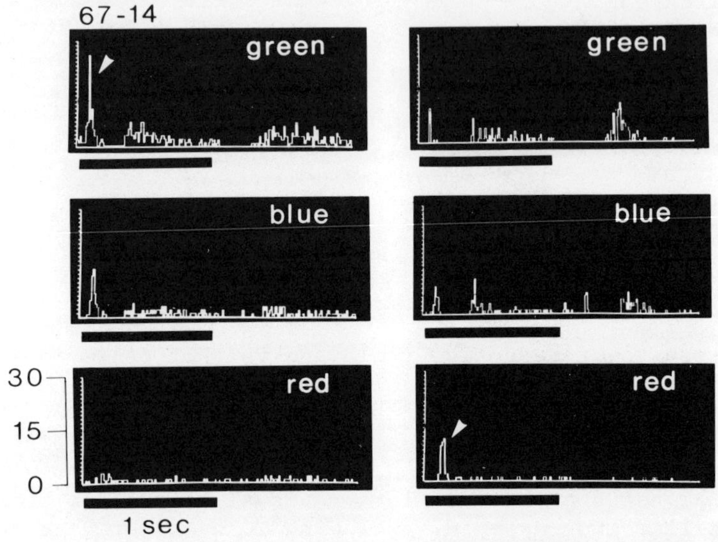

Figure 18. Opponent color responses from a P cell in the geniculate of an animal housed in cyclic light. The three PSTHs on the left were compiled from responses to a 4° spot projected on the center of the cell's receptive field for one second (note the black time bar below each PSTH). This cell gave a strong "on" response to green and a less vigorous response to blue. A red stimulus in this position had no effect on the cell. The PSTHs on the right show the responses from the cell's receptive field periphery. The response to green changed from a strong "on" response to a response that was difficult to classify as being strictly "on" or "off". The response to red, however, was a strong "on" response. Thus, this cell can be said to have red-green opponent color properties.

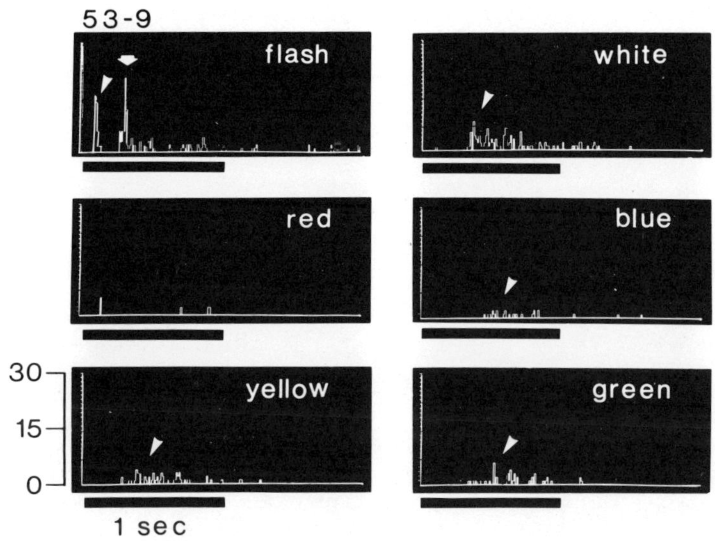

Figure 19. Responses to color from a P cell in the geniculate of an animal exposed to 16 weeks of constant light. The top left PSTH shows the response to a flash of white light, and the remaining PSTHs show the response to a 30° spot of colored light kept on for one second. The responses to different colors were not qualitatively different, with the exception of a lack of response to red. This latter observation was representative of all cells in light-exposed animals.

The spectral sensitivity of cells in the geniculate were examined in a separate series of experiments. In two control animals 10 cells were studied completely, while in 2 rats exposed to constant light for 12 weeks, 12 cells were examined. Ten cells were studied in 3 rats exposed to constant light for 24 weeks. Figure 20 shows examples of the spectral properties of the cells that were studied. Absorption curves for rhodopsin are included in Fig. 20 for comparison. The spectral sensitivity of each cell was compared to the absorption curves of rhodopsin having peak absorptions at 450, 480, 500, 520 and 550 nanometers, using a least-squares method. These calculations showed that a rhodopsin curve having an absorption peak at 500 nanometers provided the best fit to the data from 19 out of 29 cells. Five cells had spectral sensitivities that were best fit by a rhodopsin curve with a peak at 480 nm and six others most closely resembled a rhodopsin curve with a peak at 520 nanometers. Absorption curves at 450 and 550 nanometers were not best-fits to the data from any of the cells. There were no obvious differences between the spectral sensitivities of cells in control animals and those in light-exposed animals. In each of the three groups of animals, at least one cell was found with spectral sensitivities that had a small side peak at 600 nanometers associated with the larger peaks at 500 nanometers. Examples of this are shown in Fig. 20.

Figure 21 illustrates the features of light-induced retinal degeneration produced following exposure to 1, 4, 10, 16 and 24 weeks of continuous illumination at 440 lux. The sections are from plastic-embedded retinas and frame A is from a control retina. The histological details of retinal changes after exposure to various durations of constant light have been presented elsewhere in this book, authored by Anderson and Lemmon.

DISCUSSION

Taken together, the electrophysiological studies performed in our laboratory have clearly shown that the visual system of rats, exposed to extended periods of constant light, is still capable of processing perceptual information. It is important to note, however, that the conclusions one reaches depend heavily upon the specific recording techniques one uses to evaluate the integrity of the visual system. For example, our studies with ERGs, spontaneous, rhythmic potentials, and evoked slow-wave potentials all showed that there were few, if any, signs of neurophysiological activity in various parts of the rat visual system after 20-30 days of exposure to constant light. By contrast, the single-cell experiments clearly revealed that individual neurons in the LGN and VC were responsive to various kinds of visual stimuli, even after 24 weeks of exposure, at a time when the retina had undergone massive degeneration in the photoreceptor cell layer. In fact,

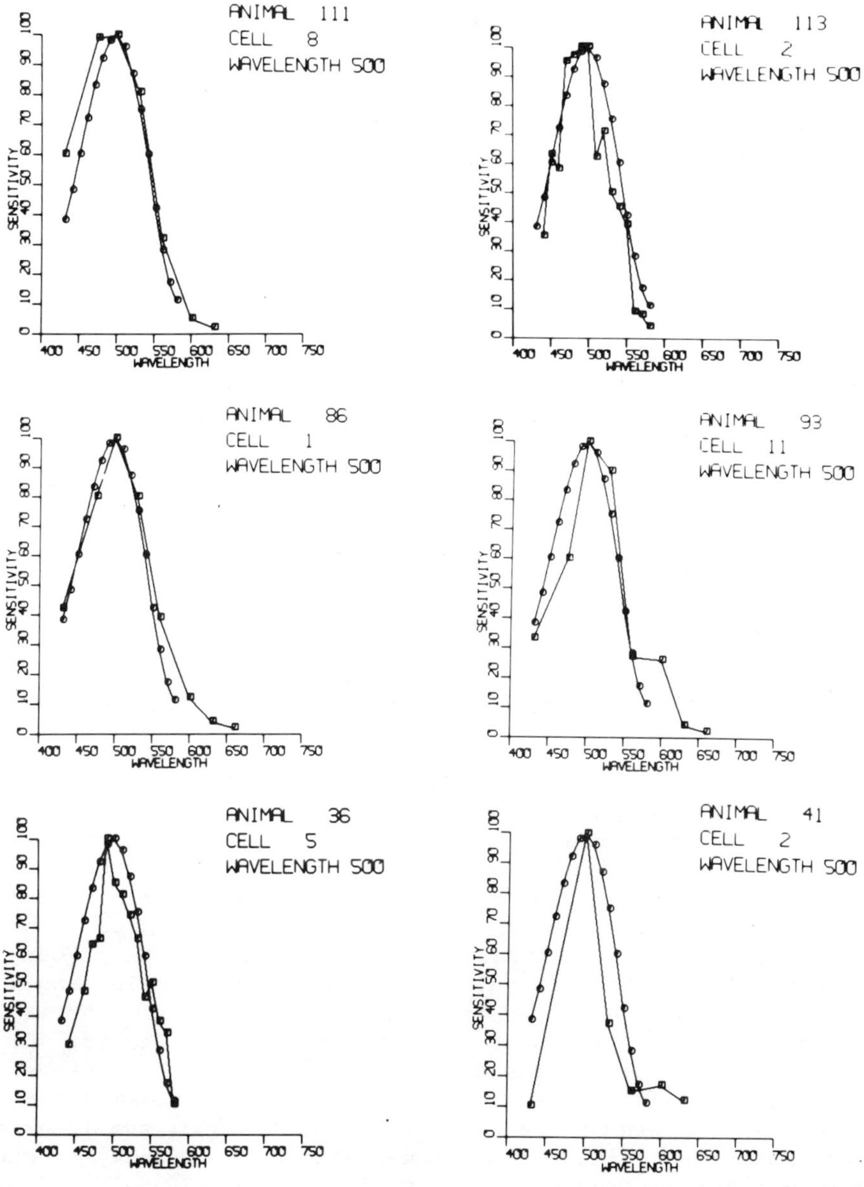

Figure 20. Spectral sensitivity curves of cells in the LGN, □, and the absorption curve for rhodopsin having a peak at 500 nm, O. The wavelength in nanometers is plotted on the X-axis. The normalized relative threshold for LGN cells and the absorption, as a percentage of the maximum, for rhodopsin are plotted on the Y-axis. The top two frames show data from cells in normal animals, and the middle and bottom frames are from animals exposed to 12 and 24 weeks of constant light, respectively. Note the small side peaks at 600 nm in cells 93-11 and 41-2.

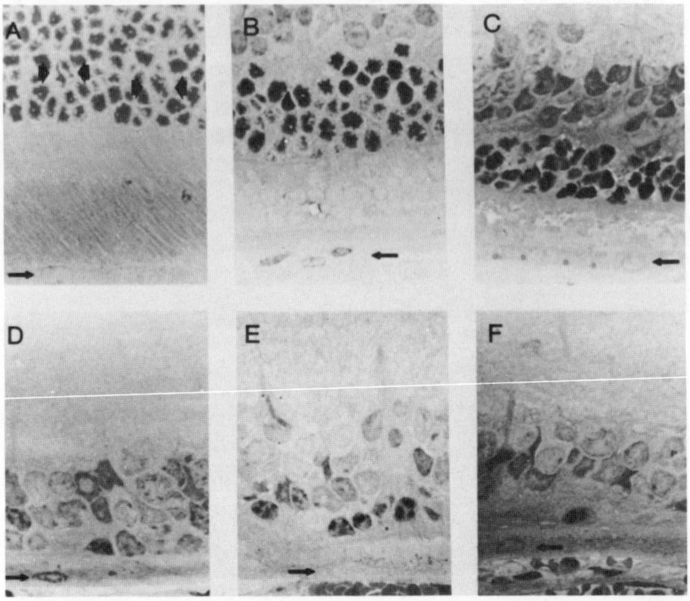

Figure 21. Sections from plastic-embedded retinas. Frames A-F show selected regions of retinas from a control rat and from rats exposed to 1, 4, 10, 16 and 24 weeks of constant light, respectively. The pictures were taken at a magnification of 250X and show sections of retina approximately one millimeter from the ciliary body. The frames were selected to show examples of photoreceptor nuclei at each exposure duration. The long, narrow, black arrows indicate the pigmented epithelium (PE). A: the bold, short arrows indicate two cone nuclei. Note the oval shape of these two nuclei and that they have more than one clump of densely staining heterochromatin. These two nuclei are also oriented perpendicular to the PE. B: the disruption of the inner and outer segments and the decrease in the distance between the PE and the photoreceptor nuclei can be seen. C: an example of the extensive disruption of the inner and outer segments in retinas from animals exposed to 4 weeks of constant light. D: one photoreceptor nucleus is seen immediately adjacent to the PE. E and F: nuclei with more than one clump of heterochromatin. Note, however, that these nuclei are not oval, nor are they oriented perpendicular to the PE, as are the cone nuclei in control animals (compare with frame A).

even the inner nuclear and ganglion cell populations show significant degenerative changes in animals exposed to 24 or more weeks of continuous illumination (20, 24). While there is no clear answer to explain why the results in slow-wave _vs_ single-unit studies should be so different, the answer could be related to the fundamental differences in the two techniques themselves. That is, one needs to keep in mind what it is that is actually recorded in macroelectrode _vs_ microelectrode methods. The potentials observed with macroelectrodes reflect a kind of summed activity consisting of the various types of electrical potentials produced by the individual neurons, such as unitary potentials, excitatory postsynaptic potentials, and even inhibitory postsynaptic potentials. Thus, it is difficult, and generally impossible, to determine the state of excitability in individual neurons by measuring changes in slow-wave activity alone. On the other hand, microelectrode techniques permit one to study neurons individually, or in small clusters, and, thus, enable one to determine the functional integrity of single neurons within a given sensory or motor system. In short, then, it is our impression that microelectrode techniques are more sensitive than macroelectrode techniques in the study of information processing in light-exposed animals.

In the animals exposed to constant light there were E, S and E-S cells at all exposure times, with the exception of 24 weeks, at which time no S cells were observed. Many of the cells in animals exposed to 10 or more weeks of constant light had extremely low levels of spontaneous activity. As a result, some cells that were suppressed following a flash may have been classified inadvertently as E cells, because it would have been impossible to detect a suppression in the absence of a significant level of spontaneous activity. The I cells observed in animals exposed to 1, 10 and 16 weeks of light had response patterns comparable to those of I cells in the control animals. That response patterns similar to those of P and I cells in control rats were found in the light-exposed rats implies that no major changes occurred in the functional connections of cells in the LGN as a result of exposure of the rats to constant light.

The response latencies to a flash of both geniculate and cortical cells increased with increasing constant light exposure durations. Evidence from cats (25, 26) suggests that the firing of one ganglion cell can result in the production of an action potential in a geniculate cell. Also, very few ganglion cells (6 or less) are thought to send excitatory input to one geniculate cell (25, 26). If these results can be extrapolated to rats, then temporal or spatial summation of inputs from ganglion cells is not necessary to produce an action potential in a geniculate P cell. Therefore, the increased response latencies of geniuclate cells is most likely due to an increase in the response latencies of retinal ganglion cells. However, a decrease in the functional convergence on LGN cells may also account for this finding. It is not surprising that the latencies of the cells in the VC increase with exposure to constant light, because the LGN cells that provide the primary input to the cortex themselves have longer response latencies. If one compares the minimum latencies of geniculate cells with the minimum latencies of the cortical cells at each exposure time, one finds that the LGN latencies are shorter than the cortical latencies at all exposure times, with the exception of 4 weeks, where the minimum latency in the cortex was 50 msec and the minimum LGN latency was 55 msec.

The receptive field studies of geniculate cells in control animals showed that, in 7 out of the 10 cells studied, there were no antagonistic surrounds, while the remaining 3 cells had antagonistic peripheries. Other workers (27, 28) have found that about ½ of the P cells had receptive fields with antagonistic surrounds and two to three percent of their cells were directionally selective. The remaining P cells had receptive fields with a dominant center and no antagonistic surround. No geniculate cells were found in the present study that had confirmed, directionally-selective properties, which is most likely due to the fact that only a very small percentage of geniculate cells have this type of receptive field organization. After 4 or more weeks of constant light, no cells having antagonistic surrounds were found and very large stimuli were required to produce a response from geniculate cells. These changes in the receptive field properties could be due to a major reorganization of the synaptic input to the geniculate cells or they could be due to changes in the response properties of the retinal ganglion cells. Of the two possibilities, the latter is much more likely, because of the changes that have been reported previously in the inner nuclear layer and the inner plexiform layer of the retina (29).

During the receptive field analyses of cortical cells in control animals, there were a significant number of cells (39%) for which no receptive field could be found. There are several possible explanations for this finding. One obvious possibility is that the visual stimuli were grossly out of focus on the animal's retinas. This is unlikely, because the tangent screen was 33 cm from each rat's eye in the experiments, and Hughes (30) has convincingly shown that the near point for rats is about 30 cm and that rats have a large depth of field. The most likely explanation is that either the proper position in the visual field was not examined for a particular cell or that stimuli having the proper temporal and spatial characteristics were not used to test the cell. In any case, the results reported here are not inconsistent with those of other investigators (31, 32), who reported finding significant numbers of cells in the rat cortex for which they were unable to identify receptive fields.

The results of our behavioral experiments (4, 18, 20), taken together with those of the electrophysiological experiments, suggest a striking relationship between the receptive field properties of visual cortex cells and the ability of animals to perform pattern discriminations. The results of the single-cell experiments revealed that there was a decrease in the number of cells having complex receptive fields as a result of exposure to constant light, so that following 16 or more weeks in constant light no complex cells were found. No simple cells were found in rats exposed to constant light for 24 weeks. This is similar to the time course seen for the decrease in the ability of the rats to perform pattern discriminations. In contrast to these results, light-exposed rats performed intensity discriminations and color discriminations throughout the 24 weeks in constant light. Similarly, cells responsive to light were found in the LGN of rats exposed to 24 weeks of constant light. Another important observation was that, during the first week in constant light, rats showed a small, but significant, impairment on pattern tasks that recovered during the second week. This was paralleled by an extreme difficulty in finding any visually responsive cells in the cortex of rats exposed to one week of constant light.

The question of color vision in control rats and light-exposed rats is an important one. Color vision in rats has been examined by previous investigators and it has been known for at least 40 years that rats can discriminate between blue and red and between green and red on a basis of wavelength (33-36). It has also been shown that, while rats can learn to discriminate between red and yellow and between blue and yellow, they only do so after an extremely large number of trials (> 500) (36). Rats have not been trained to discriminate between blue and green or between green and yellow, despite the fact that different investigators have attempted to do so (36, 37). The work of the latter investigators shows that while the rat has color vision, this ability is quite limited when compared to that of primates, or even squirrels. In order for a rat to discriminate two colors on a basis of hue, the two colors must have significantly different dominant wavelengths. In view of these observations, it seems extremely unlikely that rats have 3 different cone mechanisms having peak sensitivities at 450, 520 and 560 nm, as proposed by Cicerone (38). If rats had these cone mechanisms, then one would expect rats to be able to discriminate between blue (about 450 nm) and green (about 520 nm) or between green and yellow (about 560 nm). If, however, rats use a rod pigment (498 nm) and a single cone pigment (about 600 nm) in their color system, then one would predict that rats would be able to discriminate between colors only if their peak wavelengths were relatively far apart, as has been observed in several behavioral experiments (33-37).

In our behavioral experiments it was found that rats exposed to constant light for 24 weeks were still able to discriminate between blue and red. This capability would seem to require at least two different visual pigments. The results of the single-cell experiments on light-exposed animals indicated that, while all of the LGN cells were most sensitive at the point predicted by the rat rhodopsin absorption curve (39), some cells had spectral sensitivity curves that could not be explained simply on a basis of a rhodopsin absorption spectrum. The threshold at 600 nm for these cells was from 400-800% higher than the threshold expected from a rhodopsin absorption mechanism. These latter findings are interesting, because Granit (40) obtained spectral sensitivity curves with side peaks at 600 nm for rat retinal ganglion cells that were strikingly similar to those determined for geniculate cells in our laboratory. Granit concluded that this sensitivity to red was due to a cone mechanism and not to a rod mechanism or an error in measurement, because the two peaks of the spectral sensitivity curves had different recovery rates from light adaptation. If this is the case, then it is possible that both a cone pigment and rhodopsin are involved in the light transduction process in animals exposed to constant light and, thereby, provide a mechanism for color vision in the light-exposed rats.

Of special importance in our discussion is a very recent study on hereditary retinal degeneration in mice (41), in which the receptive field properties of cells in the visual cortex and superior colliculi were examined. Of particular interest was the fact that no electrophysiological evidence was found for cones being the primary light-transducing element in mice with retinal degeneration. Neither a Purkinje shift nor a rod-cone break in the dark-adaptation curve was found in the tectal cells studied. These findings are important because cones have been thought to be the primary remaining photoreceptor cell type in mice with hereditary retinal degeneration. Because both the evidence from mice and the evidence from our laboratory indicates that a rhodopsin mechanism and _not_ a cone mechanism is the primary light transducing mechanism in rodents with retinal degeneration, it seems that accurate and reliable criteria for distinguishing between residual rods and cones are essential.

Our behavioral studies showed that there was a decrease in the ability of rats to perform pattern vision, which was followed, much later, by a decrease in the ability of the rats to perform intensity discriminations. The single-cell experiments showed that there were important changes in the response latencies and receptive field properties of geniculate and cortical cells as a result of exposure of the rats to constant light. In view of the fact that there were photoreceptor nuclei (but no intact receptor cells) present in the retinas of the light-exposed animals that could perform visual discriminations and in the retinas of the animals with visually responsive geniculate and cortical cells, it is possible that these residual photoreceptor nuclei might be involved in the transduction of light and the transmission of information to the cells of the inner nuclear layer of the retina. Because the spectral sensitivity measurements suggest that the light transducing molecules are the same in control animals and light-exposed animals, this seems particularly likely. Therefore, there appears to be a correlation between the presence of residual photoreceptor nuclei and the ability of rats to perform visually guided tasks. The data produced by our experiments do not eliminate the possibility, however, that some novel retinal element is also involved in the transduction of light in animals exposed to long-term, constant light. Nevertheless, the observations concerning color vision and the spectral sensitivity of the light transducing element would seem to argue against the latter notion, at least at the present time.

REFERENCES

1. Anderson, K. V. and O'Steen, W. K. (1971) Expl. Neurol. 30, 555-564.
2. Anderson, K. V. and O'Steen, W. K. (1971) Expl. Neurol. 32, 502-505.
3. Anderson, K. V., Coyle, F. P. and O'Steen, W. K. (1972) Expl. Neurol. 35, 233-238.
4. Anderson, K. V. and O'Steen, W. K. (1972) Expl. Neurol. 34, 446-454.
5. O'Steen, W. K. and Anderson, K. V. (1971) Expl. Neurol. 30, 525-534.
6. O'Steen, W. K. and Anderson, K. V. (1971) Z. Zellforsch. 127, 306-313.
7. Bennett, M. H., Dyer, R. F. and Dunn, J. D. (1972) Expl. Neurol. 34, 434-445.
8. Bennett, M. H., Dyer, R. F. and Dunn, J. D. (1973) Expl. Neurol. 38, 80-89.
9. Schimmerl, G. and Stumpf, C. (1958) Arch. Exptl. Path. Pharmak. 235, 33-40.
10. Gogolak, G., Liebeswar, G. and Stumpf, C. (1969) Electroenceph. Clin. Neurophysiol. 27, 296-303.
11. Starzl, T. E., Taylor, C. W. and Magoun, H. W. (1951) J. Neurophysiol. 14, 479-496.
12. Gahm, N. H. and Sutin, J. (1968) Brain Res. 11, 507-521.
13. Trembly, B. and Sutin, J. (1962) Expl. Neurol. 5, 120-130.
14. Wells, J. and Sutin, J. (1963) Expl. Neurol. 7, 355-365.
15. Bishop, P. O. and Davis, R. (1960) J. Physiol. 154, 514-546.
16. Doty, R. W. and Kimura, D. S. (1963) J. Physiol. 168, 205-211.
17. Graham, L. T. and Pong, S. F. (1972) Expl. Neurol. 36, 399-403.
18. Anderson, K. V. and O'Steen, W. K. (1975) J. Neurosci. Res. 1, 343-351.
19. Anderson, K. V. and O'Steen, W. K. (1974) Physiol. and Behav. 12, 633-637.

20. Lemmon, V. and Anderson, K. V. (1979) Expl. Neurol. 63, 35-49.
21. Bennet, M. H., Dyer, R. F. and Dunn, J. D. (1973) Expl. Neurol. 40, 652-660.
22. Anderson, K. V., Lemmon, V. and Rosing, H. S. (1977) J. Neurosci. Res. 3, 143-152.
23. Hubel, D. H. and Wiesel, T. N. (1962) J. Physiol. 160, 106-154.
24. Lemmon, V. and Anderson, K. V. (1979) Expl. Neurol. 63, 50-75.
25. Cleland, B. G., Dubin, M. W. and Levick, W. R. (1971) J. Physiol. 217, 473-496.
26. Levick, W. R., Cleland, B. G. and Dubin, M. W. (1972) Invest. Ophthalmol. 11, 302-311.
27. Montero, V. M. and Brugge, J. F. (1969) Vis. Res. 9, 71-88.
28. Sefton, A. J. and Bruce, I. S. C. (1971) Vis. Res. Supplement no. 3, 239-252.
29. Chernenko, G. A. and West, R. W. (1976) J. Comp. Neurol. 167, 49-62.
30. Hughes, A. (1977) Vis. Res. 17, 927-939.
31. Shaw, C., Yinon, U. and Auerbach, E. (1975) Vis. Res. 15, 203-208.
32. Wiesenfeld, Z. and Kornel, E. E. (1975) Brain Res. 94, 401-412.
33. Muenzinger, K. G. and Reynolds, H. E. (1936) J. Genet. Psychol. 48, 58-71.
34. Munn, N. L. and Collins, M. (1936) J. Genet. Psychol. 48, 49, 72-81.
35. Walton, W. E. and Bornemeier, R. W. (1938) J. Genet. Psychol. 52, 165-181.
36. Walton, W. E. and Bornemeier, R. W. (1939) J. Comp. Psychol. 28, 417-436.
37. Munn, N. L. (1932) J. Genet. Psychol. 40, 351-362.
38. Cicerone, C. M. (1976) Science 194, 1183-1185.
39. Bridges, C. D. B. (1959) Nature 184, 1727-1728.
40. Granit, R. (1942) Acta Physiolog. Scand. 2, 93-109.
41. Drager, U. C. and Hubel, D. H. (1978) J. Comp. Neurol. 180, 85-114.

A PARAMETRIC STUDY OF RETINAL LIGHT DAMAGE IN ALBINO AND PIGMENTED RATS

L. M. Rapp and T. P. Williams

Department of Biological Sciences
Florida State University
Tallahassee, FL 32306

Although light serves as the adequate stimulus for vision, it can, under certain conditions, have a deleterious effect on the retina. Using rats as experimental animals, Noell et al. (1) were the first to demonstrate that morphological and functional damage to the retina can result from constant light exposures of only a few hours. The manifestations of damage to the retina included the deterioration and death of the photoreceptors with concurrent losses in the amplitude of the electroretinogram (ERG). Moderate light intensity (1,200 to 2,500 lux), several log-units lower than the levels found necessary to produce threshold retinal burns, was used to produce damage. The action spectrum of the damage corresponded to the absorption spectrum of rhodopsin, the rat's visual pigment, indicating that damage is initiated by the same primary photochemical event as is vision.

The phenomenon of retinal light damage has been subsequently observed in a variety of animal species including monkeys (2, 3), pigeons (4), and rabbits (5). Albino strains appear to be particularly susceptible to light damage (6, 7), but the role of melanin in the iris and pigment epithelium of pigmented animals in protecting against damage remains in question. Some insight, regarding the role of the iris, is presented in this chapter.

In order to produce retinal light damage, the experimental animal is either exposed directly to a light source (e.g. ophthalmoscope, laser) or allowed to move freely in diffusely illuminated cages. In the former case, measures are taken to anesthetize the animal, dilate its pupil, and prevent eyelid closure. For diffuse light exposure of unanesthetized animals, the eye open-time per exposure period is assumed to be relatively constant among the experimental animals.

Experimental conditions affecting retinal light damage

The spectral content of lighting used for damaging exposures has been manipulated in some experiments so that it includes only monochromatic light or a narrow band of wavelengths. In general, these studies show that the efficiency of light to cause retinal damage is directly related to its ability to be absorbed by visual pigment. Noell et al. (1) assessed the amount of damage in rats exposed to monochromatic lights in the red, green and blue regions of the spectrum. When these lights were adjusted to be of equal energy, the amount of retinal damage produced by each light corresponded to its effectiveness for visual excitation. In accord with this finding, monochromatic lights which differed in energy but were of equal efficiency in producing a criterion ERG were shown to have similar damaging capacities. Lawwill (5) obtained results which are consistent with Noell's by comparing the threshold intensities for damage of a broad spectrum "white" light and a monochromatic (514.5 nm) exposure light. The ratio of these thresholds was found to be approximately equal to the relative visual efficiency of the two sources as measured by the ERG. Gorn and Kuwabara (8) altered exposure light with colored plexiglass filters and also found that the degree of damage depended on the relative "bleaching power" of the different colored lights.

Effects of temperature are convoluted with other parameters in light-damage. Noell et al (1) have shown that increased body temperatures were associated with an increase in the extent and rate of damage. This effect was observed within the range of body temperatures extending from 37.5 to 39°C.

The high infrared output of incandescent bulbs can lead to increased temperatures during constant light exposures. Kuwabara (9) found that retinal damage from incandescent light was more severe and less reversible as compared to damage caused by fluorescent light of a comparable illuminance. Since the ambient temperature in the case of incandescent exposure rose to 48°C, the observed differences were attributed to the "thermal effect of light".

Numerous studies have shown that increasing either the intensity or duration of the exposure light will cause greater retinal damage. An inverse relationship between these two variables to give a constant amount of damage might therefore be expected, at least within limits. However, very few investigations have dealt with this problem specifically. Lawwill (5) found that in rabbits, there is no apparent damage when the threshold intensity for damage is doubled and the exposure time is halved. Although the relationship, (intensity time) = (a constant amount of damage), did not hold in the range observed, Lawwill suggested that "longer exposures of much less intensity might be expected to cause damage".

Morphological Sites of Damage

The majority of studies have shown that the primary site of retinal light damage is in the photoreceptor outer segment (4, 6, 10, 11). The membranous outer segment discs become disrupted showing separations and the formation of tubules and vesicles. Edema of the outer segment and nuclear pyknosis are also usually evident at early stages of damage.

As damage progresses with prolonged exposure, the lamellar arrangement of the outer segment discs becomes disoriented. The outer segments fragment and eventually undergo complete degeneration. Phagocytic cells invade the area of outer segment debris. The inner segments shrink when the outer segment is no longer present. Finally, photoreceptor cell death is evidenced by the disappearance of their cell bodies. In cases of extensive damage, the entire receptor layer may be missing. Microvilli from the pigment epithelium (if not also destroyed by damage) and Müller cells then interdigitate to form a tight adhesion (6). It is interesting to note that the inner layers of the retina are rarely affected in these extreme instances of light damage. Furthermore, behavioral studies indicate that visually mediated tasks can still be performed even when no photoreceptors appear to be remaining (12, 13).

Damage to the pigment epithelium has varied among different studies using constant-light exposure. A major influencing factor seems to be ambient or body temperature during exposure. Complete destruction of the pigment epithelium is usually observed when temperatures were above normal (1, 9, 14, 15). When temperature is controlled so that it remains at normal levels (10, 16) the pigment epithelium appears unaffected even after the photoreceptors have severely degenerated.

Purpose

The purpose of our experiments is to help characterize the conditions of light exposure which are damaging to the retinas of albino and pigmented rats. In particular, attention will be focused on two parameters of the exposure condition in relation to resulting damage to the retina. These are: 1) the intensity of illumination in the exposure chambers; and 2) the steady-state level of visual pigment in the retina during exposure.

In most of the previously reported studies on constant-light exposure of the rat retina, only a single intensity of light is used to produce damage. Furthermore, the level of this intensity is usually much greater than what a rat might experience in its natural environment. In this work, a range of light intensities which span more than one log-unit are used to cause damage. Also, the lowest intensity is an order of magnitude lower than any previously examined (cf., 14).

The second parameter of relevance to the light-damaging condition is the steady-state level of visual pigment in the retina during exposure. In the rat retina, the predominate visual pigment is rhodopsin which has been suggested by Noell et al. (1) to be the primary absorber of light initiating damage in the retina. A brief description of the manner by which steady-state rhodopsin levels are reached in the retina is as follows. Rhodopsin molecules, upon absorbing light, undergo a reaction sequence called bleaching which results in the formation of a colorless product. Light reaching the photoreceptors thus causes a reduction in the amount of rhodopsin present in the retina. This bleaching of rhodopsin in the retina is counteracted by the regeneration of the molecule and within a short period after the onset of light, a steady-state level of rhodopsin is reached. Raising the intensity of light causes a lowering of the steady-state rhodopsin level and also a quicker onset of damage to the retina.

What insights can be gained by observing and quantifying the progression of damage to the retina as a function of the steady-state rhodopsin level in the retina? Firstly it relates the severity of light damage to normal processes that occur in the retina in response to light rather than to arbitrary units of light intensity. Secondly, it could serve as a basis for comparing albino and pigmented rats with regard to the susceptibility of their <u>retinas</u> to light damage. This is because steady-state rhodopsin levels are indicative of how much light actually reaches and is absorbed by the retina. If the only protection afforded by melanin in pigmented rats is to prevent light from reaching the retina, then albino and pigmented rats should show a similar time course of damage when their steady-state rhodopsin level is lowered by the same amount.

MATERIALS AND METHODS

Experimental animals

Breeding pairs of rats were obtained from Southern Animal Farm, Prattville, Alabama. All experimental rats were raised from birth in the Institute of Molecular Biophysics animal colony. The photoperiod in this room was 12 h light, 12 h dark with a cage illumination of 5-10 lux.

Two strains of rats were used for experimentation. The Sprague-Dawley strain are albinos which completely lack the brownish-black pigment melanin. The other strain of rats used was Long-Evans (hooded) which have melanin distributed in their hair, skin and eyes. Within the eyes of Long-Evans rats, melanin is concentrated in the iris, pigment epithelium and choroid.

Male and female rats were used in equal numbers. At the onset of each experiment the rats ranged in age from 80 to 100 days.

Light exposure

The cages used for constant-light exposure were constructed of clear plexiglass which has been sandblasted in order to evenly diffuse incoming light. These cages were situated within a large box with white inside walls. Fluorescent fixtures were mounted on the inside walls, floor and ceiling of the box. Illumination was provided by "cool white" fluorescent bulbs. Inside the cages, the rats moved freely on a hardware cloth floor underneath which a sheet of plexiglass was positioned at an angle so that the animal's droppings were removed from the rats field of view. The plexiglass sheet and the cage walls served as illuminated surfaces which completely surrounded the animal. The arrangement of the fixtures provided a rather even level of illumination in the cages which did not vary in any direction by more than 20% from the mean value. Fans circulated air into and out of the box. This served to keep the temperature in the cages wihin 2°C of the room temperature and to provide fresh air.

Light intensity was controlled by metal screens that were painted flat black and placed over the fixtures. The illuminance in the cages was measured perpendicularly to the walls, ceiling and floor using a Tektronix J16 Digital Photometer with a J6501 illuminance probe. The readout of this probe was in units of lux.

Light exposure of the rats began one hour after the onset of the light cycle in the animal room. Food and water were provided ad. lib.

Measurements of light intensity made with the photometer served as the basis for adjusting to each new level of illumination. The "effective" intensity of each light level relative to the others was also measured indirectly with rhodopsin serving as an actinometer. At each illuminance, the rate of bleaching for a solution of cattle rhodopsin and for rhodopsin in enucleated rats' eyes was determined. The methodology for these experiments was as follows. Bleaching of cattle rhodopsin took place in a cuvette with a short (1 mm) light path. The cuvette was placed in a dental wax mold so that only one face was exposed to light. At various intervals during exposure to light in the experimental cages, the cuvette was removed and transferred (in the dark) to the Gilford spectrophotometer and optical density of the rhodopsin solution was measured. After several such measurements, the rhodopsin was completely bleached with bright light and optical density was again measured. The amount of rhodopsin remaining after each exposure could then be calculated and expressed as a percentage of the initial amount of rhodopsin. This procedure was repeated at different light intensities which were to be used for constant light exposure of rats. For the bleaching of excised rats' eyes dental wax was again utilized. Half-spheres were drilled in the wax that were intended to approximate the rats' eye cavity in size and absorbance properties (the wax was dark red in color). The temperature of eyes placed in this holder was kept at 18-20°C (slightly below room temperature) with ice placed beneath the holder. Bleaching again took place in the exposure cages. At various intervals during exposure, two eyes were removed and placed in cold buffer so that rhodopsin assay could be performed (see "Rhodopsin Assay" in Methods section). Bleaching in rats' eyes was measured at several light intensities including those used to bleach cattle rhodopsin.

Rhodopsin Assay

Enucleated eyes were hemisected at the ora serrata using a specialized "razor blade guillotine". The posterior half containing the retina was washed in phosphate buffer (0.067 M, pH 6.5). Using fine forceps, the retina was then carefully teased away from the eyecup and its connection at the optic nerve was pinched off. The retinas from the left and right eyes of each rat were placed in centrifuge tubes containing phosphate buffer and spun at 9000g for 15-20 minutes. Extraction of rhodopsin followed in 0.25 ml of 15% Triton X-100 per retina for a period of three hours. After extraction, 1 M hydroxylamine was added in an amount that gave a final concentration of 0.1 M. This was done to insure that photoproducts were rendered colorless. Centrifugation was again performed to separate the insoluble material from the rhodopsin extract. A 180 microliter aliquot of the extract was transferred to a cuvette that had a 1 cm light path. All of the preceding steps took place in either dim red light or darkness. A model 2000 Gilford spectrophotometer was used to measure the optical density of the extract at 500 nm before and after exposure to 10 min of bright light which completely bleached the visual pigment. The concentration of rhodopsin was calculated from the Beer-Lambert law assuming a molar absorbance of 4000 m^2/mole.

Histology

Following a 24 h period of dark adaptation, the rat was anesthetized (in dim red light) by the intraperitoneal injection of sodium pentobarbital (50 mg/kg body weight). Fixation of the eye was accomplished by the vascular perfusion of a mixture of 2.0% glutaraldehyde and 0.6% formaldehyde (prepared from paraformaldehyde) in 0.1 M cacodylate buffer, pH 7.35, which contained 0.038% calcium chloride. Adequate fixation was attained with the perfusion of 100-150 ml of the fixative for a period of at least 5 minutes. Immediately following the perfusion, the eyes were dissected out and placed in the same fixative for one hour. The cornea and lens were then removed and the eyecup trimmed to the desired size.

The fixed tissue was rinsed several times in cacodylate buffer over a period of 2-3 hours. A gradual series of ethanol was used to dehydrate the tissue. The tissue was then cleared in toluene and embedded in Paraplast Plus (manufactured by Lancer, Saint Louis, MO).

Orientation of the eye was accomplished by injecting a small amount of hematoxylin dye into the extraocular musculature at the temporal corner of the eye just prior to fixation. The placement of this mark was later checked relative to a large blood vessel running across the back of the eye in the nasal-temporal direction. These landmarks were used to orient the eye during embedding into the parafin block. Orientation in the block was such that sectioning proceeded from nasal to temporal portions of the eye with each section being divided equally into superior and inferior regions. Sections passing through the optic nerve head were thus roughly in the verticle meridian.

Sections were serially cut at 7 microns and mounted on glass slides. Hematoxylin eosin was used for staining. Measurements of the outer nuclear layer thickness were made at a magnification of 400 x with an eyepiece micrometer calibrated in divisions of 4 microns.

RESULTS

The Regional Distribution of Damage

A primary consideration to the investigation of the damaging effects of light on the retina was the choice of an appropriate measure to follow the course of damage. Although light damage is manifest both morphologically and functionally in the retina it seemed logical that the most direct information could be obtained by histological observation. Several investigators have shown that as the retina becomes damaged there is a decrease in the number of photoreceptor cells present due to their death and phagocytosis (1, 10). The outer nuclear layer, which is comprised of the cell bodies of the photoreceptors, becomes reduced in thickness in parallel to cell loss. The thickness of the outer nuclear layer is therefore an appropriate measure of the severity of light damage to the retina, however; preliminary experiments indicated that the distribution of photoreceptor cell destruction was not uniform throughout the retina. By orienting the eye during histological processing, a study was undertaken to determine the area in the retina showing the most severe damage. Previous studies concur that there are relatively fewer photoreceptors lost in the far periphery as compared to more central regions of the retina (1, 14, 17, 18). However, a more precise localization of the spatial distribution of damage in the retina was not attempted by these investigators. We have found that the region of most pronounced damage in the retina was not centered at the posterior pole. Instead, sections taken in the verticel meridian showed the most severe damage to be in an area superior to the optic nerve head. (Note: the optic nerve head is located close to the midpoint in the verticle meridian although this is not true for other planes of section.) The distribution of light damage in the superior and inferior portions of the verticle meridian can be seen in the micrographs of Fig. 1. This retina was taken from an albino rat exposed to 100 lux for 24 hours. The low power micrograph on the top passes through the optic nerve head with inferior retina displayed toward the left and superior toward the right. The inferior retina appears to have very little loss of photoreceptor cells since the thickness of the outer nuclear layer is approximately the same as that found in a control retina. In the superior retina, though, the outer nuclear layer thickness begins to taper off at about 500 microns superior to the optic nerve head and shows a rather uniform reduction until around 2000 microns in the superior direction. At this point the outer nuclear layer gradually returns to its nearly normal thickness. The micrographs at the bottom of Fig. 1 are higher magnifications of representative areas along the verticle meridian at equal distances from the optic nerve head but in opposite directions. The inferior retina, shown on the left, appears normal with the exception of what may be early stages of outer segment degeneration. In contrast, the superior retina on the right shows highly disorganized outer segments, edema throughout the retina, and a substantial decrease in the outer nuclear layer of about 40%.

Figure 2 is a graphic representation of the distribution of damage along the verticle meridian in the retinas of albino rats exposed constantly to 40 lux. Outer nuclear layer thickness is plotted as a function of distance along the verticle meridian. At four days exposure, the area of greatest damage in the verticle meridian was more clearly delineated. An 85%, or greater, reduction in the outer nuclear layer was seen in the region extending from 800 to 2000 microns superior to the optic nerve head. This corresponded to the survival of 1-2 rows of photoreceptor cell nuclei. In comparison, no fewer than 8-9 rows of nuclei remained in the inferior retina (one row of nuclei equals approximately 4 microns of outer nuclear layer thickness).

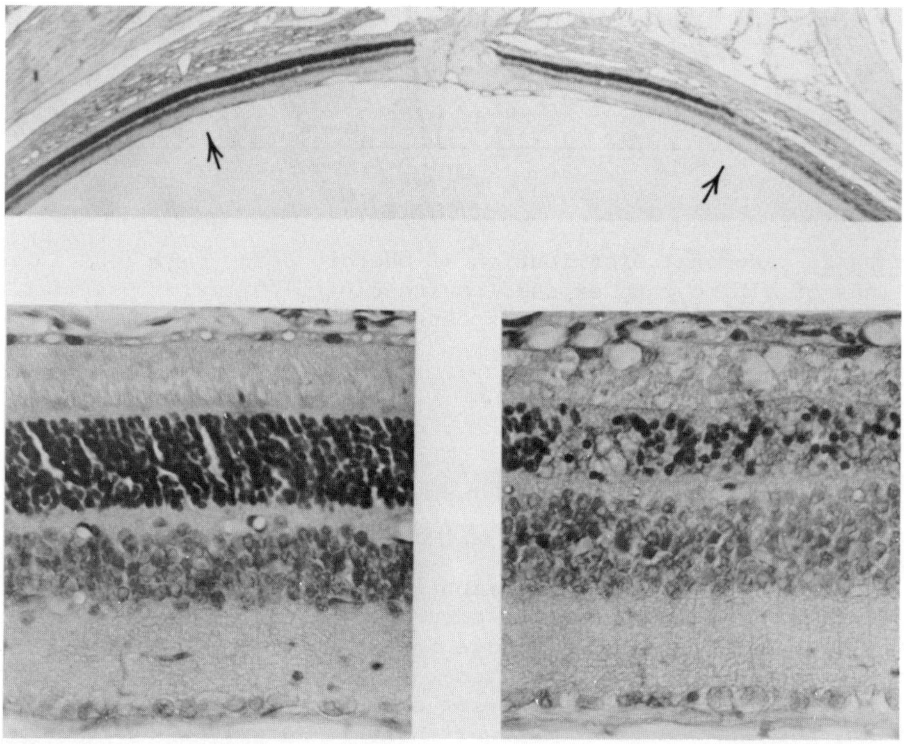

Figure 1. Top: Micrograph of an albino rat exposed to 100 lux for 1d. The optic nerve head is centered with superior retina shown to the right and inferior to the left. Arrows point to regions 800 microns in the peripheral direction from the optic nerve head. x25. Higher magnifications of the regions indicated by the arrows are shown below. x400. Bottom left: Relatively unaffected region in the inferior retina. Bottom right: Lesioned area in the superior retina.

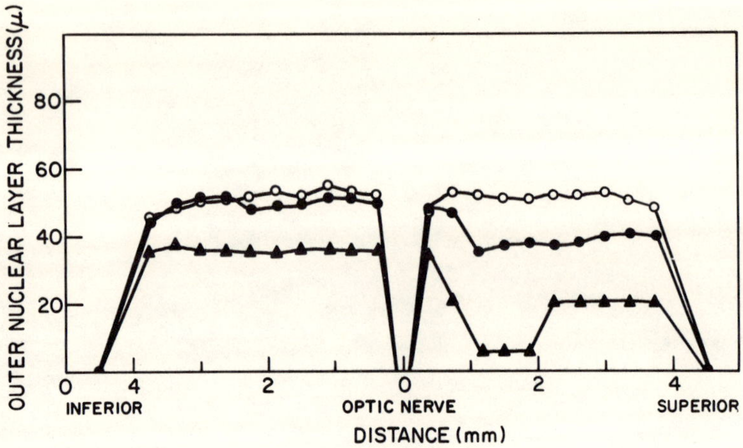

Figure 2. Regional distribution of photoreceptor cell loss in the retinas of albino rats exposed constantly to 40 lux. Open circles: control; open triangles: 4d exposure.

The same approximate distribution of photoreceptor cell loss as depicted above was found to occur at all of the light intensities investigated that produced damage to the retina. This was also true for the retinas of pigmented rats (Fig. 3). The consistency in observing the superior-inferior distribution of damage was apparent, but the causes of this effect were largely unknown. One possibility was that a higher irradiance existed at the superior retina in comparison to the inferior retina. This was not thought to be likely since measures were taken to provide uniform illumination in the exposure cages. Despite this precaution, the possible influence of lighting gradients that may have existed in the exposure cages was not ruled out. Experiments were conducted in which the distribution of damage in the retina was assessed following exposure to exaggerated light gradients. Light was provided either only from above or only from below the rat's cages. The average of the illuminance measurements taken in all directions was the same for both conditions but there was a 2 log-unit difference between the illumination in the up and down directions. The results of these experiments showed that lighting from above (which would supposedly cause a greater irradiance of the inferior retina as compared to the superior) was effective in only partially equalizing the severity of damage in the inferior and superior retina. The

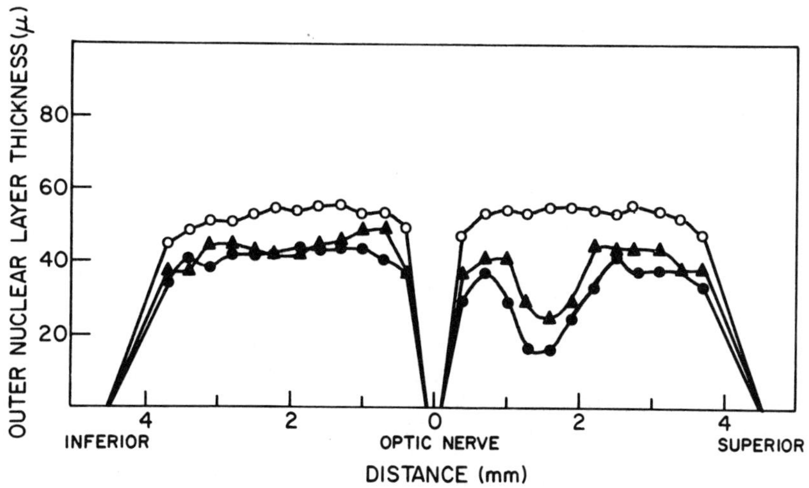

Figure 3. Regional distribution of photoreceptor cell loss in the retinas of pigmented rats exposed constantly to 40 lux. Closed circles: control; closed triangles: 2d exposure.

average thickness of the outer nuclear layer in the superior retina was still about one half that of the inferior retina. When lighting was provided from only below the superior retina was readily damaged, but the inferior portion was spared. With this lighting condition, as compared with lighting from only above, an exposure duration of only about one fifth as long was required to produce the required reduction in the superior outer nuclear layer.

Another control was performed which took into account the possible influences of a constricted pupil, the shape and position of the eyelids and the rat's behavior during exposure on the distribution of retinal light damage. In this experiment, collimated light from a xenon light source was directed on the eyes of anesthetized albino rats with their pupils dilated and eyelids retracted. The light passed through a green interference filter (Corning 4-67) and a heat filter (Corning 1-75). The duration of the exposure was four hours with the body temperature of the rat maintained at 38°C. With these exposure conditions, damage was again found to be most pronounced in the superior retina. A localized region in which almost all of the photoreceptor cells had been destroyed was found in the superior retina. This "lesion" in the retina, where only scattered photoreceptor cell bodies could be found again extended from around 700 to 2500 microns superior to the optic nerve head. All other areas along the verticle meridian showed only about a 50% reduction in the outer nuclear layer. Destruction of pigment epithelial cells was also found with xenon exposure. This damage was also localized to the region associated with the severe photoreceptor loss.

As mentioned earlier, it was intended that damage to the retina would be monitored by reductions in the outer nuclear layer. Since the above findings show that this measure is confounded by position in the retina, it was decided that only a specific area of the retina would be used for making comparisons among retinas with regard to light damage. The region of the superior retina which consistently showed the most severe damage was chosen for this purpose. Six measurements of the outer nuclear layer thickness were routinely taken at 250 micron intervals in the region extending from 750 to 2000 microns superior to the optic nerve head in the verticle meridian. The average of these 6 measurements will, hereafter, be reported as "outer nuclear layer thickness".

An important dependent variable measured in this work was the duration of exposure necessary to produce a criterion amount of damage. The criterion chosen for this purpose was a 35% reduction of the outer nuclear layer thickness. Damage at this criterion level (or of greater severity) was typically associated with disorganized outer segments and phagocytic cells in the area of photoreceptor degeneration.

Damage as a Function of Light Intensity

The experiments using rhodopsin actinometers confirmed that the effective intensity of light in the exposure cages varied in proportion to the measured illuminance. The first preparation utilized to address this point was cattle rhodopsin in solution. Within experimental error, the bleaching of cattle rhodopsin was found to be first order with intensity (Fig. 4). That is to say, the rate of bleaching was directly proportional to rhodopsin concentration and light intensity and thus followed the equation

$$-dR/dt = kIR \tag{1}$$

where R is the fraction of unbleached rhodopsin, k is the bleaching rate constant, and I is the flux density of the light incident on the retina.

The above measurements on cattle rhodopsin were sufficient to determine the effectiveness of the exposure lights on the bleaching of this visual pigment. But in order to directly relate the kinetics of rhodopsin to processes occurring in the rat's eye during light exposure, it was necessary to conduct similar experiments on rat rhodopsin *in situ*, (i.e. in eyes from enucleated rats). Previous studies indicate that no rhodopsin regeneration occurs in isolated rat eyes (19). If in addition no photoreversal occurred with the

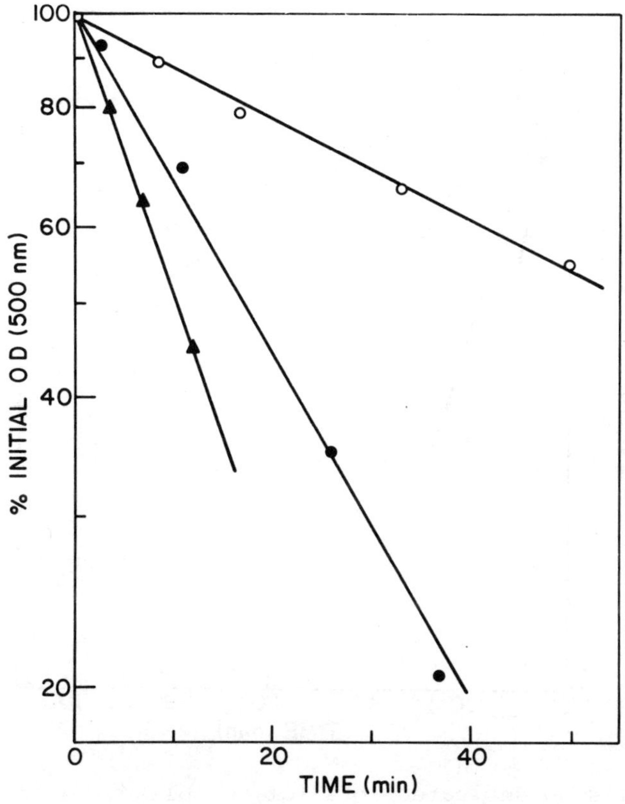

Figure 4. First-order rate plots for the bleaching of cattle rhodopsin solutions in the light exposure cages. Open circles: I = 5 lux; closed circles: I = 10 lux; closed triangles: I = 20 lux.

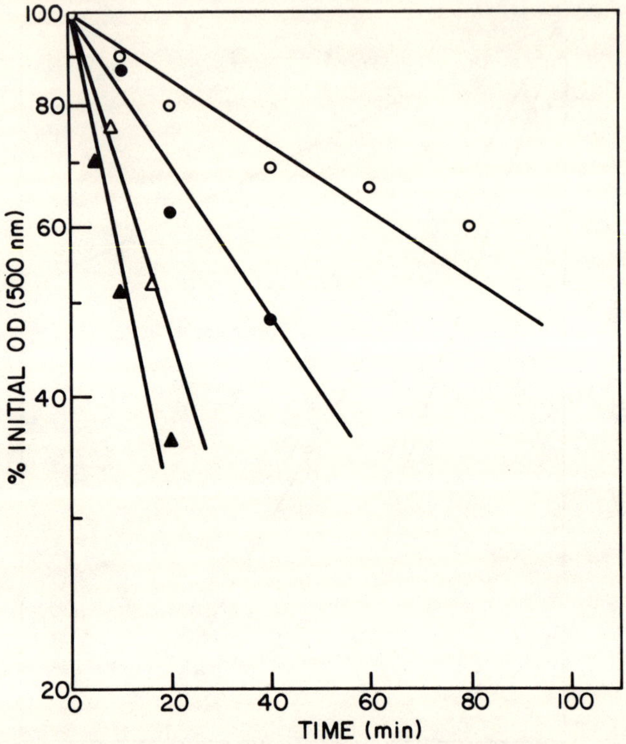

Figure 5. First-order rate plots for the bleaching of rhodopsin in excised albino rats' eyes by light in the exposure cages. Temperature of the eyes was maintained at 18-20°C. Open circles: I = 5 lux; closed circles; I = 10 lux; closed triangles: I = 40 lux.

exposure light used, bleaching in rat eyes should also be first-order with intensity. This result was in fact obtained for the bleaching of rhodopsin in enucleated eyes of albino rats (Fig. 5). It should be pointed out that bleaching of cattle rhodopsin _in vitro_ was about twice as fast as rat rhodopsin _in situ_. A possible explanation for this could be the absorption and/or scattering of light by the ocular media of the rat's eye.

Fig. 6 shows the time course of photoreceptor destruction in albino rats with exposure to light in the range of 10-160 lux. Five lux was also examined but no damage was observed with 16 days of exposure. Decreases in outer nuclear layer thickness did not occur gradually from the onset of light exposure. Instead, the control thickness was maintained for a length of time followed by a relatively rapid decrease. The initial period when no apparent damage occurred was longer for the lower illuminances. The latency for damage-onset decreased with increasing intensity. The interrupted line in Fig. 6 represents the criterion level of damage in the retina. The relationship between illuminance in the exposure

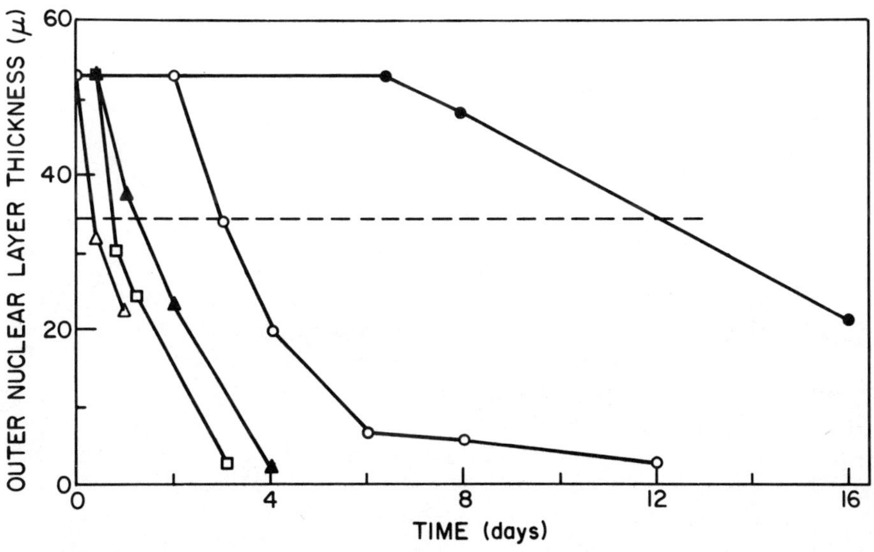

Figure 6. Time course of photoreceptor cell loss in albino rats exposed to constant light. Closed circles: I = 10 lux; open circles: I = 20 lux; open squares: I = 80 lux; open triangles: I = 160 lux. The interrupted line intersects the curves at a 35% decrease in the outer nuclear layer thickness which was the criterion level of damage.

cages and the time required to produce a criterion amount of damage is examined later.

Pigmented rats were also examined for retinal damage at several light intensities. But with the same range of light intensities used for exposure of albino rats, it was found that no damage to the retina occurred. This was also true for light intensities an order of magnitude higher. It was postulated that the opaque iris of pigmented rats was effective in lowering the retinal irradiance to non-damaging levels. An attempt was made to eliminate the screening effect of the iris by maximally dilating the pupil with drops of atropine sulfate (2%) on the eye. With just a single application the pupil remained dilated for several days. Under these conditions, the retinas of pigmented rats showed damage by constant light. The experimental results reported on pigmented rats will therefore only pertain to the light exposure of these rats when their pupils were maximally dilated.

Sufficient information was obtained for comparing albino and pigmented rats by examining the latter at only three light intensities. Figure 7 shows the time-course of retinal light damage for pigmented rats exposed to 40, 80 and 160 lux. As with albino rats,

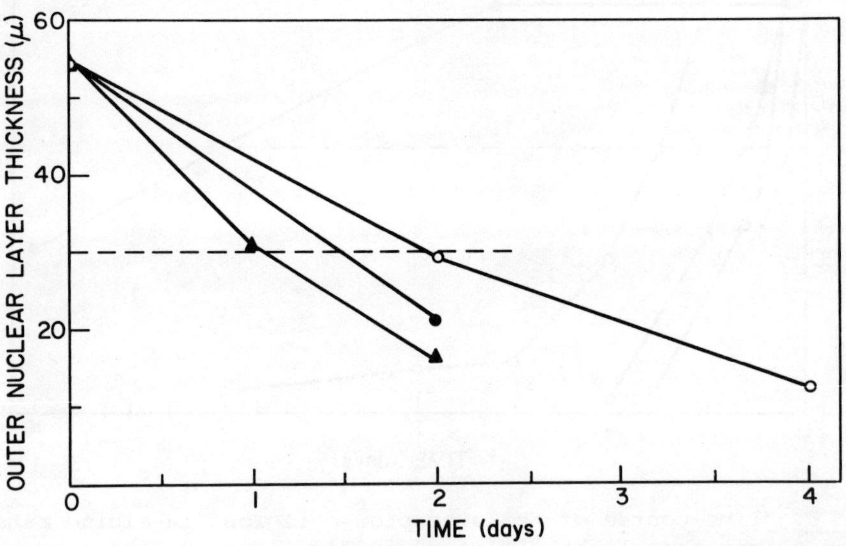

Figure 7. Time course of photoreceptor cell loss in pigmented rats exposed to constant light. Open circles: I = 40 lux; closed circles: I = 80 lux; closed triangles: I = 160 lux. The criterion level of damage is again represented by the interrupted line.

the plotted curves are non-overlapping with the highest light intensity causing damage with the shortest duration exposures. Lack of sufficient sampling at early time points does not allow a description of the initial plateau in Fig. 7. The criterion level of damage is again indicated by the interrupted line.

In Fig. 8, the exposure time required to produce a criterion amount of damage is plotted as a function of illuminance in the exposure cages for albino and pigmented rats. The points for the pigmented rats lie above those for the albinos at any given illuminance indicating the longer exposure times necessary to produce criterion damage in pigmented rats. The difference between albino and pigmented rats in this regard is rather small in comparison to the differences observed when the pupils of the pigmented rats were not dilated. In the case where chemical dilation of the pupil was not performed, no damage was observed even with sixteen days of exposure to the highest light intensity used on albinos. When the pupils of pigmented rats were dilated (Fig. 8), only about twice as much time was required to produce criterion damage in these rats as compared to albinos.

Damage to the Retina as a Function of Steady-State Rhodopsin Levels

The basic procedure used for measuring the rhodopsin content of the retina was as follows. The rat was anesthetized in the light exposure chamber and then immediately transferred to dim red light

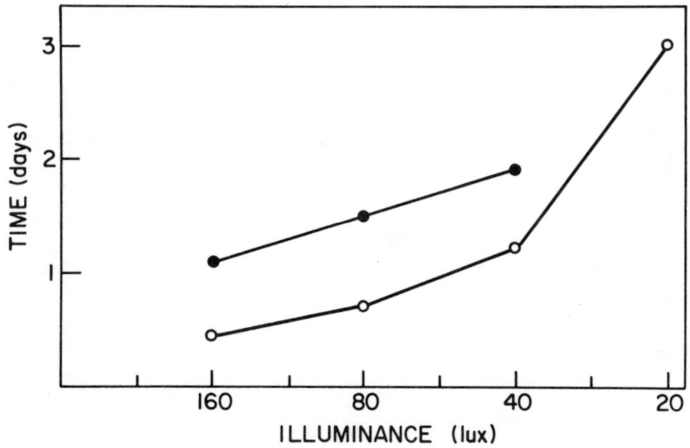

Figure 8. Time required to produce criterion damage as a function of illuminance in the exposure cages. Open circles: albino rats; closed circles: pigmented rats.

for dissection. The eyes were quickly removed and transferred to cold buffer (1-2°C) in the dark. The amount of rhodopsin in the retinas of these eyes could then be measured as described in a previous section. The amount of rhodopsin in retinas stored for several hours at 1-2°C in the dark did not change appreciably with time so it was assumed that negligible regeneration or thermal decomposition of rhodopsin occurred under these conditions. All values of the amount of rhodopsin in the rat retina are expressed as a percentage of the amount found in control rats. The controls were taken from the light cycle in the animal room and dark adapted overnight (14-16 h) before rhodopsin assay.

The existence of a steady state for rhodopsin bleaching in the rat retina is shown in Fig. 9. Rhodopsin levels in the retina were measured in albino rats at various intervals of exposure to constant light of 10 and 20 lux. At both light intensities the level of rhodopsin decreased rapidly during the first 30 min of exposure. Although the value of this initial reduction differed for the two intensities, there was, in common, a leveling-off of the rhodopsin content of the retina during the next 30 min of exposure (points not shown). A relatively constant level of rhodopsin was observed during the period from 1 h to at least 24 h after the onset of exposure. A secondary drop in the rhodopsin content of the retina occurred after a few days of constant-light exposure. The onset of this decline occurred sooner for the higher intensity and further experimentation showed that these (lowered) rhodopsin levels did not return to the control value with prolonged dark adaptation. It is therefore likely that this secondary decline of rhodopsin is a

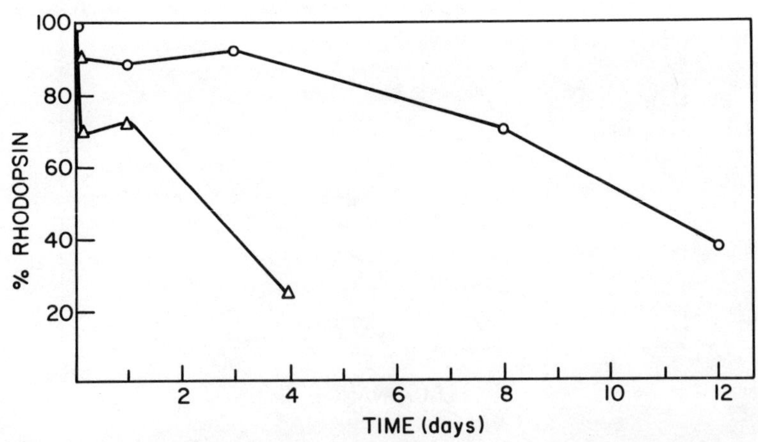

Figure 9. Rhodopsin levels in the retinas of albino rats during constant light exposure. Open circles: I = 10 lux; open triangles: I = 20 lux.

further manifestation of retinal light damage, i.e. it merely reflects the loss of photoreceptor cells. In any case, it was desirable to choose an exposure duration at which to measure steady state rhodopsin levels and not have these measurements confounded with cell death. On the basis of the data in Fig. 9, an exposure duration of 2 h was chosen for this purpose.

Steady state rhodopsin levels, as measured at 2 h after the onset of constant-light exposure, are plotted in Fig. 10 as a function of the illuminance during exposure for albino and pigmented rats. The curve for albino rats was found to be s-shaped but could not be fit by the hyperbolic function that would obtain from simple bleaching and regeneration kinetics. The curve for pigmented rats was not fully described since only three experimental points were obtained for these rats. It was clear, however, that in the range that was examined, there was greater steady-state bleaching (i.e. a lower steady-state level of rhodopsin) for albino rats than for pigmented rats at any particular light intensity. Looking at the data from a different perspective, it can be seen that the curve for pigmented rats is displaced horizontally to higher intensities

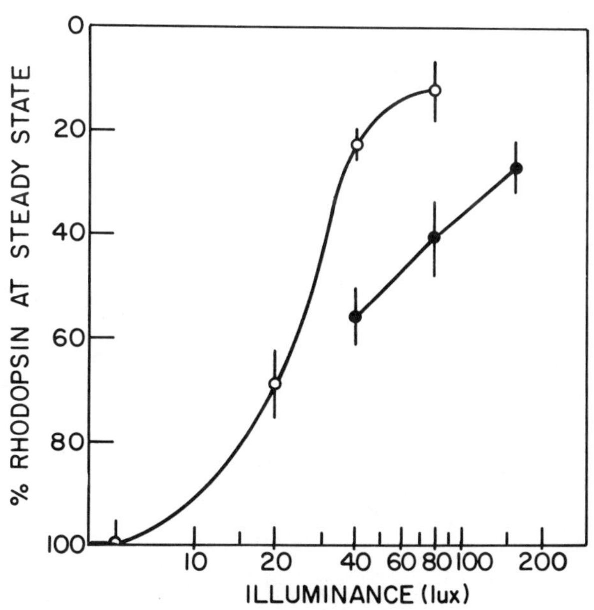

Figure 10. Steady-state levels of rhodopsin in the retina as a function of the illuminance in the exposure cages. Open circles: albino rats; closed circles: pigmented rats. Vertical bars denote the standard deviations of groups of two or three rats.

(relative to albino rats) indicating the higher intensity of light necessary to produce any particular level of steady state bleaching in pigmented rats.

The exposure time required to produce a criterion amount of light damage as a function of the steady-state rhodopsin level in the retina is one indication of the sensitivity of the retina to damage. With respect to the relationship between these two variables albino and pigmented rats show virtually no difference (Fig. 11). The points for pigmented rats in Fig. 11 lie slightly below those for albino rats, (i.e. at slightly shorter times to criterion damage) suggesting that if any difference does exist between these two strains it is in the direction that would indicate a greater sensitivity to light damage of the pigmented rat's retina.

DISCUSSION

The morphological manifestations of retinal light damage observed in this work were essentially the same as those which have been reported by other investigators who also used light microsocpy to examine this phenomenon (1, 20). The initial signs of damage to the retina included disorientation and fragmentation of the photoreceptor outer segments accompanied by the presence of invading phagocytic cells and edema. Degeneration of photoreceptor cells led to their death and disappearance from the retina. The pigment epithelial cells were not visibly affected by constant light

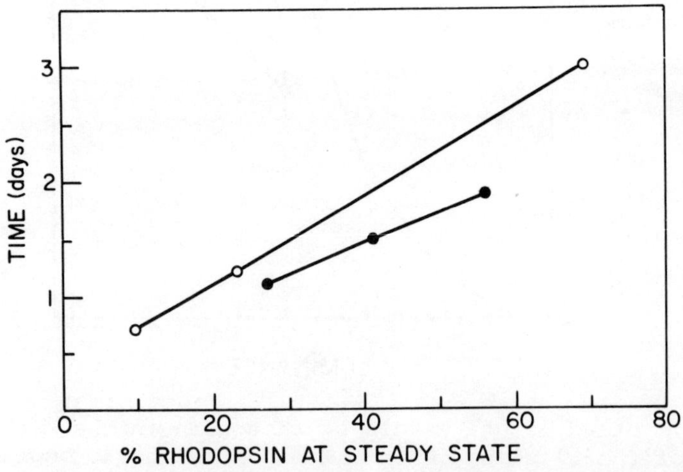

Figure 11. Time required to produce criterion damage as a function of the steady-state rhodopsin level in the retina. Open circles: albino rats; closed circles: pigmented rats.

exposure of low intensity (10-160 lux). However, the high-intensity xenon exposure did cause the destruction of pigment epithelial cells. Damage to pigment epithelial cells may thus have a higher threshold intensity than the photoreceptors or may be related to increases in temperature in the microenvironment of these cells.

The observation that photoreceptor cell death by constant light exposure was greater in the superior versus the inferior retina was a new and relevant finding. A direct application of this finding was made in this work by restricting the measurement of the outer nuclear layer thickness to a localized portion of the superior retina. In previous investigations on retinal light damage (21, 22), histological sections passing through the optic nerve head but without regard to orientation were obtained for the purpose of outer nuclear layer measurements. Although this procedure may have been adequate for the assessment of extensive damage in groups of several animals, it was not likely to be as sensitive or as consistent a type of measure as the one employed in the present study.

The experiments using exaggerated lighting gradients in part showed that damage to the superior retina was most effectively produced by light from below the rats. But regardless of whether the brightest light was below or above the rats, damage was more severe in the superior retina as compared to the inferior. Furthermore, the manipulation of other external conditions which were thought to be potentially related to the regional distribution of damage was in no case sufficient in reversing or equalizing damage to the superior and inferior retina. These altered conditions involved dilation of the pupil, retraction of the eyelids and anesthesia to eliminate behavioral effects. The implication of these control experiments was that the primary reason for the observed distribution of damage lies in intrinsic differences in the susceptibility of the photoreceptors, themselves, to damage by light.

A more specific understanding of the basis for superior-inferior differences in the retina comes from studies which have examined photoreceptor structure and composition in these two regions. In frogs, it has been shown that the superior region of the retina contains a higher concentration of visual pigment (23) and longer outer segments (24) as compared to the inferior region. Battelle and LaVail (25) found that both pigmented and dark-reared albino rats have longer outer segments in the superior versus the inferior retina. This difference was not observed in albino rats maintained in cyclic light. Although the rats used for experimentation in the present study were also maintained in cyclic light prior to constant light exposure, the morphology of their photoreceptors might have more closely resembled those in the dark-reared condition in Battelle and LaVail's study. This could have been since very low intensity light (5-10 lux) was used in this work during the light portion of the cycle which resulted in no measurable bleaching of rhodopsin at steady state. In any case the observations of Battelle and LaVail suggest that the two parts of the retina are different metabolically.

A study by Organisciak and Noell (26) suggests a possible link between the morphology/composition of rod outer segments and the susceptibility of the retina to light damage. They found that albino rats, maintained in complete darkness for at least two weeks, had rod outer segments that were longer and contained relatively higher concentrations of rhodopsin. A previous study by Noell and Albrecht (27) showed that dark-maintained rats are more vulnerable to light damage. Thus, the connection between vulnerability and receptor morphology/composition might be a metabolic state which is influenced by both darkness and position in the retina.

It was shown that in pigmented rats a considerable degree of protection from retinal light damage comes from the attenuation of light by the heavily pigmented iris. The pupils of these rats were observed to be as small as 0.1 mm in diameter during light exposure. This would reduce the retinal irradiance by over two log-units relative to the maximally dilated condition. Indeed, the dilation of pigmented rat's pupils with atropine sulfate reduced the threshold intensity for damage by more than two log-units. The possibility had to be considered, though, that atropine sulfate either caused damage directly to the retina or interacted in such a way as to cause an increased susceptibility of the retina to light damage. Control experiments were therefore performed in which damage to the retina was assessed for both the chemically dilated and normal condition in pigmented rats kept in darkness and in albino rats exposed to constant light. With these two situations there was no difference between the dilated and non-dilated conditions so it was assumed that the effect of this drug relevant to retinal light damage was merely the elimination of the screening of light by the iris of the pigmented animals.

Pigmented rats with chemically dilated pupils still required somewhat longer exposure durations to produce retinal damage as compared to albinos (Fig. 8). This may have been due to the ability of their pigment epithelium to absorb light as compared to the same structure in albino rats which completely lacks melanin and might thus permit the scattering of light back to the photoreceptors. A further possibility is that the aperture provided by dilation of the pigmented rat's pupil might still cause less light to reach the retina as compared to the non-dilated pupil of albino rats. This is because the entire eye-ball of an albino rat is somewhat translucent and can admit some of the light which strikes any part of its surface.

There was another possible explanation for the differences observed in Fig. 8 that was, however, discounted by further experimentation. This explanation was that, perhaps, the retina of albino rats was, in some way, inherently more susceptible to light damage than the pigmented rat retina. Experiments which determined the time to criterion damage as a function of steady-state rhodopsin levels in these two strains of rats contradicted this hypothesis. In fact, pigmented rats damaged slightly faster than albinos as a function of steady-state rhodopsin levels possibly indicating a greater susceptibility of their retinas to light damage. One reason for this difference could be higher concentrations of rhodopsin in the retinas of pigmented rats. The dark-adapted level of rhodopsin has been found by other investigators to be slightly higher in pigmented rats than in albinos (25).

SUMMARY

The major conclusions which can be drawn from this work are summarized in the following paragraphs.

1) The damaging effects of constant, low intensity light. Constant-light exposure of albino rats leads to a pathological condition in the retina characterized by the degeneration of photoreceptor cells. These cells eventually die and undergo phagocytosis. Other cells in the retina do not appear to be damaged by constant, low intensity light. Damage to the pigment epithelium does occur however when light exposure is associated with high light intensity.

The light used for the constant exposure of rats in this work could be considered to be of low intensity for two reasons. Firstly, the lowest intensity used to damage the retina was considerably dimmer than any previously reported. This light was on the order of a log-unit lower in intensity than what is typically found in indoor environments and laboratory animal colonies. Secondly, rhodopsin in the retina was not completely bleached at these intensities even though damage was caused and even though complete bleaching of the albino rat retina does readily occur at slightly higher intensities. Within the limit of a maximum of 16d for exposure, the lowest light intensity to cause damage to the albino rat retina only bleached about 10% of the rhodopsin at steady state.

2) Superior vs inferior retina. The superior region of the retina was more severely damaged by constant-light exposure than the inferior region. The region extending from around 700-2000 microns superior to the optic nerve head along the verticle meridian would typically show up to a 40% reduction in the outer nuclear layer while the inferior retina showed no apparent loss of cells. Control experiments, in which various external conditions were manipulated, indicated that the primary cause of the observed regional distribution of light damage in the retina was probably related to intrinsic differences in the susceptibility of the photoreceptor cells to damage.

3) Pigmented vs albino rats. The threshold intensity for causing damage to the retina is considerably higher (over two log-units) in pigmented rats than in albinos. The protection afforded by pigmented rats is primarily due to the screening of light by melanin in the iris. This can be surmised since maximal dilation of the pupil caused the retinas of pigmented rats to be damaged with exposure durations of only about twice as long as those which caused similar damage in albino rats. This relatively small difference between albino rats and pigmented rats with dilated pupils is not due to a differential susceptibility of their retinas to light damage. Experiments which measured steady-state rhodopsin levels showed that when exposure light is equated in terms of its "effectiveness" for bleaching rhodopsin, then damage occurs at about the same rate in these two strains of rats. Damage to the retinas of pigmented rats shows the same approximate regional distribution as in albino rats.

ACKNOWLEDGEMENTS

The authors thank Ms. Beverly Tranchand for her patience and skill in preparing this manuscript, Mr. B. Williams for drafting and photographing the figures and Mr. W. Howell for technical assistance with the histology. This work was supported by NEI grant 5R01EY02250.

REFERENCES

1. Noell, W. K., Walker, V. S., Kang, B. S. and Berman, S. (1966) Invest. Ophthal. 5, 450-473.
2. Adams, D. O., Beatrice, E. S. and Bedell, R. B. (1972) Science 177, 58-60.
3. Ts'o, M. O. M. (1973) Invest. Ophthal. 12, 17-34.
4. Marshall, J., Mellerio, J. and Palmer, D. A. (1972) Expt'l. Eye Res. 14, 164-169.
5. Lawwill, T. (1973) Invest. Ophthal. 12, 45-51.
6. Kuwabara, T. and Gorn, R. A. (1968) Archs. Ophthal. 79, 69-78.
7. Reuter, J. H. and Hobbelen, J. F. (1977) Physiol. Behav. 18, 939-944.
8. Gorn, R. A. and Kuwabara, T. (1967) Arch. Ophthal. 77, 115-118.
9. Kuwabara, T. (1970) Amer. J. Ophthal. 70, 187-198.
10. O'Steen, W. K., Shear, C. R. and Anderson, K. V. (1972) J. Anat. 134, 5-22.
11. Hannson, H. A. (1970) Virchows Archives B, 6, 247-262.
12. O'Steen, W. K. and Anderson, K. V. (1972) Expt'l. Neurology 34, 446-454.
13. Bennett, M. H., Dyer, R. F. and Dunn, J. D. (1973) Expt'l Neurology 38, 80-89.

14. Grignolo, A., Orzalesi, N., Castellazzo, R. and Vittone, P. (1969) Opthalmologica 157, 43-59.
15. Friedman, E. and Kuwabara, T. (1968) Archs. Opthal. 80, 265-280.
16. Rapp, L. M. and Williams, T. P. (1977) Nature 267, 835-836.
17. O'Steen, W. K. and Kraeer, S. L. (1977) Invest. Ophthal. 16, 940-946.
18. LaVail, M. M. (1976) Invest. Ophthal. 15, 64-70.
19. Cone, R. A. and Cobb, W. H., III. (1969) Nature 221, 820-822.
20. O'Steen, W. K. and Lytle, R. B. (1971) Amer. J. Anat. 130, 227-234.
21. O'Steen, W. K., Anderson, K. V. and Shear, C. R. (1974) Invest. Ophthal. 13, 334-341.
22. Olafson, R. P. and O'Steen, W. K. (1976) Invest. Ophthal. 15, 869-874.
23. Reuter, T. E., White, R. H. and Wald, G. (1971) J. Gen. Physiol. 58, 351-371.
24. Besharse, J. C., Hollyfield, J. G. and Rayborn, M. E. (1977) J. Cell Biol. 75, 507-527.
25. Battelle, B. and LaVail, M. M. (1978) Exp. Eye Res. 26, 487-497.
26. Organisciak, D. T. and Noell, W. K. (1977) Invest. Ophthal. 16, 188-190.
27. Noell, W. K. and Albrecht, R. (1971) Science 172, 76-80.

THE NATURE OF CHRONIC LIGHT DAMAGE TO THE RETINA

Theodore Lawwill, M.D., R. S. Crockett, Ph.D. and G. Currier
University of Louisville
School of Medicine
Louisville, Kentucky 40202

In 1966, Noell et al. (1), found that rats exposed to constant light with effective retinal irradiance of 11μW/cm^2 for a one week period had irreversible damage to their retinas. This occurred only if the animal had a pre-exposure dark adaptation period.

Since Noell's work, Kuwabara (2) has exposed a monkey fundus to 0.27 W/cm^2 from an indirect ophthalmoscope for 15 minutes and reported severe destruction seen histologically. They assumed an ocular transmission of only 0.36.

Lawwill has published thresholds for retinal damage in the rabbit (3) and monkey (4, 5) for four-hour exposures to broad band white light and to several lines of the argon laser.

There is no question that the susceptibility of the diurnal animals which have been tested is considerably less than that of the rat. There are also differences in the distribution of damage in the fundi of different species. The relative response to the blue end of the visible spectrum is also different between the rat and the diurnal animals tested. In the rat, Noell did not find the increased sensitivity in the blue found by Ham (6) and Lawwill (7) in monkeys.

In the studies reported here, we have used a Maxwellian view system, concentrating the incident beam at the iris plane. The beam is diverging from this point and evenly illuminates 50° of the retina. The intensity of the field varies no more than 2 to 1 from center to edge with the center area being most intense. The standard duration of exposure is 4 hours.

The first characteristic of importance is the distribution of visible damage in the fundus. In dutch belted rabbits, the area affected by threshold intensity was an oblong area parallel to the visual streak (Fig. 1). The second area affected was a similarly shaped area superior to the myelinated nerve fibers. This distribution is not correlated with any inhomogeneities in the exposing field since they would be concentric in distribution. Obviously, this specific area of the retina in rabbits is more sensitive. Furthermore, in eyes with varigated fundi, there was no obvious threshold difference between areas located over deeply pigmented retinal pigment epithelium and those over lightly pigmented retinal pigment epithelium. The transition in pigmentation was abrupt and often cut through the line of damage, but no difference could be seen.

In the monkey the area of greatest damage is perimacular with sparing of about 2° in the central macula. The damage usually extends out to the temporal vessels. The fluorescein angiographic sequence in Fig. 2 shows early filling defects in the choriocapillaris followed by window type defects of the pigment epithelium and then late staining of the retina-proper with stain pooling in the area just around the macula. A similar sequence is seen in Fig. 3 where retinal staining is more extensive.

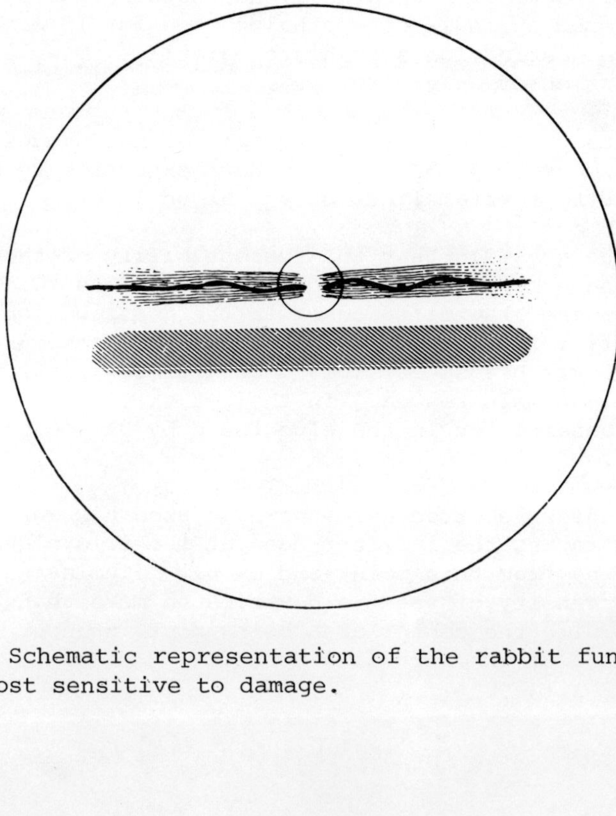

Figure 1. Schematic representation of the rabbit fundus showing the area most sensitive to damage.

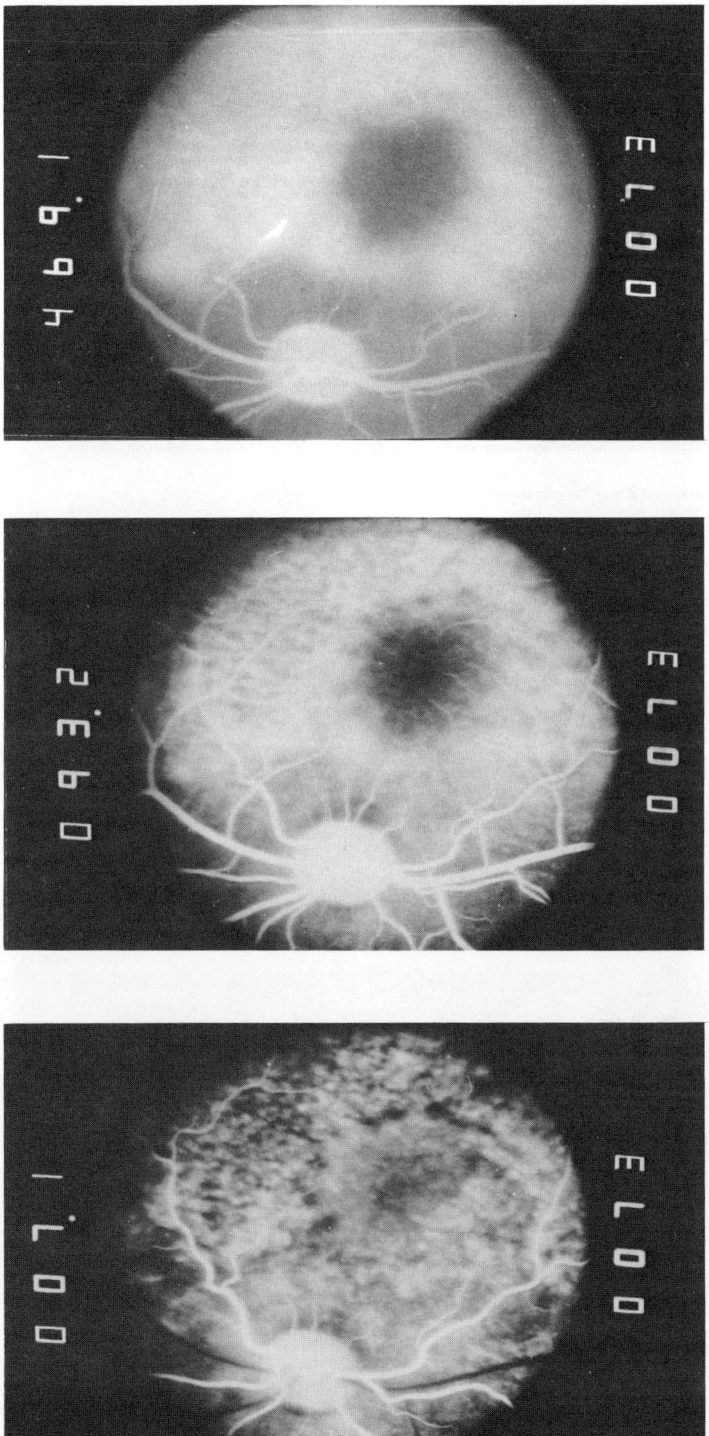

Figure 2. Fluorescein angiography of monkey fundus damaged by light. The first frame shows an early phase where choroidal filling is irregular. The second frame shows window defects in the retinal pigment epithelium. The third frame shows late retinal staining in the area of severe damage.

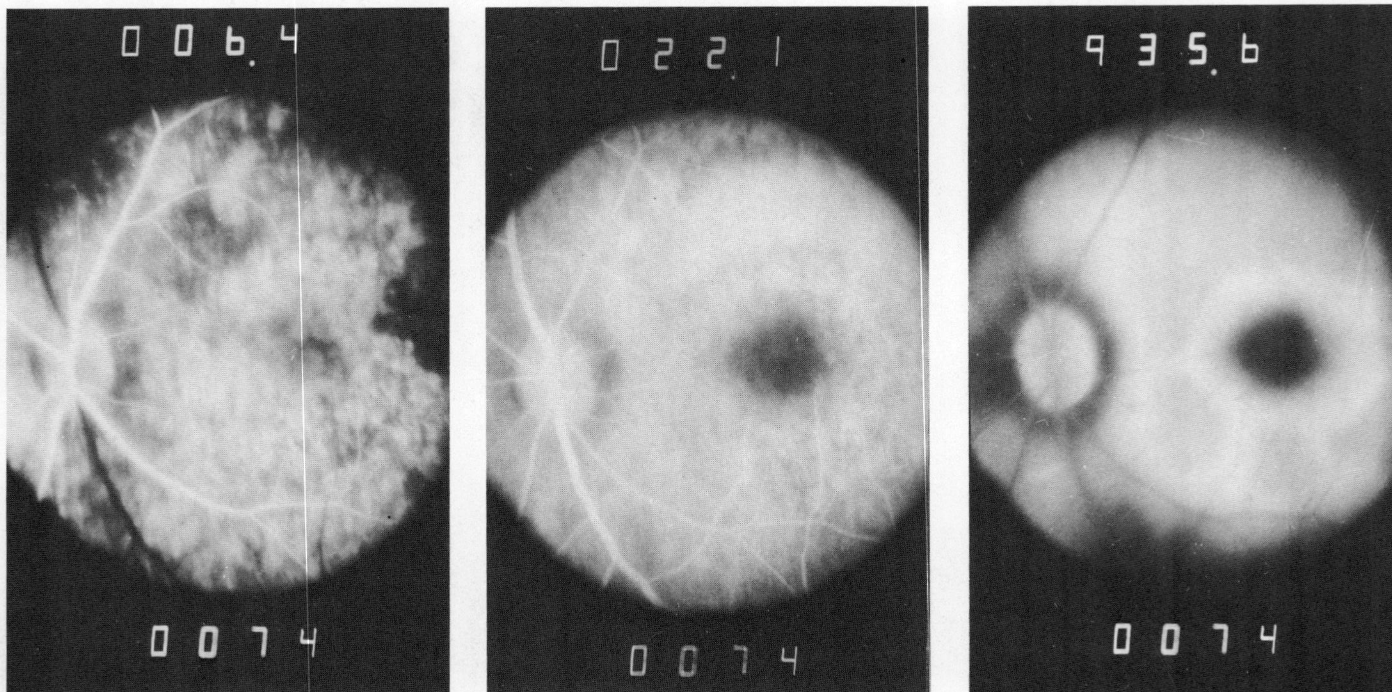

Figure 3. A sequence similar to Fig. 2 where irregular choroidal filling is more evident and late staining markedly limited to the parafoveal area.

We have typically followed the functional state of the retina after exposure with flash electroretinography. The amplitudes of the a- and b-waves of the dark-adapted ERG decrease immediately after exposure and continue to be depressed for at least two weeks and then begin to return toward normal (Fig. 4). In the most severely damaged eyes (Fig. 5) the levels may not return to normal for 30 days. Even in these severely damaged eyes there is almost always significant recovery.

The histological evaluation of the damaged retinas is carried out approximately two weeks after exposure. After fixation, the eye is carefully sectioned into 2 mm blocks and the position of each block in the fundus is recorded. The plastic-embedded blocks are cut one micron thick and stained with toluidine blue for light microscopy. Grossly, the area of greatest damage correlates with what is seen funduscopically. However, the damage is often patchy, not only as to the area of retina (adjacent areas differ sharply in the amount of damage), but also as to which layer of retina sustains the most damage (pigment epithelium versus nuclear layers).

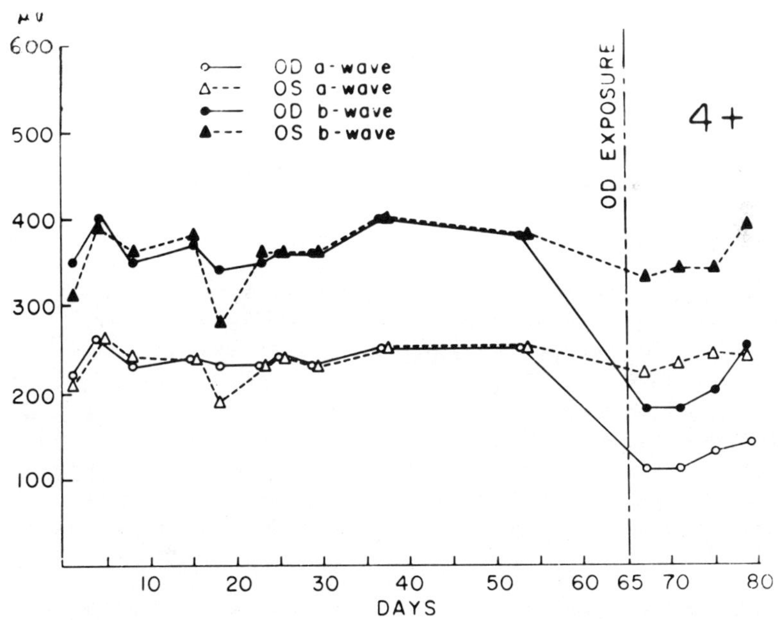

Figure 4. The flash ERG a- and b-wave amplitudes followed over time. The decrease shown here is a 4+ change and is about the most severe seen.

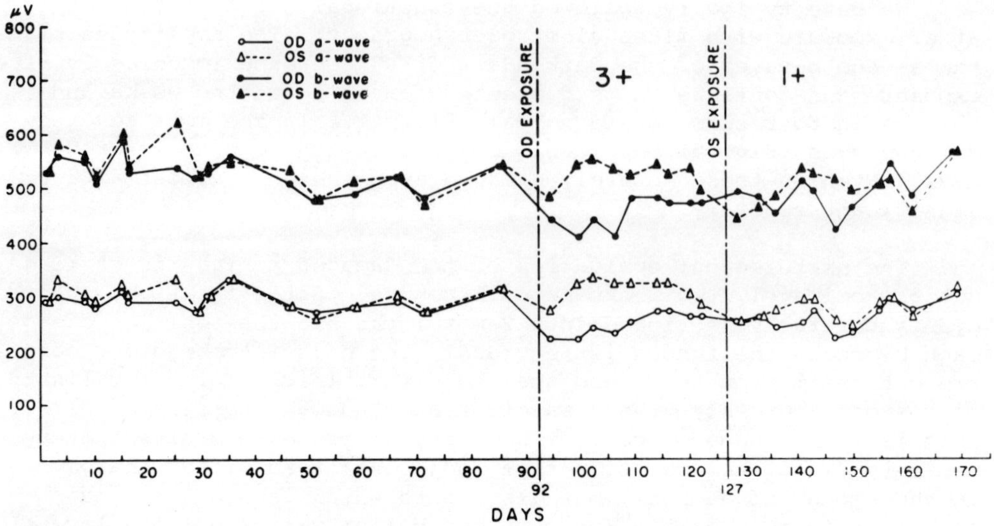

Figure 5. The flash ERG a- and b-wave amplitudes following exposure. In this case 30 days were required for the amplitude to return to normal.

A sharp transition from damaged to undamaged area is seen in Fig. 6. In the left two-thirds of this section, the nuclear layers are thinned, the pigment epithelium is totally destroyed, and few remnants of the outer segments remain. In the right third, the nuclear layers are thicker and there are more residua of the outer segments of the receptors. This is a typical sharp transition in degree of damage often seen within a homogenously irradiated area. There are areas where the pigment epithelium and outer segments are severely affected while the outer nuclear layer is well preserved. Also the pigment epithelium may be well preserved when there is fairly widespread loss of nuclei in the outer nuclear layer. Electronmicroscopically, one can see single necrotic cells adjacent to reasonably normal cells (Fig. 7). In several instances we have seen inclusions in the myoid which resemble outer segment discs (Fig. 8). We also see an increased incidence of migration of the nuclei into the inner segments of the receptor cells (Fig. 9). There is accumulated glycogen in the cytoplasm near the nuclei in the outer nuclear layers (Fig. 10) and in the cell processes (Fig. 11).

We use several techniques to look for pathological changes caused by the light exposures: opthalmoscopy, fluorescein angiography, electroretinography, light and electron microscopy. In order to record our results, we grade the degree of damage seen on a zero to 4+ scale. In each technique we use the grade of 1+ to

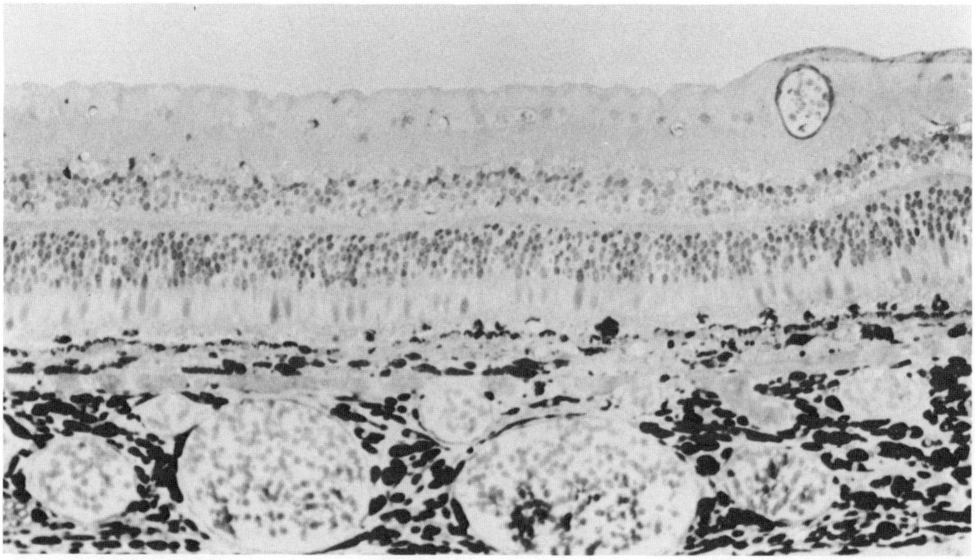

Figure 6. Retinal section showing transition from severely to less severely damaged areas. (x215).
(Reduced 10% for purposes of reproduction.)

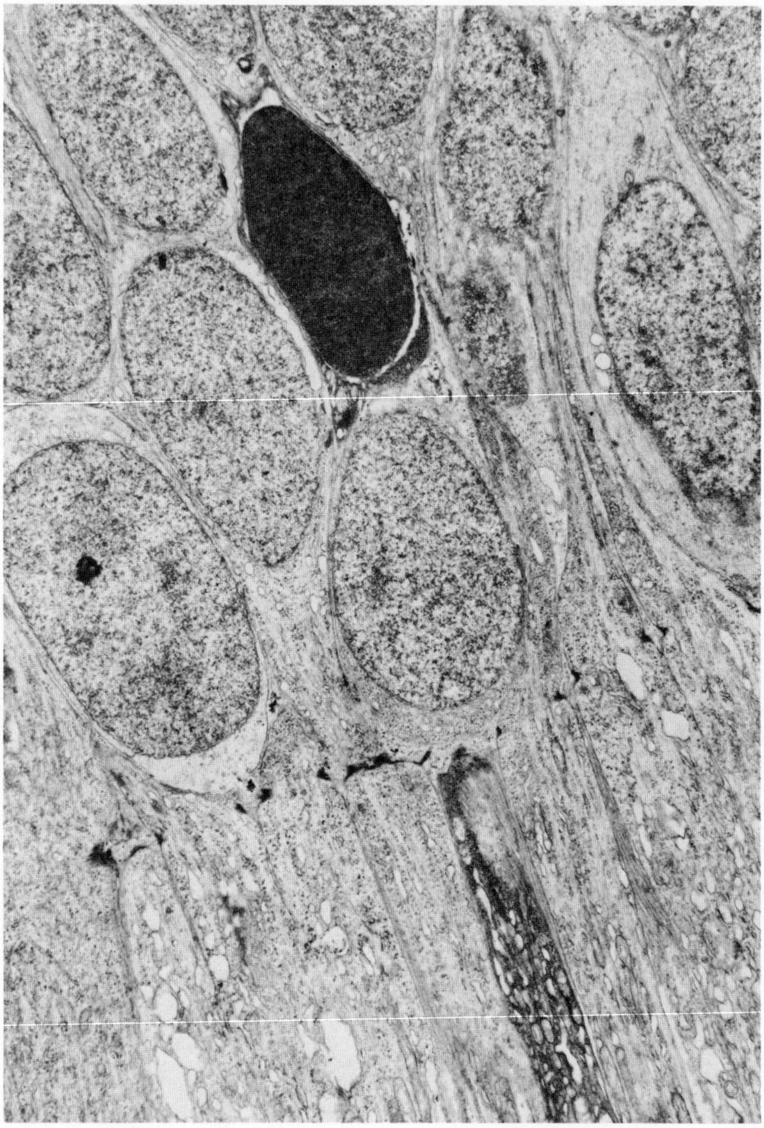

Figure 7. Electron micrograph of junction between outer nuclear layer and inner segments showing an isolated damaged cell. (x4,600).

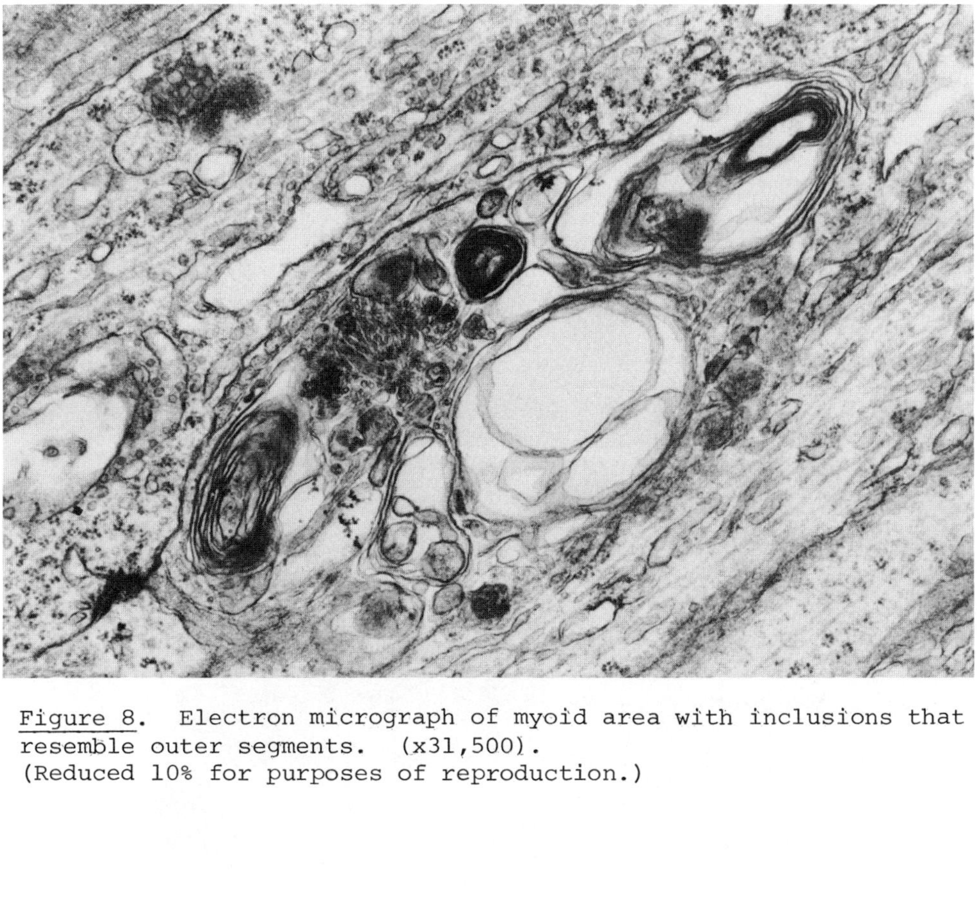

Figure 8. Electron micrograph of myoid area with inclusions that resemble outer segments. (x31,500).
(Reduced 10% for purposes of reproduction.)

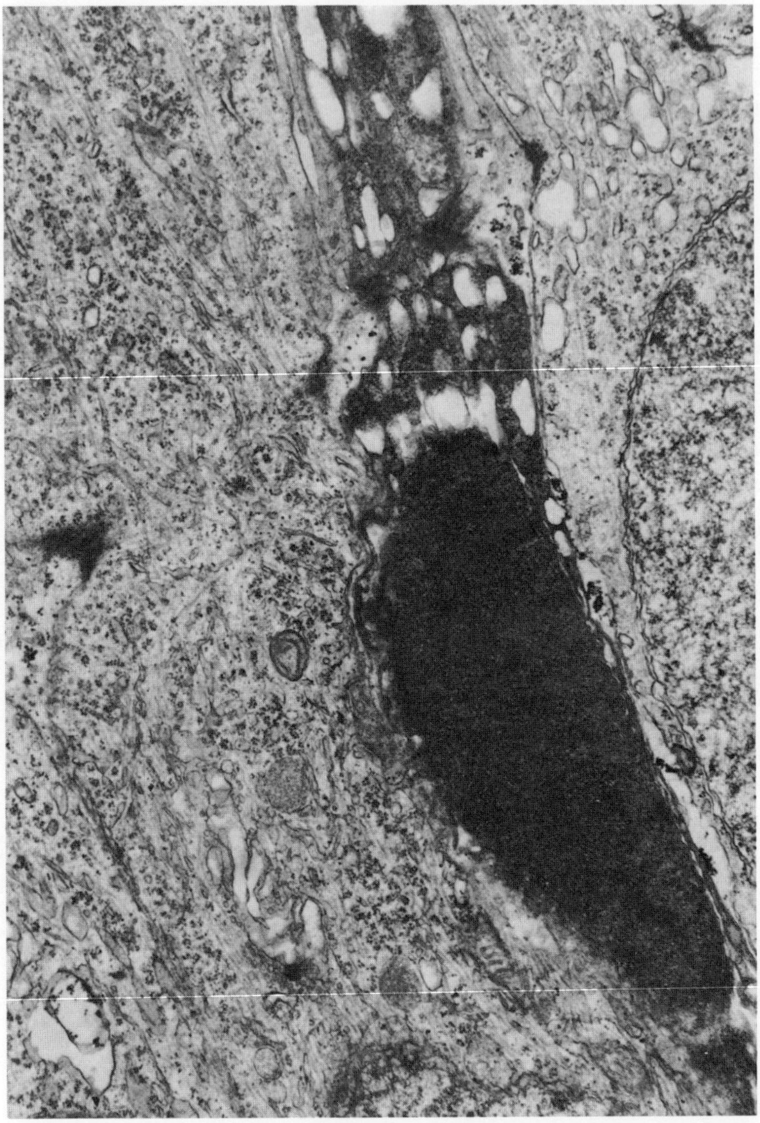

Figure 9. Electron micrograph showing nucleus displaced into inner segment. (x19,800).

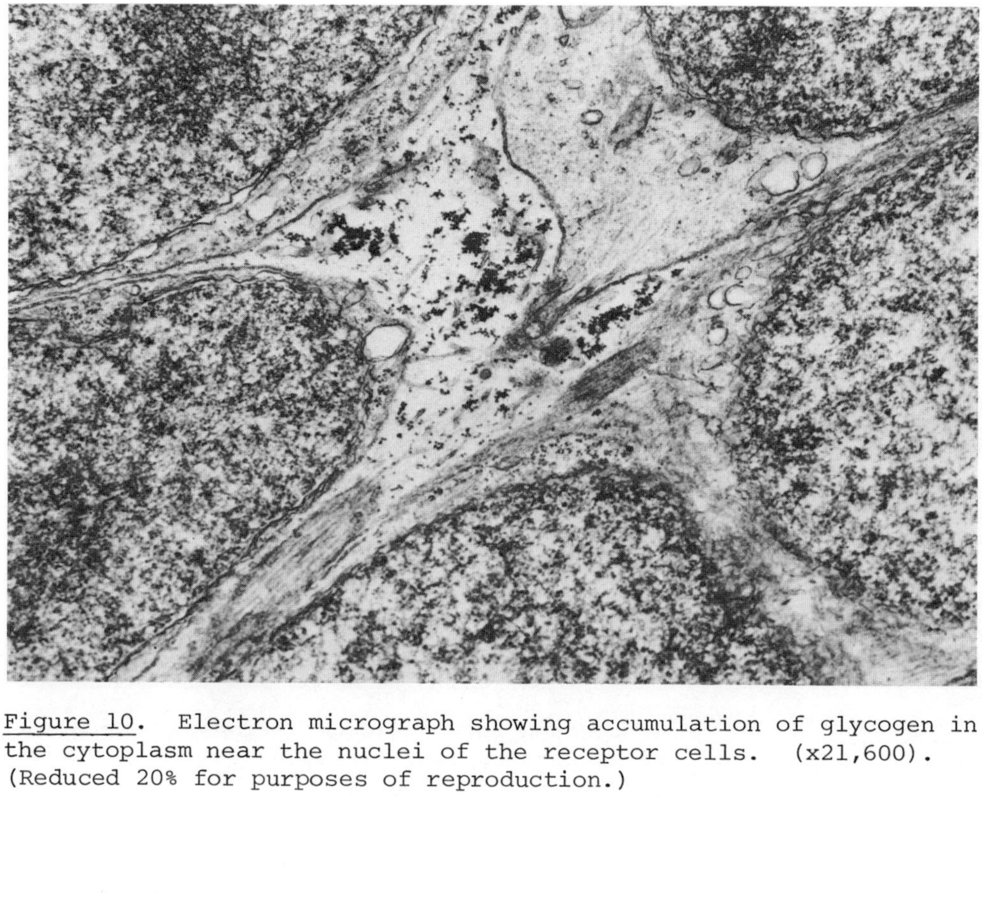

Figure 10. Electron micrograph showing accumulation of glycogen in the cytoplasm near the nuclei of the receptor cells. (x21,600). (Reduced 20% for purposes of reproduction.)

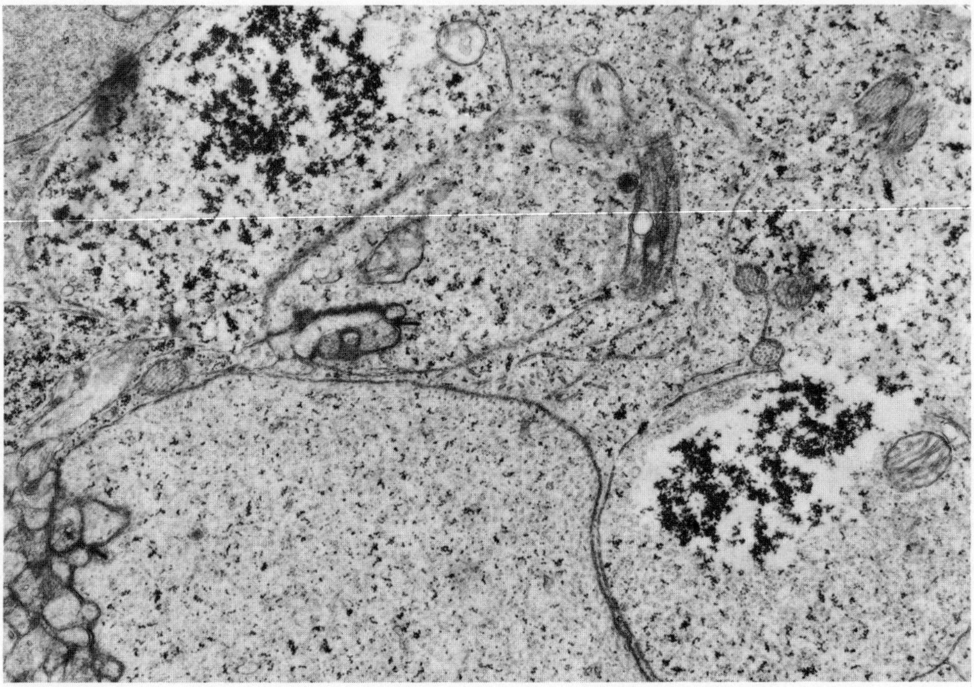

Figure 11. Electron micrograph showing accumulation of glycogen in the processes of the receptor cells. (x19,800). (Reduced 10% for purposes of reproduction.)

mean that we are better than 95% sure that we are seeing actual damage and not artifact. The grading in each case is done without knowing the intensity of the exposure. We have published (5) the criteria for our several grading systems.

We see excellent correlation between the grading systems for the several techniques. But, because a different technique may pick up damage earliest in any specific case, we use the grade for all techniques for each eye exposed. Our results are shown, as in Fig. 12, as a scattergram giving the grades for each test on each eye plotted with the abscissa being the intensity of exposure for the four hour exposures. In this case the damage scores for the 488 nm argon laser exposures are given along with the ERG amplitudes produced by 20 msec flashes of the same source plotted on the same intensity scale. This correlation with the ERG gives us another way to compare our measured intensities to a biological standard.

The importance of light damage to the retina must be made in relationship to every day experience and possible hazards. To do this, measures must relate to known light sources. Since the

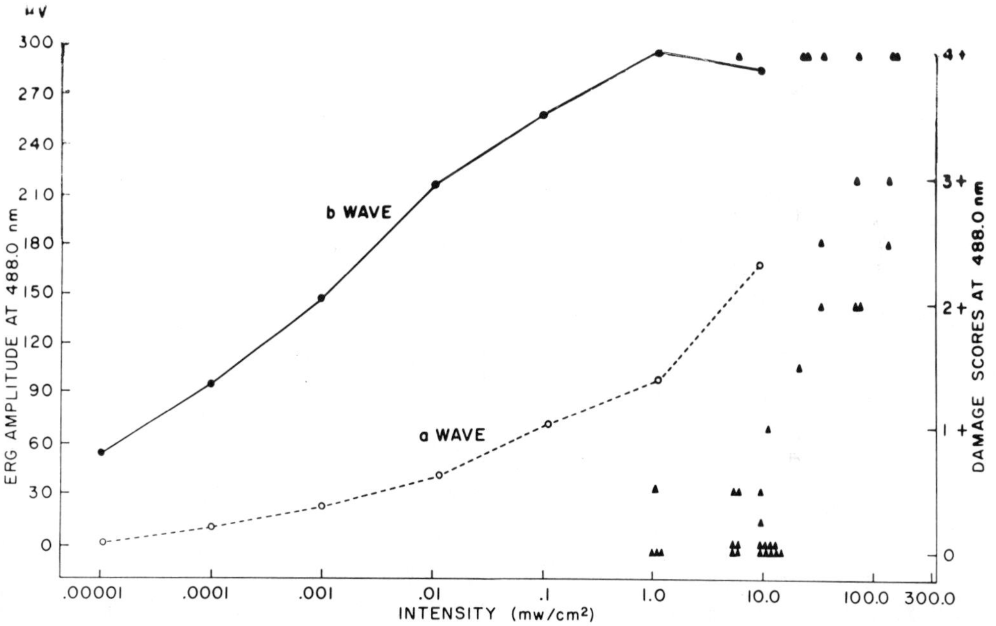

Figure 12. The damage grades for eyes exposed to 488 nm light at various intensities along with the ERG amplitude produced by 20 msec flashes from the same source plotted on the same intensity scale.

experimental laser sources are unlike most broad band sources, we have chosen a luminance or brightness scale for comparison (Fig. 13). There are certain defects in this approach, since the threshold for damage decreases in the blue end of the spectrum while visual sensitivity peaks in the blue-green portion. For the same physical intensity expressed in W/cm^2, the highly chromatic blue lines of the argon laser are more damaging than, but less bright than, the green lines of the argon laser or broad band white light.

By measuring the intensity at each visible wave-length for common ophthalmic diagnostic instruments one can estimate the equivalent damaging effect of that light using known data from the laser sources. We have measured the intensity of three common clinical instruments: the direct ophthalmoscope, the indirect ophthalmoscope (with 20D. lens) and the slit lamp with Hruby lens. None of the instruments were at their most intense setting but provided sufficient brightness for reasonable ophthalmic examination. However, these instruments are often used at higher intensities in the clinic.

In Fig. 14, we have given the curve for the retinal irradiance produced in the human eye by an old AO slit lamp microscope using the integral Hruby lens in the mode for directly examining the retina. On one scale, the intensity is given in $mW/cm^2/nm$. At 30 nm intervals the power is integrated and the total power for that 30 nm in mW/cm^2 is written. The total integrated power, beginning at the blue and going to the red end of the spectrum, is plotted. And, the thresholds for damage in monkeys at 457.9, 514 and 590 nm are plotted and connected by a dotted line. It is obvious that the retinal damage threshold for four-hour exposures in monkeys is at all times below the power available from the examining instrument.

While the indirect ophthalmoscope had only 70% as much power averaged across the field, there were differences in intensity of greater than 10 to 1 across the field at the plane of the retina. This suggests that hot spots may approach 10 times the average power. Thus, some retinal areas could be exposed to considerably higher intensity.

SUMMARY

The nature of retinal light damage from specific experimental light sources has been reviewed. Techniques of evaluating damage included ophthalmoscopy, fluorescein angiography, light and electron microscopy, and electroretinography. The variability of threshold in different areas of the fundus is noted. A comparison is drawn between the threshold intensity of light causing damage at specific wave-lengths and the intensity and wave-length distribution of light from commonly used ophthalmic diagnostic instruments.

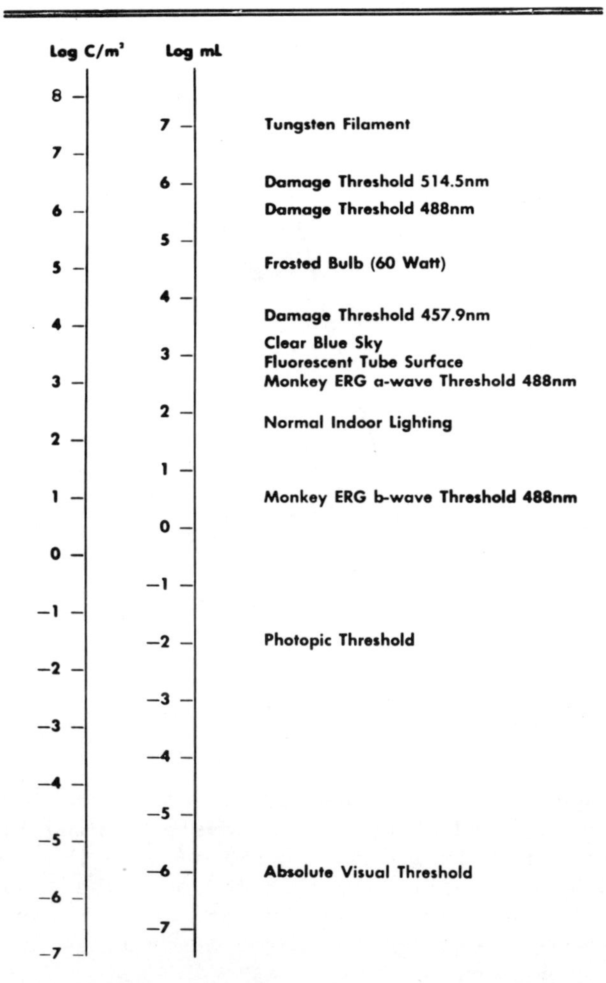

Figure 13. Common light sources and damage thresholds for three laser sources on a luminance scale. The comparison is distorted because the scale is based on visual sensitivity and not intensity.

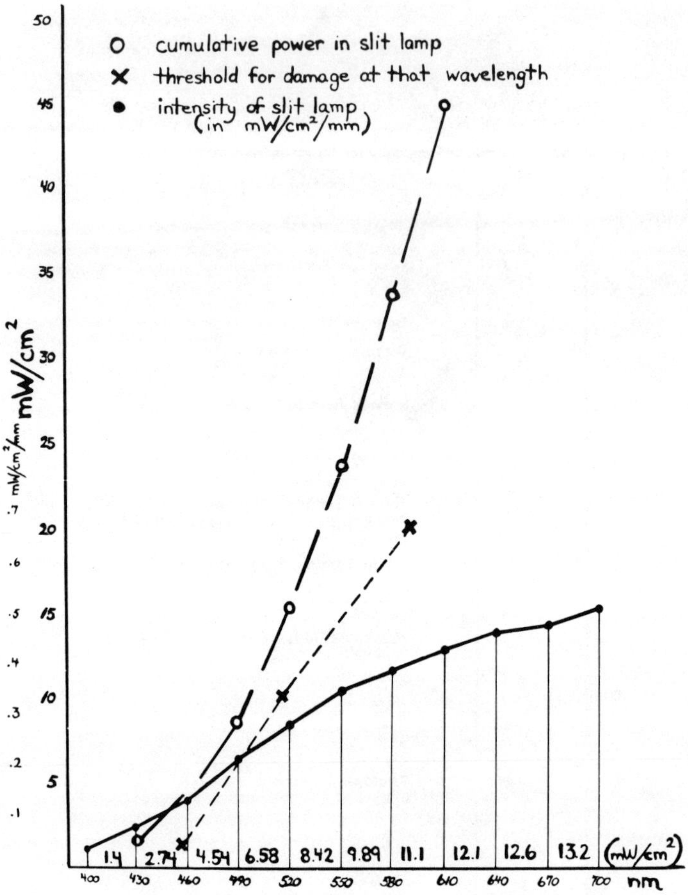

Figure 14. Graph showing retinal irradiance in the human eye produced by a slit lamp and Hruby lens fundus examining instrument. The spectral intensity is plotted in $mW/cm^2/nm$. The cumulative power for each 30 nm band is given in mW/cm^2. The cumulative power from this source beginning in the blue and going toward the red is shown. The damage thresholds for three laser lines are plotted and connected by a line.

The data from light damage studies suggests that the light intensity at which our retinas might be damaged in chronic viewing are perilously close to many every day environmental situations, but common sense and pain threshold are generally sufficient to keep us from injuring ourselves. There are a few situations where common sense gives way to interest such as during a solar eclipse when people do injure their eyes. However, diagnostic instruments are submitted to in the interest of good health and they must be carefully evaluated to be certain they provide more help than harm.

REFERENCES

1. Noell, W. K., Walker, V. S., Kang, B. S. and Berman, S. (1966) Invest. Ophthalmol. 5, (5), 450-473.
2. Friedman, E. and Kuwabara, T. (1968) Arch. Ophthalmol. 80, 265-279.
3. Lawwill, T. (1973) Invest. Ophthalmol. 12, 45-51.
4. Lawwill, T. (1972) Documenta Ophthalmologica Proceedings Series 2, 65-76.
5. Lawwill, T., Crockett, S. and Currier, G. (1977) Documenta Ophthalmologica Proceedings Series 15, 285-295.
6. Ham, W. T., Jr., Mueller, H. A. and Sliney, D. H. (1976) Nature 260, 153-155.
7. Lawwill, T., Crockett, S. and Currier, G. (1977) Documenta Ophthalmologica 44, (2), 379-402.

LIGHT DAMAGE IN DYSTROPHIC AND NORMAL RATS

M. Kaitz and E. Auerbach

Vision Research Laboratory
Hadassah University Hospital
Jerusalem, Israel

Albino rats of the Royal College of Surgeon (RCS) strain have a single gene mutation that causes a progressive retinal degeneration (1). The rate of degeneration is accelerated by exposure to light (2). Because the pathology of RCS rats resembles that of retinitis pigmentosa, these rats have been widely used as a model of the human condition. Over the last decade, it has become clear that light stress can also induce retinal damage in nondystrophic albino rats and that the histological picture of light-induced degeneration in RCS and nondystrophic albino rats is qualitatively similar (3) (see Discussion). In this paper, we shall review our own data on behavioral and electrophysiological measures of retinal dystrophy in normal and dystrophic albino rats. These studies together with the histopathological data suggest that one of the effects of the RCS mutation is to exaggerate the normal retinal sensitivity to light stress.

MATERIALS AND METHODS

Animals

The RCS rats used in these studies were pink-eyed, beige-hooded animals descended from the original RCS strain (4). Nondystrophic albino rats used for comparison in the behavioral studies were of the Fisher strain born to parents originally obtained from Charles River Laboratory (Boston, Mass.). In the electrophysiological studies, Sabra albino rats were used for control.

Rearing Conditions

Animals used in the behavioral experiments were reared from birth in cyclic light (12 hr light/12 hr dark) or in constant darkness. Those reared in cyclic light were housed in steel mesh cages. Illumination at the center of the cages was 194 ± 21 meter candles (mc) during the light period. The dark reared animals were housed in plastic cages kept in a lightproof chamber. When transferred to continuous light conditions, animals were exposed to either 430 ± 53 mc ("high") or 108 ± 21 mc ("low") illumination for a one month duration (Exp. 1).

For the electrophysiological experiments, animals were reared in one of three compartments of a lightproof chamber. Each compartment was equipped with one or two fluorescent lamps connected to a timer which provided the 12 hr light/12 hr dark cycle. The intensity and color spectrum of the illumination within each compartment was controlled by covering the lights with layers of metal screening and/or large colored gelatin filters. For Exp. 2, animals were raised in one of three levels of white illumination: 2239 ± 220 mc (high intensity), 391 ± 73 mc (low intensity) and 0 mc (darkness). For exps. 3a, b rats were reared under red, green or blue chromatic illumination with intensity adjusted to equate either energy flux (Exp. 3a) or to evoke the same ERG positive amplitude (Exp. 3b). In the former condition, radiant power at the center of the cages was approximately 3.2 watts/cm^2 and luminance levels in the red, green, and blue chambers were 40.9, 1560.7 and 120.6 candles/meter^2 (cdm^2), respectively. For the latter condition, the red chamber was illuminated at 40.9 cdm^2 and the blue and green chambers were dimmed to this value by a factor of 1.35 and 1.85 log units, respectively. (For spectral distribution of light energy within these housing quarters (see Fig. 1 in [5].) Animals in Exp. 4 were raised in high intensity illumination conditions (2239 ± 220 mc) of 12 hrs light/12 hrs dark or 6 hrs light/6 hrs dark/6 hrs light/ 6 hrs dark. Finally, for Exp. 5, illumination by white light measured at the center of the cage was 160 ± 67 mc.

In all animal quarters, the temperature was maintained at 22-24°C and the rats were fed ordinary rat chow and water throughout the experiments.

Behavioral Method

The conditioned suppression paradigm was used to measure brightness detection and spectral sensitivity thresholds in normal and RCS rats (6). The behavioral method has been described in detail in a previous publication (7). Briefly, an animal was engaged in a key pressing response at a high baseline rate (VR 96). At random intervals (VI 5 min), a light which warned of an impending footshock was presented. One could assess whether the animal detected the light by observing the presence or absence of response suppression during the stimulus-shock interval (1 min). To obtain brightness detection thresholds to white or colored lights, the stimulus light was modified by introducing neutral density or interference filters into the optical path. (For spectral characteristics of the stimulus light, see Fig. 1 in [8].) Determination of sensitivity were made at least four times weekly over a period of six months or longer.

Electrophysiological Method

Prior to the recording session, dark and light reared animals were light adapted (160 ± 67 mc) for 5 min and subsequently dark adapted for 3-4 hours. Following pretreatment with atropine sulfate (.05 mg/kg, subcu), the animals were anesthetized with sodium pentobarbital 40-60 mg/kg, ip). The left pupil was dilated with atropine sulphate drops (1%). The electrical responses were recorded through silver-silver chloride electrodes placed on the left cornea. The right and left ear served as ground and reference, respectively. Responses were amplified and recorded on an x-y plotter. The stimulus used to evoke the electroretinogram (ERG) was a 10 μsec flash of white light delivered from a Grass PS2 photostimulator positioned 10 cm from the eye.

The amplitude of the b-wave was measured from the trough of the a-wave to the peak of the b-wave. Responses less than 20 μV in amplitude were considered below threshold.

RESULTS

Intensity Dependence of Light Induced Effects

Light accelerates degeneration in RCS rats. From 2.75 to 6.0 months of age, dark reared RCS rats had brightness detection thresholds of 0.38 ± .03 millicandles/meter^2 (mc/m^2). Thresholds of rats raised in cyclic light were 1.90 ± 0.86 mc/m^2 from at least 2.5 to 7.0 months of age (7). Similarly, the electroretinographic responses of dark reared rats were of larger amplitude than those exposed to cyclic light conditions until at least 90 days of age (Fig. 1a). Nondystrophic albino rats, on the other hand, showed similar ERGs and brightness thresholds whether reared in cyclic light or darkness (0.34 ± 0.16 and 0.27 ± 0.19 mc/m^2, respectively) (Fig. 2b) (7).

However, the brightness detection thresholds of both RCS and albino rats increased upon exposure to continuous light (Exp. 1).

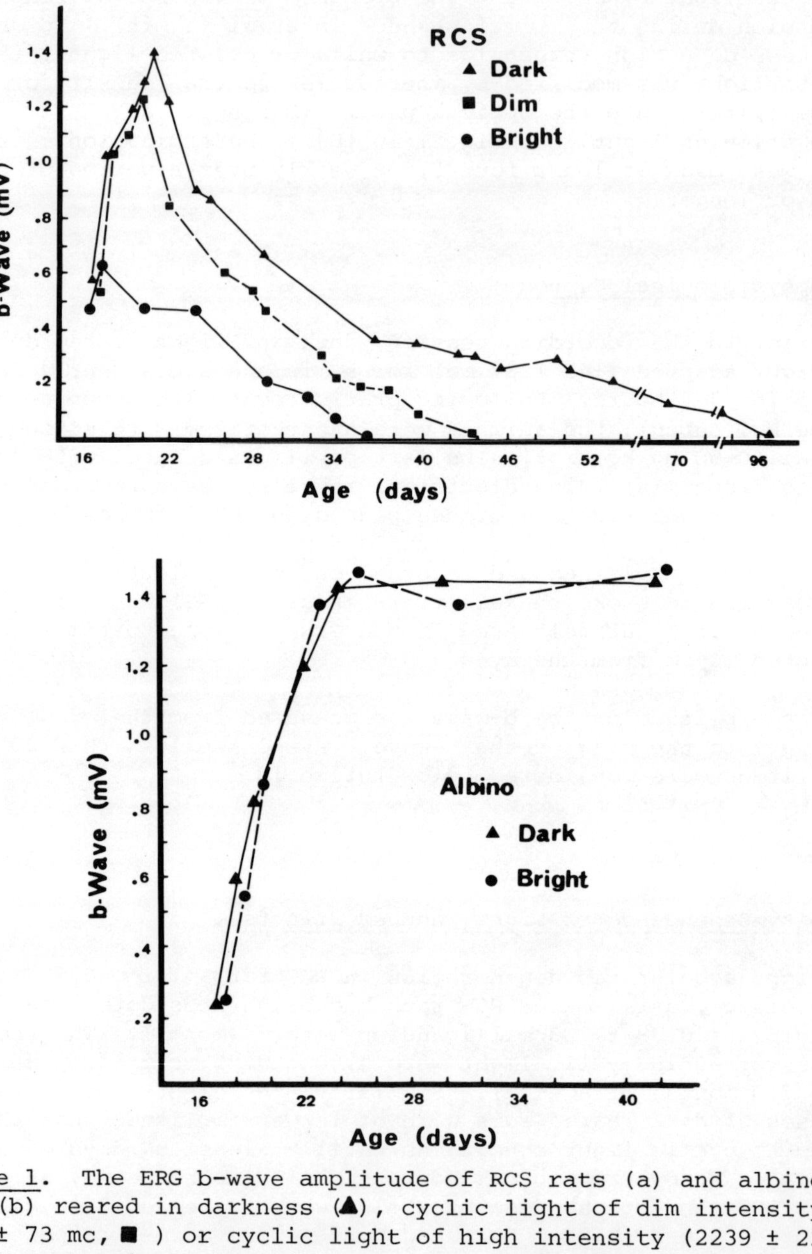

Figure 1. The ERG b-wave amplitude of RCS rats (a) and albino rats (b) reared in darkness (▲), cyclic light of dim intensity (391 ± 73 mc, ■) or cyclic light of high intensity (2239 ± 220 mc, ●). Each point constitutes the average ERG amplitude obtained from 2-4 animals. For all points, the standard deviation was less than 50 µV. (From Kaitz and Auerbach, "Retinal Degeneration in RSC Rats Raised under Ambient Light Levels," in Vision Res. 19, 79-81, 1979)

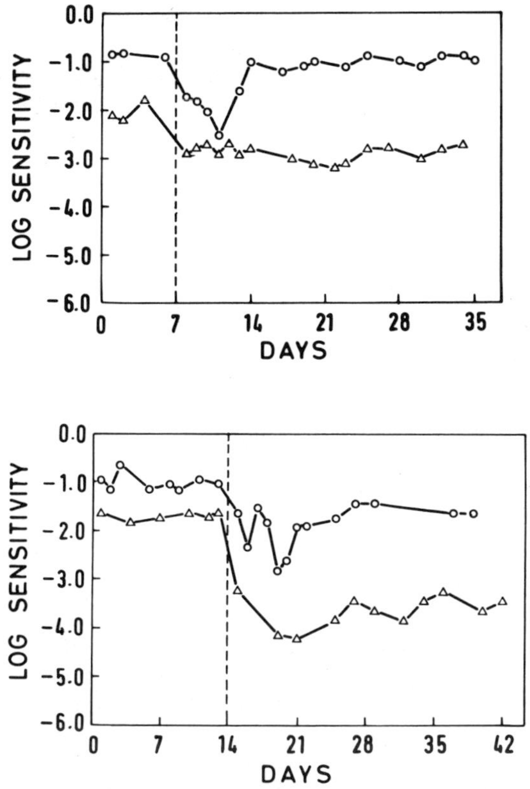

Figure 2. The effect of continuous illumination at 108 mc (a) or 430 mc (b) on the brightness sensitivity thresholds of RCS (Δ) and nondystrophic albino (O) rats. The rats were raised and maintained in cyclic light until day 7 (dashed line) then moved into continuous illumination until day 35. Initial variability is likely due to the upset in environmental lighting conditions.(From Kaitz, "The Effect of Light on Brightness Perception in Rats with Retinal Dystrophy," in *Vision Res.* 16, 141-148, 1976)

RCS rats reared in cyclic light and transferred to low intensity, continuous light conditions for one month showed a 1 log unit elevation in threshold (2.4-30.0 mc/m^2); exposure to high intensity illumination resulted in a 3 log unit decrement in sensitivity (1.9 mc/m^2 to 151.0 mc/m^2) (Figs. 2a, b). Albino rats, on the other hand, showed no change after a month of exposure to low level continuous illumination but a 1 log increase (.3 to 1.2 mc/m^2) in the threshold during exposure to high intensity continuous light, (Figs. 2a, b). Thus, both groups of rats are sensitive to light stress but the RCS rats are the more sensitive.

Electrophysiological data plotted in Fig. 1 illustrate the results from experiments in which albino and RCS rats were reared in darkness or cyclic light environments differing in the intensity of the illumination during the light cycle (Exp. 2). Starting at 22 days of age the ERGs (b-wave amplitude) of dystrophic rats reared in darkness declined from nonlinearly until at approximately 100 days of age, a criterion b-wave could not be evoked by the photostimulator flash. RCS rats reared in the dim cyclic light conditions showed an earlier onset of decline and a faster deterioration of the ERG b-wave than those reared in total darkness. When the level of illumination was raised, the dystrophic rats showed a still earlier onset and more drastic decline in the ERG amplitude. The nondystrophic albino rats were unaffected by low or high intensity cyclic illumination. However, if the light period is extended to 24 hrs/day, albino rats do show deficits which increase with increasing levels of illumination (3, 8, 9).

Time Course of the Light Induced Effects

RCS rats reared in cyclic light from birth and transferred to conditions of low intensity continuous illumination showed a drop in behaviorally measured brightness sensitivity during 24 hrs of light exposure. During the remaining 29 days of continuous light, their thresholds did not change (Fig. 2a). When exposed to high intensity light, the threshold changes, while more drastic, followed the same time course (Fig. 2b). Nondystrophic albino rats were unaffected by continuous low level illumination. However, they showed reduced sensitivity following the first day of exposure to continuous light of high intensity. As with the dystrophic rats the total loss in sensitivity occurred during the first period of extended light and thresholds stabilized after this (Figs. 2a, b).

Spectral Sensitivity Following Light Effects

Figs. 3a, b illustrate the change in behaviorally determined threshold of dystrophic (RCS) and nondystrophic rats before and after one month exposure to continuous high intensity white light illumination (Exp. 1). While the normal albino rats showed less overall decrement than do RCS rats, both strains showed a more profound deficit in detecting short wavelength light following continuous light exposure than exposure to medium or long wavelength lights.

Correspondingly, the action spectrum of the damaging effect of light follows the rhodopsin absorption curve (Exps. 4a, b). When the animals were raised in quarters illuminated by chromatic light equal in total energy flux, those maintained under red lights showed a slower decline in ERG amplitude than those reared

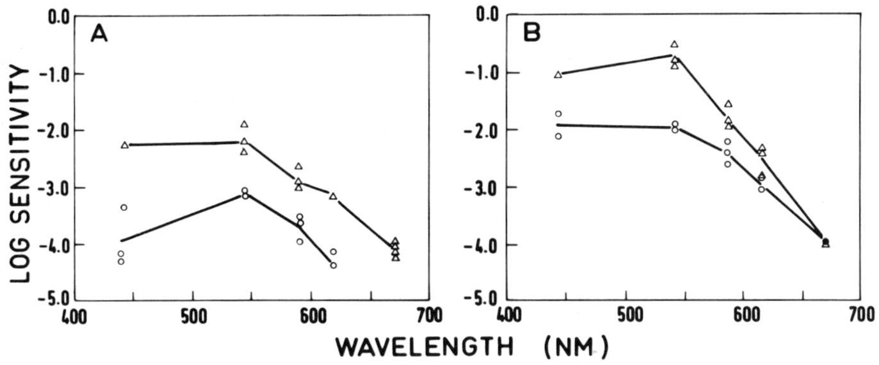

Figure 3. Spectral sensitivity of RCS (a) and nondystrophic albino (b) rats raised and maintained in cyclic light before (Δ) and after (O) four weeks exposure to continuous illumination to white light at 430 mc. (From Kaitz, "The Effect of Light on Brightness Perception in Rats with Retinal Dystrophy," Vision Res. 16, 141-148, 1976)

under the green or blue light conditions (Fig. 4a). When the colored lights were matched for equal energy absorption by rhodopsin, the ERG decline of all rats was identical (Fig. 4b). The nondystrophic animals exposed to the same chromatic illumination showed no decline in ERG amplitude at the intensity levels used regardless of the energy spectrum (Figs. 5a, b). However, it has been reported that if the light period is extended to 24 hrs/day, the exposure to green light results in a larger diminution in the a- and b-wave of the ERG than equal energy illumination at shorter or longer wavelengths. As in the present experiment, when chromatic

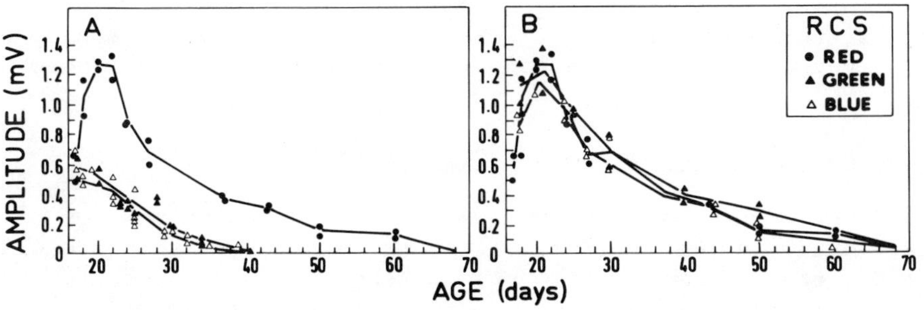

Figure 4. The time course of ERG (b-wave) amplitude decline of RCS rats reared under red (range: 590-690 nm), green (range: 480-565 nm) or blue (410-560 nm) lights of equal energy (a) or matched in luminance to be equally absorbed by rhodopsin (b). (From Kaitz and Auerbach, "Action Spectrum for Light Induced Retinal Degenerat in Dystrophic Rats," Vision Res. 19, 1041-1044, 1979)

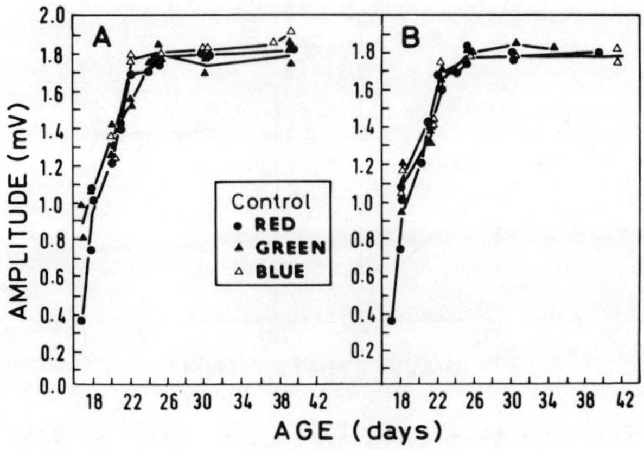

Figure 5. The ERG (b-wave) amplitude of nondystrophic albino rats reared under red, green or blue lights of equal energy (a) or matched luminance to be equally absorbed by rhodopsin (b). (From Kaitz and Auerbach, "Action Spectrum for Light Induced Retinal Degeneration in Dystrophic Rats," *Vision Res.* 19, 1041-1044, 1979)

light intensity was matched for scotopic efficiency, ERG reduction was the same regardless of wavelength (3, 9, 10).

Temporal Distribution of Cyclic Light

In Exp. 4, a comparison was made of the effects of cyclic light broken into 6 hr periods (i.e., 6 hr light/6 hr dark/6 hr light/6 hr dark) and cyclic light of 12 hr light/12 hr dark. It was found that at 30 days of age the ERGs of dystrophic rats reared in the former condition were one half the amplitude of rats reared in the latter dark condition. The ERG of albino rats was also affected by the fractionalization of light/dark intervals (Table I).

Age Dependence of Light Induced Degeneration

The effect of light on RCS rats is dependent on the age of the animals during exposure (Exp. 5). Rats born and raised in cyclic light and then transferred to the dark before 30 days of age actually improved in retinal response such that at 40 days of age, their ERG was equal in amplitude to that of animals reared in darkness from birth. However, animals maintained in light conditions longer than 30 days of age and then transferred to darkness show a permanent diminution in response amplitude and never equal to dark reared controls (Table II).

TABLE I

ERG a and b wave amplitudes of RCS and normal albino rats reared in cyclic light of 12 hrs light/12 hrs dark or 6 hrs light/6 hrs dark/ 6 hrs light/6 hrs dark. (From Kaitz and Auerbach, "Effect of Early and Late Light Exposure on Inherited Retinal Dystrophy in Rats." Exp. Eye. Res. 26, 699-704, 1978)

STRAIN	AGE (DAYS)	ERG AMPLITUDE (μV, $\overline{X}$) 12 HRS L/12 HRS D			6 HRS L/6 HRS D/ 6 HRS L/6 HRS D		
		N	A	B	N	A	
		3	36.9	342.7	3	42.8	177.0*
RCS	31	3	8.3	245.7	3	23.8	108.5*
AL	25	3	899.1	1731.6	2	600.0	1394.1*
AL	30	3	853.0	1639.7	2	366.3	1265.0*

*Significantly different from 12 hrs light/12 hrs dark condition, P < .05.

TABLE II

The mean ERG (b-wave) amplitude ± 1 S.D. of 40, 60 and 80-day-old RCS rats which had been transferred from cyclic light (12/12) to darkness (0/24) at various ages (days). (Kaitz and Auerbach, Ibid.)

RCS Age at transfer from 12/12 to 0/24 (N=)	ERG AMPLITUDE (μv) DAY 40	DAY 60	DAY 80
0 (0/24 controls)(2)	465.83 ± 65*	217.5 ± 35*	110 ± 03*
15-19 (5)	471.0 ± 23*	273 ± 50*	121 ± 32*
20-24 (4)	479.0 ± 30*	334 ± 80*	121 ± 35*
25-29 (4)	461.5 ± 80*	302 ± 18*	111 ± 63*
30-34 (4)	371 ± 23*	335 ± 68*	93 ± 20*
35-39 (5)	333 ± 43*	208 ± 30*	25 ± 12*
40-44 (4)		89 ± 2*	<20
60-64 (4)			<20
12/12 controls (2)	122 ± 20	<20	<20

*Amplitudes significantly larger than 12/12 controls p < .05.

O'Steen, Anderson and Shear (11) have shown for nondystrophic rats that there is less retinal damage following exposure to stressful illumination in weanlings than in older animals. Rats 3-4 weeks old were apparently unaffected by continuous exposure. Slightly older rats showed damage in only the central retinal regions. Beyond 3 months of age, exposure resulted in severe degeneration of photoreceptor and outer retinal layers over a broad retinal area.

DISCUSSION

A number of lines of evidence suggest that the process whereby light exposure damages the retina in RCS and nondystrophic albino rats is fundamentally the same. The differences are in degree, not of kind. This conclusion is based on common peculiarities in the time course, age dependence, spectral dependence and histopathology of light stress in the two strains.

Psychophysical Evidence

For both dystrophic and nondystrophic animals, the results of the present study support previous reports that the increment in threshold following continuous light exposure is related to the light intensity to which the animals are exposed (3). The RCS rat is, however, affected at lower intensities and shorter durations than the nondystrophic albino.

Although affected at different levels of illumination, the time course of the threshold changes following continuous illumination is similar for the two strains. Both show a sudden decline in visual sensitivity during the first 24 hrs of continuous exposure and a subsequent stabilization in threshold. Increased light intensity after stabilization results in a second sudden step of deterioration.

Finally, for both RCS and albino rats, threshold increases were greater at the short than at the long wavelength end of the spectrum. This observation suggests that in both strains the retinae contain more than one photopigment and that the pigment with the short wavelength spectrum is more affected by constant illumination (12, 13).

Electrophysiological Evidence

The decline of ERG amplitude of both RCS and albino rats during exposure to light stress is intensity dependent (3). Curiously, the decline in ERG amplitude does not follow the time course of threshold change seen in the behavioral paradigm. Dark reared rats exposed to cyclic illumination show a sudden increase in detection threshold whereas the decline of ERG amplitude is accelerated but still gradual, extending over several weeks. This peculiarity is seen in both strains of rats. In addition, adult cyclic-light-reared RCS rats with undetectable ERG (<20 μV compared to nearly 2.0 mV in nondystrophic albinos) have detection threshold only one log unit below that of nondystrophic albinos. Even more surprising, dark reared RCS rats with extinct ERGs show the same brightness detection thresholds as the controls.

The results of rearing animals in chromatic environments matched for energy or scotopic efficiency taken with the spectral sensitivity data of both strains following exposure to white light suggests that light damage is dependent on absorption by rhodopsin.

For both strains, animals are protected from the damaging effects of light when exposure is restricted to the first month of life. We have shown that for RCS rats, early light exposure produces reversible damage. Similar exposure to animals more than 30 days of age results in more permanent effects. Previous reports by O'Steen have shown that exposure to stressful light conditions cause only minimal structural damage in young rats whereas in older animals exposure to the same conditions results in severe and widespread retinal damage (11). O'Steen suggests that trophic hormones of the pituitary or their target organs may influence the susceptibility of retinal photoreceptors to light damage in older animals (14).

Finally for both strains, maintenance in fragmented light/dark intervals resulted in a reduction in ERG amplitudes compared to animals reared in 12 hr light/12 hr dark conditions. The importance of a day/night-like cycle has been underlined by recent work showing the circadian shedding of photoreceptor disks (15). It is possible that continuous light or light and dark periods of short time intervals could upset the intrinsic retinal timing mechanism, impede normal functioning, and consequently cause retinal deterioration.

Histological Evidence

The histological picture of light induced degeneration in albino and RCS rats are qualitatively similar. In light-deprived RCS rats, the accumulation of extralamellar material between the pigment epithelium and the apical processes of the rod outer segments is followed by swelling and misalignment of photoreceptors. Eventually the rod outer segments atrophy, leaving only a scant number of intact receptors. The peripheral receptors appear to be less affected than those of central origin. Concurrent with the receptor changes, pigment epithelium cells become disorganized and their number decreases. The retina is ultimately left with few receptors and the outer retinal layers have narrowed to such an extent that the inner layers are adjacent to a thin pigment epithelial layer (2, 16-18). The retinal deterioration seen in RCS maintained in light is strikingly similar to that seen in dystrophic rats reared in darkness from birth. The time course of the retinal degeneration however, is only accelerated as a consequence of the light stimulation (2, 17).

As with the dystrophic strain, during exposure to continuous light of low intensity or high intensity for a short duration, the retinas of normal albino rats show structural edema followed by alterations in the outer segments of the receptor cells and pigment epithelium cells (3, 19-21). As with the dystrophic rats, the receptors of the central retina degenerate more quickly than those in the periphery (13). The extent of disruption seen in the more proximal layers of the retina is dependent on the intensity and duration of illumination conditions. However, the effects on inner retinal layers due to light stress are less severe and probably secondary to the photoreceptor and pigment epithelial cell damage (22).

In summary, both albino and RCS strains show the main lesion in the rod outer segments followed by a secondary effect in the pigment epithelium. Receptors in the central regions are more light sensitive than those in the periphery. Further, the inner layers remain relatively normal in appearance even following complete destruction of the photoreceptor cells.

The Common Mechanism

The underlying basis for the light induced strain is unknown. While hypotheses have been suggested, no one proposal can explain all the available data. For instance, Noell and Albrecht have suggested that light damage may be due to a toxic photoproduct (23). Since vitamin A deficiency was shown to protect the retina from light damage, retinol was proposed as a candidate. The hypothesis has been generally disregarded because of its inability to explain several experimental results (see (24) for complete discussion). However, in agreement with predictions of the hypothesis it has recently been confirmed that excessive retinol can have deleterious effects on retinal structure (25, 26). Furthermore, the proposal has the virtue that it explains the hypersensitivity of RCS rats to light. Since the rhodopsin content of both light and dark reared RCS rats is higher than that of normal albinos, retinol concentration presumably exceeds normal levels following light exposure of high intensity. On the other hand, dark-reared RCS rats with twice the retinal rhodopsin of animals reared in cyclic light are no more effected by continuous light exposure than cyclic light reared RCS rats (2, 27, 28). In fact, the brightness thresholds of RCS rats reared in darkness and transferred to continuous light conditions for one month were lower than that of RCS rats reared in cyclic light from birth (29). In addition, administration of high levels of vitamin A (retinyl acetate) does not affect the time course of ERG decline in light-reared RCS rats (Kaitz, unpublished). Thus, for both RCS and albino rats, retinol concentration during light exposure is apparently not the critical factor in determining light induced damage.

Another suggestion is that light may disrupt cellular metabolism directly (22). In accordance with this hypothesis, Haanson has found depressed oxidative enzyme activity in the retina following continuous light exposure and a failure of the epithelial cells to transport nutrients to the outer segments (30). Insufficient data on light induced damage in RCS rats makes a comparison to the albino rat impossible. If, however, it is shown that light does disturb the metabolic transport from the pigment epithelium to retinal cells in the dystrophic strain, then light would only aggravate the already pathological relationship between pigment epithelium and receptor cells. Even in dark reared rats, an extralamellar barrier is seen interposed between the photoreceptor and pigment epithelial cells (2, 16). This layer of debris presumably affects the flow of metabolites even without further disruption from light exposure.

Gorn and Kuwabara (9) proposed that light may damage the membrane of the outer segments directly. While the suggestion remains plausible under some conditions, the pigment epithelial cells appear to be initially affected following exposure to extreme illumination levels (30). Thus, with high intensity light, the destruction of the outer segment must be preempted by another damaging mechanism. In addition, such a proposal requires that membrane properties be different for rods and cones and also depend on retinal eccentricity since in both dystrophic and nondystrophic rats the cones are less susceptible to light damage than rods, and peripheral rods are less affected than central rods (12, 13, 18).

The hypothesis more likely, at least for the RCS rats, is that the effect of light exposure is to further upset the relationship between the synthesis of new photoreceptor disks and membranes and the disposal of the old material. The pigment epithelium cells of dystrophic rats show an inability to carry out their usual phagocytic functions (32-34). This deficit may account in part for the accumulation of extralamellar material seen in the retinae of young RCS rats independent of rearing condition (2, 17). However, since light exposure increases the rate of photoreceptor turnover, retinal deterioration in dystrophic rats should and does occur in light at a faster rate than in darkness (36).

A comparison of the effects of light damage and the basis for such damage between normal and dystrophic rats is clearly difficult, given the paucity of available data. Whatever the ultimate answer, it appears that to maintain normal visual physiology, the illumination levels to which the animals are exposed must be restricted to moderate levels at cyclic intervals approximating a day/night cycle. For those animals with degenerative retinae, even exposure to moderate light levels is damaging.

In summary, albino rats show retinal degeneration following exposure to continuous light or to high intensity illumination for short durations. The retinal dystrophy seen in RCS rats is accelerated by exposure to light. Based on electrophysiological and behavioral data, it appears that while the RCS rat has a far lower threshold for the light induced damage, the effects of damage from light stress are similar between strains. The data are related to the prevailing hypotheses on the basis of retinal damage caused by light.

ACKNOWLEDGEMENTS

We thank M. Devor, D. Lebow and I. Perlman for their help in the preparation of this manuscript.

REFERENCES

1. LaVail, M. M., Sidman, R. L. and Gerhardt, C. O. (1975) J. Heredity, 66, 242-244.
2. Dowling, J. E. and Sidman, R. C. (1962) J. Cell Biol. 14, 73-109.
3. Noell, W. K., Walker, V. S., Kang, B. S. and Berman, S. (1966) Invest. Ophthal. 5, 450-473.
4. Bourne, M. C., Campbell, D. A. and Tansley, K. (1938) Br. J. Ophthal. 22, 613-625.
5. Kaitz, M., Auerbach, E. (in press) Vis. Res.
6. Estes, W. K. and Skinner, B. F. (1941) J. Exp. Psychol. 29, 390-400.
7. Kaitz, M. (1976) Vis. Res. 16, 141-148.
8. Kaitz, M. and Auerbach, E. (1977) Ophtahl. Res. 9, 351-356.
9. Gorn, R. A. and Kuwabara, T. (1967) Archs. Ophthal. 77, 115-118.
10. Anderson, K. V., Coyle, F. P. and O'Steen, W. K. (1972) Exp. Neurol. 35, 233-238.
11. O'Steen, W. K., Anderson, K. V. and Shear, C. L. (1974) Invest. Ophthal. 13, 334-339.
12. Cicerone, C. M. (1976) Science 194, 1183-1185.
13. LaVail, M. M. (1976) Invest. Ophthal. 15, 64-70.

14. O'Steen, W. K. (1977) Exp. Eye Res. 25, 361-369.
15. LaVail, M. M. (1976) Science 194, 1071-1073.
16. LaVail, M. M., Sidman, R. L. and O'Neal, D. (1972) J. Cell Biol. 53, 185-209.
17. LaVail, M. M. and Battelle, B. A. (1975) Exp. Eye Res. 21, 167-192.
18. LaVail, M. M., Sidman, M., Rauzin, R. and Sidman, R. L. (1974) Vis. Res. 16, 693-702.
19. Kuwabara, T. and Gorn, R. A. (1968) Archs. Ophthal. 79, 69-78.
20. Grignolo, A., Orzales, N., Castellazzo, R. and Vittone, P. (1969) Ophthalmologica, 157, 43-59.
21. Shear, C. R., O'Steen, W. K. and Anderson, K. V. (1973) Am. J. Anat. 138, 127-132.
22. Parnavelas, J. G. and Globus, A. (1976) Exp. Neurol. 51, 171-187.
23. Noell, W. K., Delmelle, M. C. and Albrecht, R. (1971) Science, 172, 72-76.
24. Noell, W. K. and Albrecht, R. (1971) Science 172, 76-79.
25. Vento, R. and Cacioppo, F. (1973) Exp. Eye Res. 15, 43-49.
26. DeWar, A. J., Barron, G. and Reading, H. W. (1975) Exp. Eye Res. 20, 63-72.
27. Delmelle, M. C., Noell, W. K. and Organisciak, D. T. (1976) Exp. Eye Res. 21, 369-380.
28. Battelle, B. A. and LaVail, M. M. (1978) Exp. Eye Res. 26, 487-497.
29. Kaitz, M. (1976) Invest. Ophthal. 15, 153-156.
30. Haanson, H. A. (1971) Exp. Eye Res. 12, 270-274.
31. Kaitz, M. and Auerbach, E. (1978) Exp. Eye Res. 26, 699-704.
32. Bok, D. and Hall, M. O. (1971) J. Cell Biol. 49, 664-682.
33. Herron, W. K., Riegel, B. W., Brennan, E. and Rubin, M. L. (1974) Invest. Ophthal. 13, 87-94.
34. Edwards, R. B. and Szamier, R. B. (1977) Science 197, 1001-1003.
35. Young, R. W. (1967) J. Cell Biol. 33, 61-72.

PROLONGED INTENSE SPECTRAL LIGHT EFFECTS ON RHESUS RETINA

Harry G. Sperling

University of Texas Health Science Center
Sensory Sciences Center
Houston, Texas 77025

To understand the interactions between receptors which result in color and brightness discriminations, it would be valuable to know the spatial distribution of color receptors. It would be useful just to learn the percentage of red-, green-, and blue-sensitive cones as a function of retinal position. It would be much more useful to know the detailed classification within small regions from receptor to receptor, especially if it were true in primates, as it is in fish, that there is a regularly repeated pattern in the cone mosaic.

The following will describe a series of experiments, in which we have been engaged for about ten years, that have answered some of these questions and show promise of answering the others. These studies involve the interplay of several disciplines, ranging from psychophysics performed on humans and monkeys to anatomical and histochemical studies performed on monkeys and fish. The work grew out of our ability to obtain psychophysical increment-thresholds on rhesus monkeys which were as detailed and quantitatively accurate as those which we had been obtaining on humans. This enabled us to obtain baseline data, then induce pathology by intense light exposure, perform subsequent psychophysical measures and then histology all on the same animal.

The intense light effects which constitute one of the techniques--the results from which help us to identify the distribution of color receptors--has had another equally important goal of obtaining evidence, on a species close to man, on the spectral and intensity parameters of photic damage to cones. Fig. 1 shows the typical optical system which we have used throughout our psychophysical experiments. It consists of a three path maxwellian view

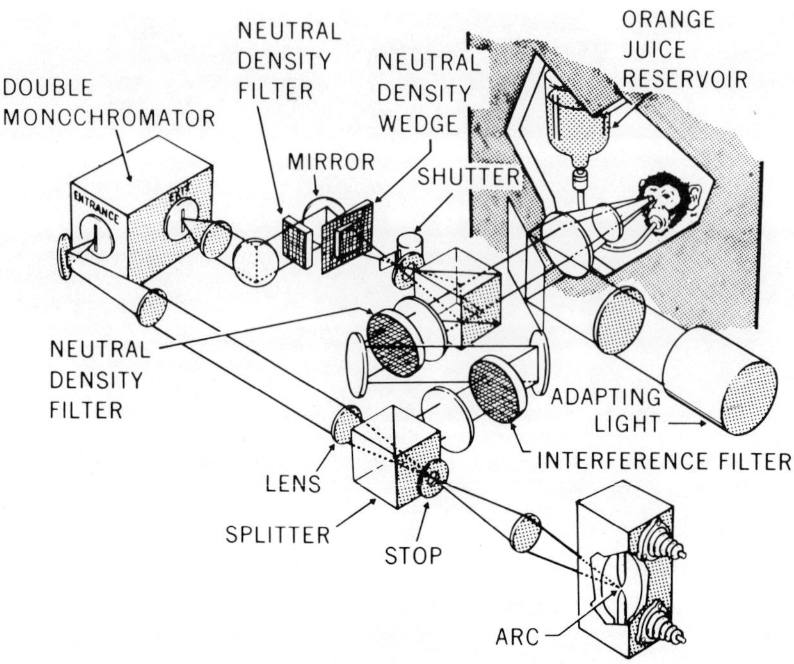

Figure 1. Diagram of apparatus for psychophysical measurement of increment thresholds on monkeys (from Ref. 2).

with a xenon arc (Osram 1600 watt) as the source for the intense spectral exposure path and for a spectral path which provides a 2° circular test light in the center of the 18° background field. The third, white background, path is from a tungsten iodide source, also aperatured to 18° and focused in the pupil of the eye. The three aperatures are in conjugate focal planes so that the viewing distance may be placed at infinity, for relaxed accommodation. The images in the pupil are smaller than the smallest natural pupil at the background intensities used. The image of the test target is largest, so that lateral movements of the head vignette the test beam, reducing its intensity, hence the probability of its detection. Since the animal works to obtain his daily fluid in the experiment, he learns to hold the beam centered in his pupil. We have demonstrated this by infrared television viewing. We have also shown that the animal places the test light on the fovea by making tiny laser burns on the retina of a trained monkey through the optical system during the course of threshold determination (1).

The procedure for obtaining thresholds is described in detail by Sidley and Sperling (2). Briefly, the animal is trained both to look for a test light flash in the center of the background field and to hold a lever down, when he hears a tone. If he releases the lever within 500 ms after the test light, he hears a beep each time and receives a portion of synthetic orange juice on a fraction of those times. The size of the portion of the fraction are arranged to obtain thresholds for twenty-two wavelengths from violet to red, once in each direction in a single session, before the monkey's thirst is slaked. In a good subject, all of the animal's daily fluid needs are met in the experimental session which may last several hours. This is especially valuable in obtaining the intense light effects which will be reported.

The threshold criterion is two consecutive (SPO) flashes with no response in a descending intensity series. The number of quanta in the first flash is taken as the threshold. Fig. 2, open circles, shows a typical increment-threshold spectral sensitivity function obtained against a 3000 Troland, 5500°K neutral background. We always plot thresholds in log (reciprocal quanta) as a function of

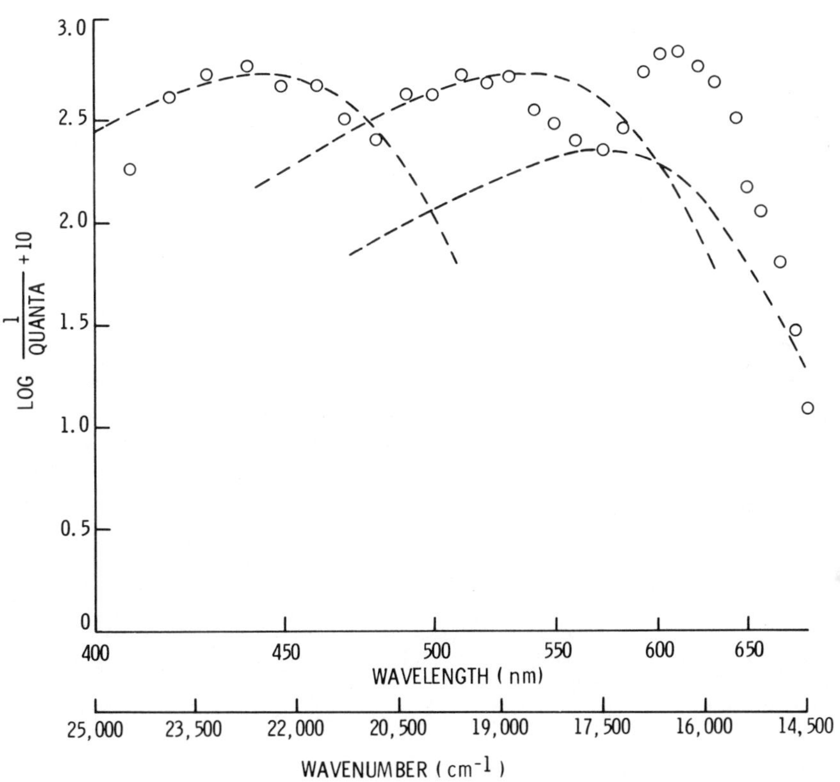

Figure 2. Mean spectral sensitivity for one monkey compared with generalized absorption spectra of the three classes of primate cones (from Ref. 4).

equal wave number intervals. Here the thresholds are compared with the abstracted absorption spectrum of the three classes of primate cones after Marks, Dobelle and MacNichol (3). A major finding in our work with monkeys and humans is that when measured against intense neutral backgrounds, we always find three peaks in the spectrum at about 440, 540 and 610 nm. The shortest wavelength peak is very well fit by the absorption spectrum of the short wavelength sensitive class of cones, corrected for pre-retinal absorptions. The middle and long wavelength peaks are too narrow to be fit by input from only one class of cones, and the 610 peak is 35-40 nm too far toward the red-end of the spectrum to be fit by red-sensitive cones which peak near 570 nm. Therefore, we have proposed, and have shown considerable data to support, a model of the increment-threshold sensitivity of primates (Table I) which assumes the three peaks to result from three independent channels. Sensitivity is determined by the most sensitive channel, acting alone, over a spectral region. This allows us to account for the sharp dips near 480 and 580 nm and the way in which they shift with changing adaptation. The narrowed peak at 540 nm, and the narrowed and displaced peak at 610 nm are accounted for by inhibitory interaction between the red- and green-sensitive cones in channels 1 and 2 of the model. We have found through a series of neutral and chromatic adaptation studies on monkeys and humans (4) that linear subtractive interaction between functions representing the absorptions of the red- and green-sensitive cones best models the inhibitory interactions in the red and green channels.

Following the statement and experimental testing of the model, we began studies (5, 6) of the effects of intense, prolonged spectral light exposures on these functions in rhesus. Fig. 3 shows results of blue light exposures on two monkeys. These exposure intensities lie approximately midway between the upper limit of light intensities encountered in nature and the threshold for gross burn damage to the retina. They were achieved by filtering the arc lamp with a 6 nm half band-width interference filter, with peak transmission at 463 nm. The sensitivity peak at 440-445 nm, which represents the

Table 1. Equations of Spectral Sensitivity (SS) Model (Sperling and Harwerth, 1971), where Q is number of quanta at threshold; R, G, B the three regions of the spectrum served by the three independent channels; α_{575}, α_{535}, α_{445} are the absorption spectra of the cones.

$$SS_\lambda = M \begin{cases} 1/Q_R = [K_1\,\alpha_{575} - K_2\,\alpha_{535}] & (1) \\ 1/Q_G = [K_3\,\alpha_{535} - K_4\,\alpha_{575}] & (2) \\ 1/Q_B = [K_5\,\alpha_{445}\qquad\qquad\;] & (3) \end{cases}$$

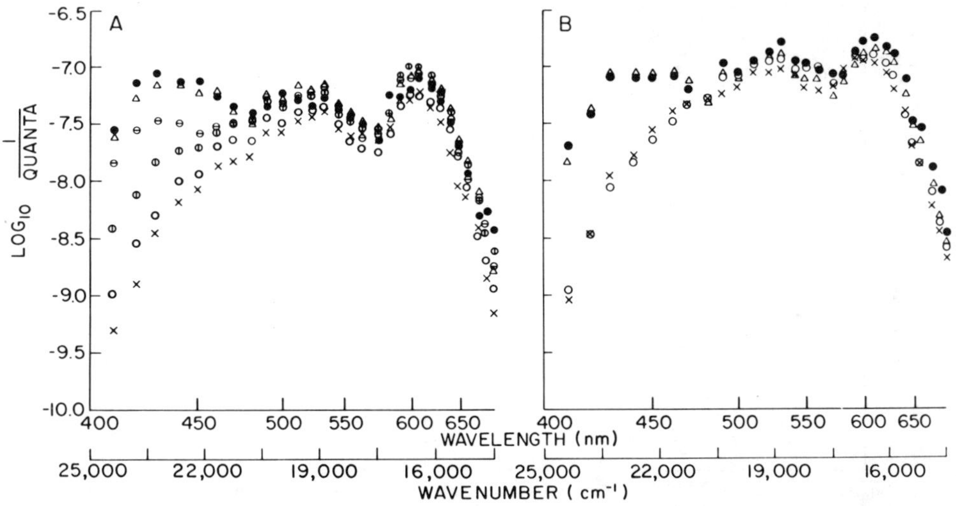

Figure 3. Increment-threshold spectral sensitivity of two rhesus monkeys before exposure (solid circles). In Part A, 6 weeks after exposure to 4.2 x 10^{-4} watts/steradian at the cornea of 463 nm blue radiation for: 560 minutes (horizontally divided circles); 120 minutes (vertically divided circles); 1680 minutes (open circles). Data taken 5 months later (crosses). In Part B, the open circles were taken 2½ weeks after 280 minutes exposure to 7.9 x 10^{-4} watts/steradian, and the crosses five months later. The open triangles are control data taken from the unexposed eye two weeks after the last exposure (from Ref. 5).

response of the blue-sensitive cones alone, was reduced in three series of seven day exposures in the monkey represented in Fig. 3A, and in one seven day series for the monkey in Fig. 3B. The exposure intensity for 3B was double that of 3A and the total exposure time was halved. The filled circles represent the pre-exposure sensitivity; the open circles post-exposure sensitivity obtained some days after each series of exposures; the crosses represent sensitivity five months later in the exposed eye; the open triangles represent the sensitivity of the unexposed eye after the entire series of exposures. It is clear from these data that repeated brief exposures for several hours each day, day after day, for a week or more reduces the sensitivity of the blue responding cones at least below that of the adjacent green responding cones, and that the green- and red-sensitive cones are unaffected by those exposures. Fig. 4 shows data on the same monkeys obtained in the exposed and unexposed eye at the end of 24 months. These data were obtained with an intense yellow background calculated to reduce the sensitivity in the green and red regions far below that

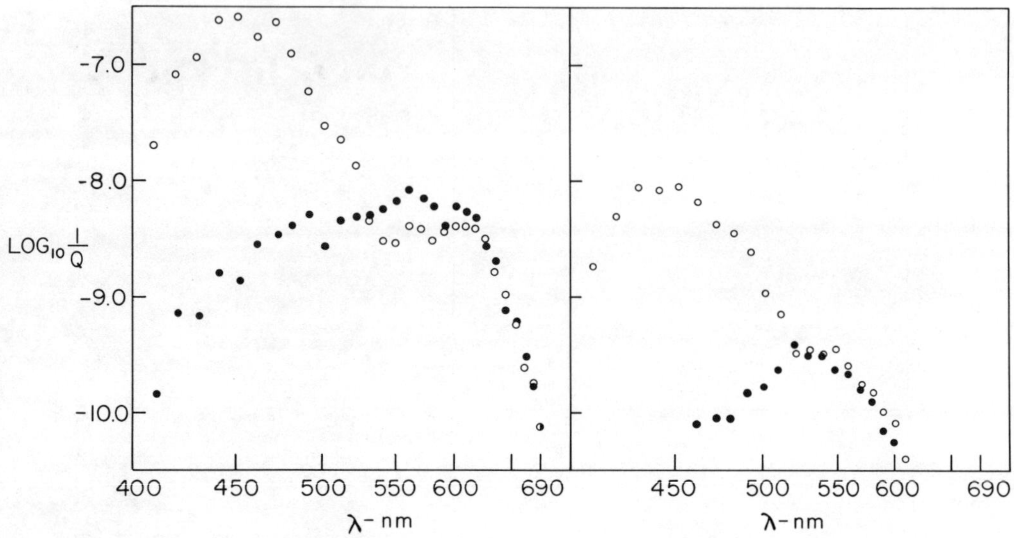

Figure 4. Increment-threshold spectral sensitivity of blue-blinded monkeys of Fig. 3. Taken against 10,000 tds of 550 nm added to 3000 td, white, 24 months after blue exposure. Open circles are from the unexposed eye. Filled circles are from the experimental eye.

in the blue, thus revealing any recovery of the blue mechanism. It is clear that no such recovery is seen, that the blue sensitivity in the unexposed eye is quite normal, and that loss of blue response is probably total.

Fig. 5 shows data obtained on one of the blue-exposed monkeys at the end of 36 months. The exposed eye is represented in the right hand column; the unexposed in the left. These data were obtained against increasing intensities of a narrow band 650 nm red background (added to our standard 3000 td white), designed to reveal the shape of the green-sensitive cone response function. Leaving aside various results with respect to the nature of green-sensitive cone response and some evidence of blue inhibition of green-sensitive cone and of rod response, it is clear that there is no sign of blue-sensitive cone recovery at the end of three years.

Fig. 6 shows data on two animals in another experiment where they were exposed to intense narrow-band green light. The full circles are pre-exposure on each of the two monkeys in Fig. 6A and 6B; the open circle data labeled 1 were taken 24 hours after the end of a series of exposures to green light. It is clear that the animals were reduced to dichromacy (deuteranopia). Their data are fit by the absorption spectrum (weighted for pre-retinal absorptions)

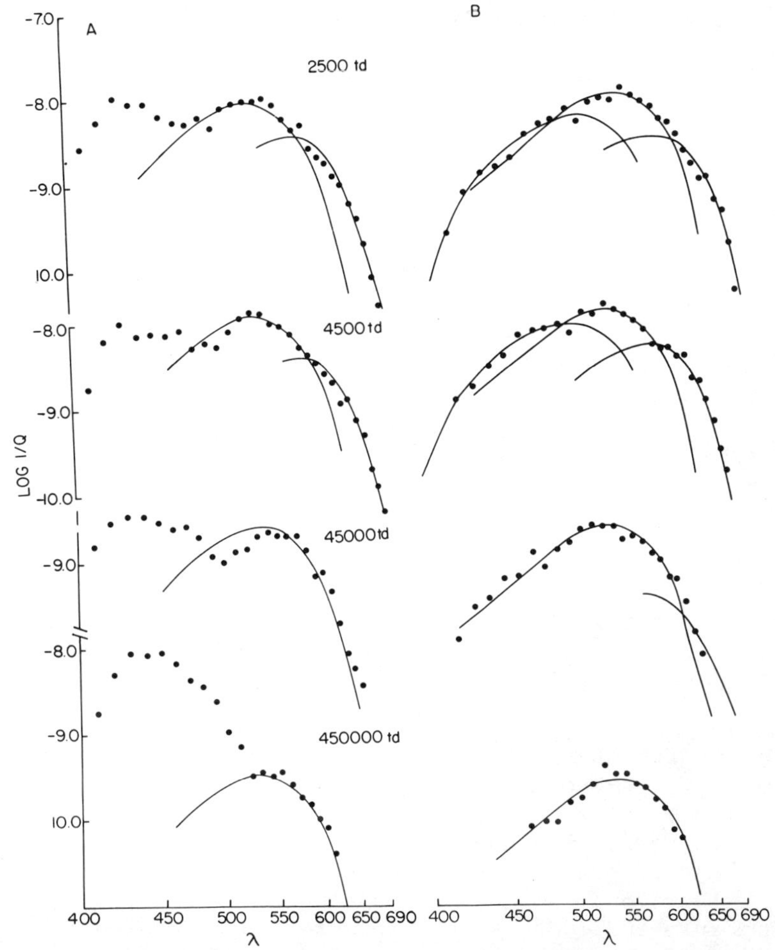

Figure 5. Comparison of "blue-blinded" eye B with unexposed eye A of a rhesus monkey 36 months after exposure. Increment-thresholds were obtained with 2° centrally fixated spectral test lights against a 3000 td, 5500° K white background plus different amounts of 650 nm, red as labeled. The solid curves fitted to the data represent rod and cone response functions as follows: the low curve with the longest wavelength peak, modeling the response of the red cones, is taken from monkey threshold studies against intense blue-green backgrounds; the mid-spectrum curve, modeling the green cone response, is the spectral sensitivity of protanopes by flicker photometry, taken in our laboratory. It agrees very closely with Stiles' $\pi 4$. The shortest wavelength is the absorption spectrum of rhodopsin weighted for pre-retinal absorptions (from Ref. 8) and models rod response.

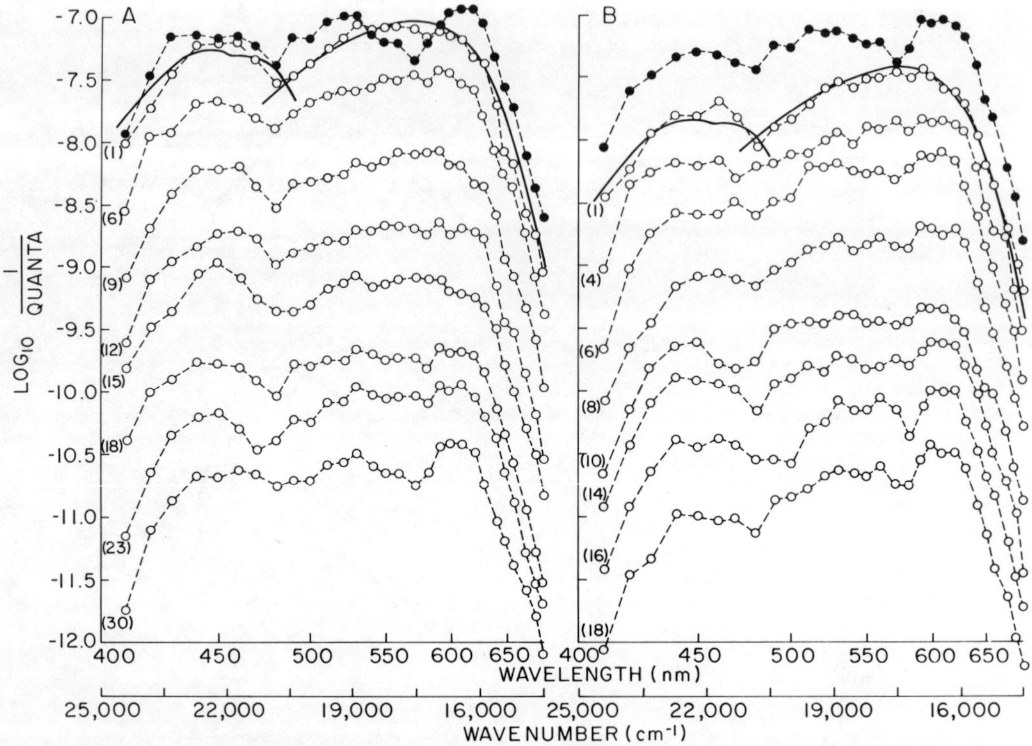

Figure 6. Increment-threshold spectral sensitivity of two monkeys before and after exposure to green light. Solid circles are data before exposure. The open circles are data obtained at different days after exposure; the number to the left of each curve gives the number of days after exposure. The solid curves are the absorption spectra of the 575 nm and 445 nm cone photopigments. The data before exposure and 1 day after exposure are on the same scale; each successive day's measurements are displaced downward by 0.5 log unit for ease of viewing (from Ref. 5).

of the blue-sensitive and red-sensitive cones, shown as the two solid curves. Then, over a period of weeks, there is gradual recovery to trichromacy. By the 18th or 30th day, (bottom curves), each monkey had recovered to very nearly the same sensitivity as pre-exposure.

A further psychophysical study examined the effect of the blue exposures on the rod photoreceptors, 36 months after exposure on the same monkey as in Fig. 5. Using a newly developed behavioral technique for holding the animal's fixation away from the center of the field (7), we were able to hold the test light at 6° from the central fovea on the horizontal meridian, in the temporal field. The animal was dark-adapted for thirty minutes, a dark background was used and thresholds were obtained in both the "blue-blinded" and the unexposed eye. The results are shown for the exposed eye in Fig. 7. The functions drawn through the data points are the absorption spectrum of rhodopsin corrected for pre-retinal absorptions (8) and part of the green-sensitive cone response function. It is clear that we did not succeed in eliminating all cone function, but that the main peak is that of rhodopsin. The sensitivity in the two eyes is nearly the same. We can conclude that if the intense light exposures had affected the rods, they would have fully recovered. Since the blue sensitivity peak was completely absent in this eye, at this time, it is clear that rods do not serve the blue sensitivity function, as was once proposed (9). It is also clear that the rods are either less susceptible to these effects or

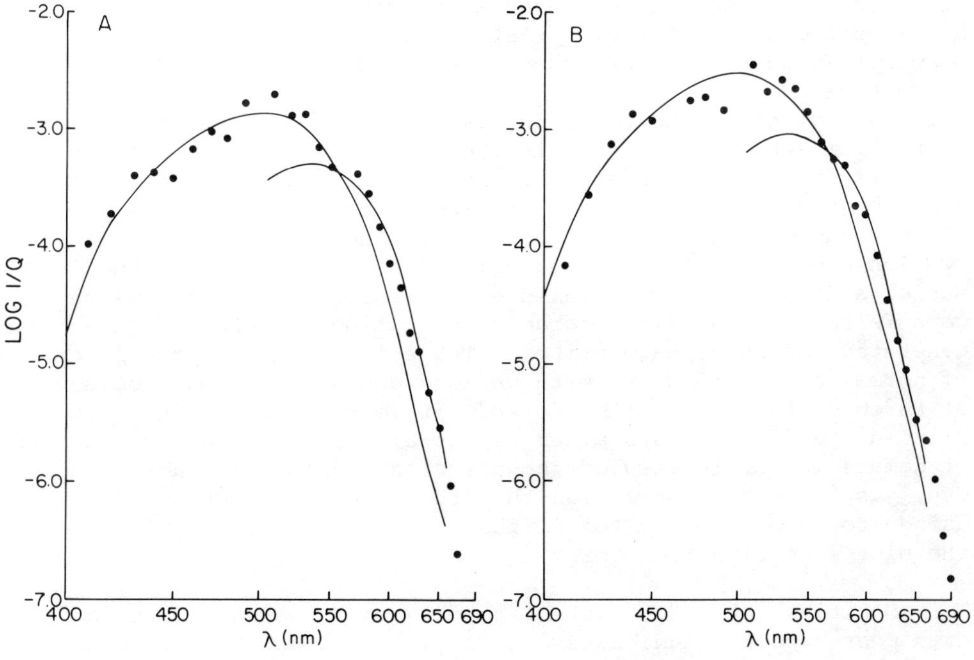

Figure 7. Increment-thresholds at 6° in peripheral retina against dark background after 30 minutes dark adaptation on monkey of Fig. 5. The solid curves model green cone response (longer wavelength peak) and rod response (shorter wavelength peak) as in Fig. 5.

recover more rapidly, despite the fact that the peak sensitivity of the rods and blue cones are close to each other in the spectrum.

These psychophysical data have shown that blue-light exposure reduces or eliminates the response of the blue mechanism without affecting the green or red mechanisms. Green light exposure selectively reduces or eliminates the response of the green mechanism for weeks with little effect on the blue and red mechanisms. This establishes a long-term, differential effect of spectral light. From this, it is quite reasonable to hypothesize that a lesion occurs _via_ absorption of light by the cone photopigments; the "green-blindness" is produced by the absorption of light by the green-sensitive cone photopigment present only in the green-sensitive class of cones; the "blue-blindness" by absorption in the photopigment of the blue-sensitive cones. Clearly this differential spectral effect, producing data which resemble the threshold sensitivities of congenital tritanopes and deuteranopes (10) is different from that produced by absorption of light in the non-selective pigment epithelium which accounts for thermal burns.

The long-term nature of these effects suggests tissue damage to the photoreceptors themselves. To test this possibility, we produced additional blue-blinded animals by the same procedures and sacrificed them for tissue study at different times after exposure. We sacrificed a blue-blinded animal 15 days after exposure. In vertical, one micron-thick sections by light microscopy through various regions in the unexposed eye, the photoreceptors and pigment epithelium appear quite normal--the ellipsoids of the cones are regularly spaced and unswollen; the cone nuclei are lightly staining and lie just above the outer limiting membrane. Fig. 8 is typical of sections of the "blue-blinded" eye. There are normal cones, swollen cones, pyknotic cones, some of which show both pyknotic nuclei and pyknotic or shrunken ellipsoids. Frequently, we see gaps in the cone mosaic, probably indicating where cones have degenerated and been phagocytized. Table II summarizes the percentage of normal cones and cones with various degrees of damage observed at three distances from the fovea. It is clear that, in a blue-blinded eye fifteen days after exposure, there is a broad spectrum of damage which, in view of the normal appearance of the unexposed eye, justifies the conclusion that the effect involves selective damage to cones, but is too diffuse to be useful in picking out the blue-sensitive receptors.

Sometime later, a "blue-blinded" animal was sacrificed at thirty days post-exposure and, as shown in Fig. 9, consistently showed a pattern of degenerating cones fairly regularly spaced among normal-looking cones. Fig. 10 shows electron microscope sections of the apparent gaps in the mosaic, establishing that these are highly degenerating photoreceptors. Serial vertical and horizontal sections were made from different parts of this retina, and the proportion

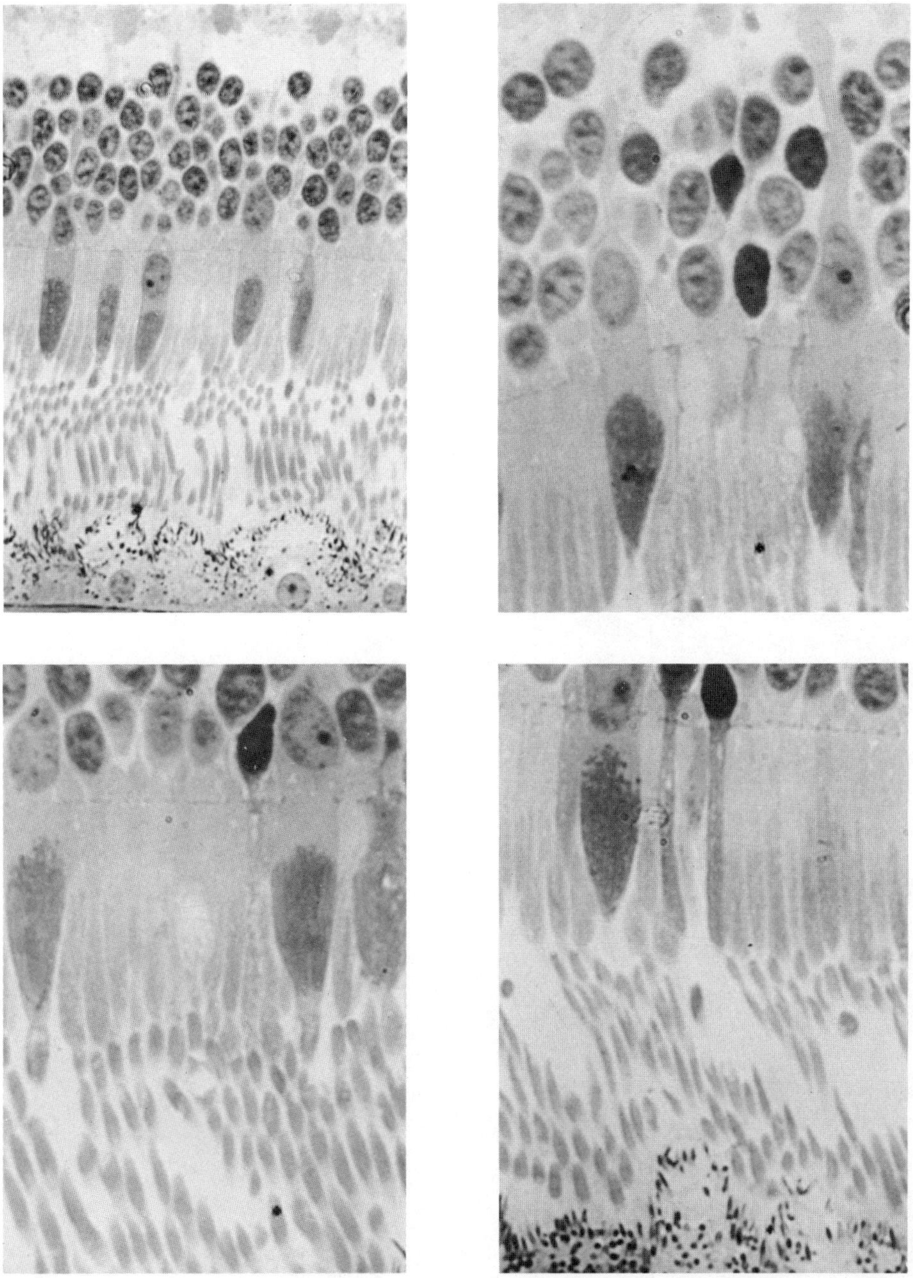

Figure 8. Light photomicrographs of vertical 1 μm thick sections from 15 day post-exposure blue-blinded eye in parafoveal locations and shows: swollen cones, pyknotic nuclei, atrophied outer segments and gaps in the cone mosaic.

TABLE II

Percentage of 5,000 Cones Per Position

Per MM	300 Microns Experimental	300 Microns Control	500 Microns Experimental	500 Microns Control	800 Microns Experimental	800 Microns Control
Normal Cones	62%	92%	41%	96%	53%	88%
Swollen Cones	17%	1%	19%	0	18%	7%
Degenerate Cones	9%	0	20%	0	18%	0.5%
Pyknotic Nuclei	2.8%	0.4%	28%	0	8%	0.2%
Displaced Nuclei	0.28%	0	0.4%	0	0.1%	0.5%

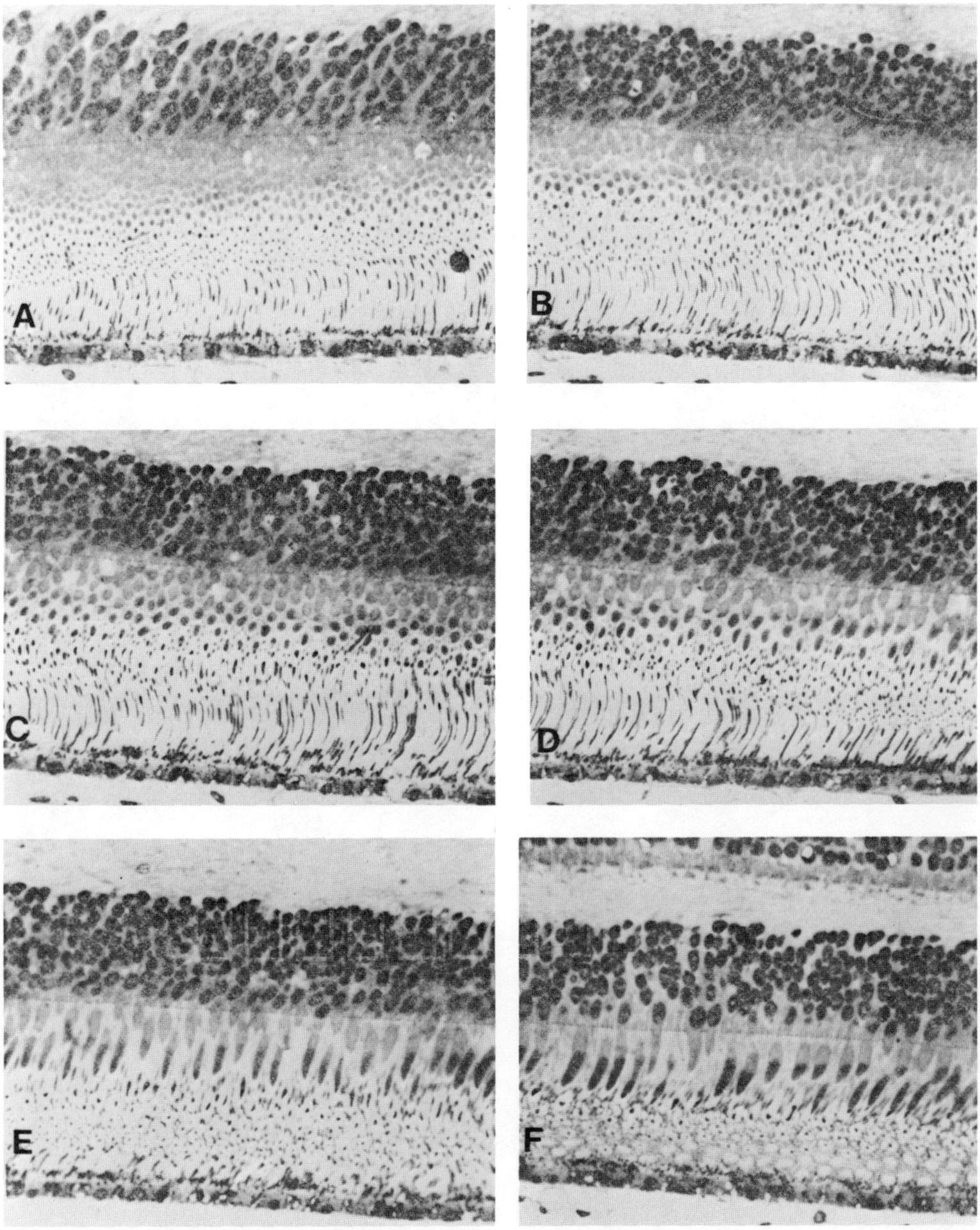

Figure 9. Photomicrographs of blue-blinded monkey retina sectioned 30 days post-exposure 1 μm thick sections at: A - .25°, B - 65°, C - 1.25°, D - 3.0°, E - 5.0° and F - 7.0° in the nasal field of the right eye. Only gross damage seen as gaps in the mosaic were counted in the data of the following figures. (Reduced 20% for purposes of reproduction.)

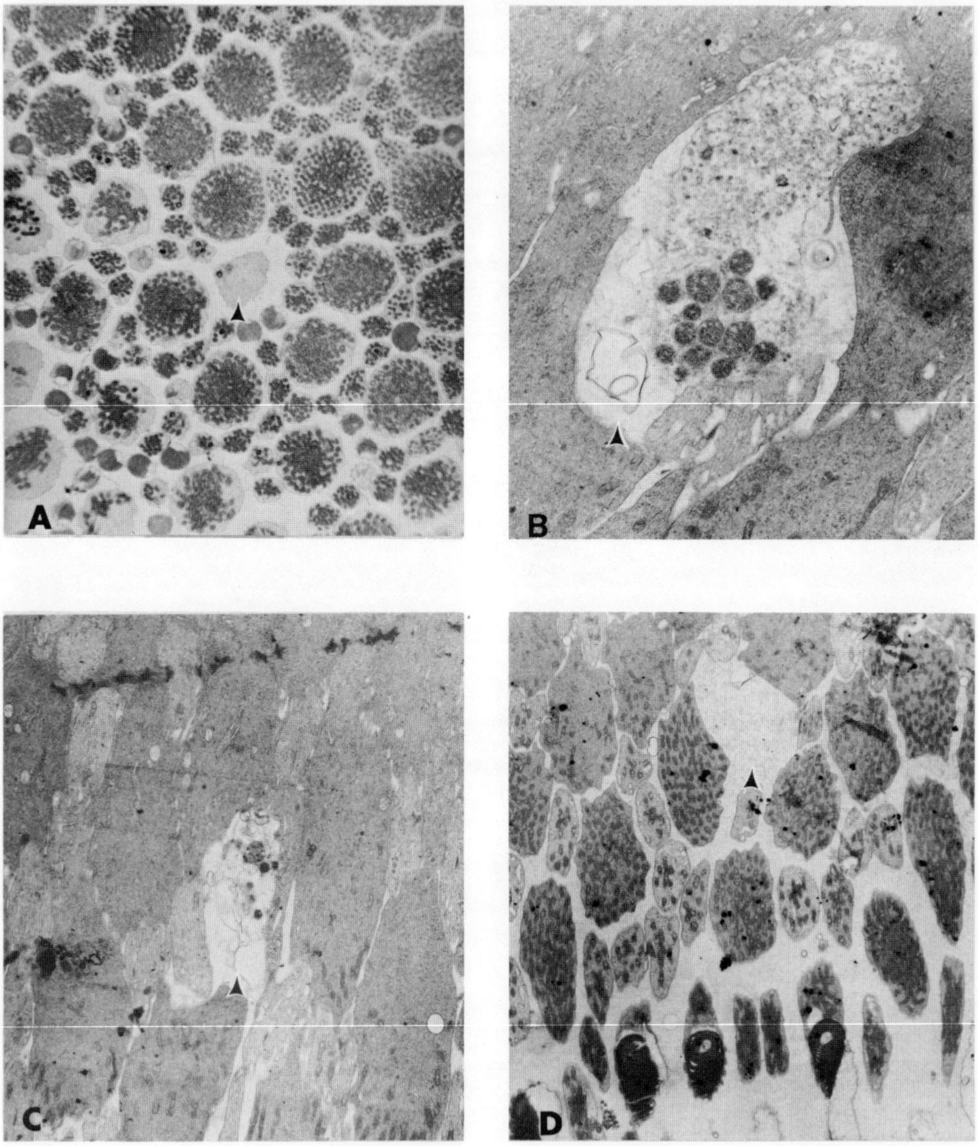

Figure 10. Electron micrographs showing gaps in the retinal mosaic. The arrows indicate degenerating photoreceptors in: A-ellipsoid level; B and C - myoid level with remains of cone debris; D - space left in the ellipsoid level. (Reduced 20% for purposes of reproduction.)

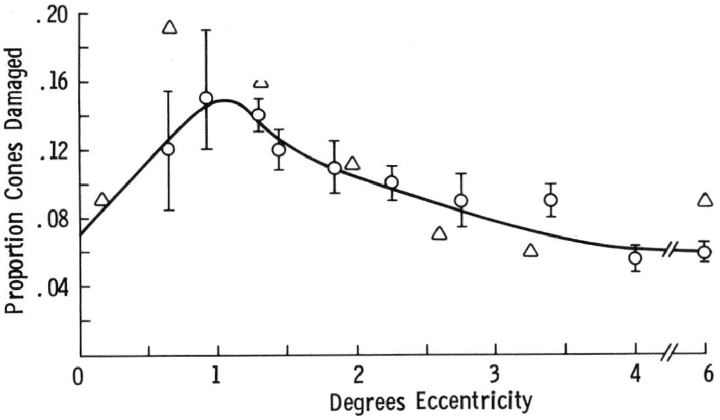

Figure 11. Proportion of grossly damaged to normal cones in sections of blue-blinded retina at 30 days post-exposure, as a function of distance from the central fovea in degrees. Open circles are mean proportions with the S.E.M. shown as vertical bars from vertical sections. Open triangles are from tangential sections on the same retina. The curve was fitted by eye.

of damaged cones as a function of position on the retina was estimated. Fig. 11 shows the mean (and standard error of the mean) of counts obtained by grouping areas at different distances from the foveola in all directions. The means from the vertical sections are shown as open circles, S.E.M. as bars. The open triangles are from counts obtained from the horizontal sections. It is clear that the distribution of damaged cones peaks a short distance from the center of the fovea in the 1.0° to 1.5° region and then falls gradually towards the periphery, appearing to asymptote at 8 to 10% of the cones at 5° to 6°. In Fig. 12, the distribution obtained from the vertical sections is compared with the percentage of blue cones determined from 0° through 6° in another approach taken in our laboratory (11, 12, 13). In the latter study, we have established the distribution of the blue-, green- and red-sensitive cones by a cytochemical technique involving the conversion of nitro-blue tetrazolium to diformazan (NBT-DF reaction) in the presence of different spectral lights. Both studies show a lower frequency of blue-sensitive cones towards the center of the fovea, a peak near 1.0° to 1.5° and a gradual decline of responding cones towards the periphery. The percentages at the peak are both about 16% of the cones. The light-damaged cones appear to fall off somewhat more rapidly toward the periphery than the proportion of NBT-stained cones. This may be accounted for by a non-uniform light distribution toward the edge of the exposure field. Taken together, these results support the conclusion that the blue-blinding effect with

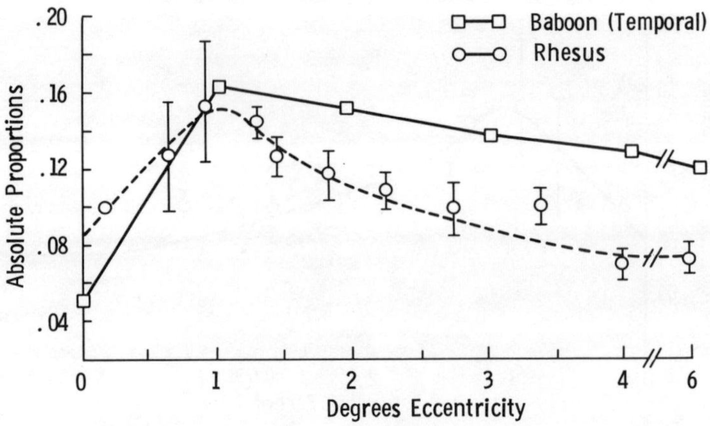

Figure 12. Comparison of proportion of damaged cones from vertical sections in blue-blinded eye at 30 days post-exposure with proportion of cones stained by the NBT-DF reaction to blue light.

repeated, intense blue light exposure is the result of damage to the blue-sensitive cones resulting from differential absorption of energy by the "blue cone" photopigment. The light damage results, in turn, add additional support to the hypothesized distribution of blue-sensitive cones based on the NBT-DF studies in primates.

We have induced irreversible tritanopia in infra-human primates with repeated brief exposures to blue lights much less intense than lights which produce thermal lesions. We have induced deuteranopia which recovers in 18 to 40 days. We show pathology which clearly points to damage at the receptor level, and we have established a correlation in the case of the "blue-blindness" between the degenerated cones and cones which respond differentially to blue light by conversion of nitro-blue tetrazolium to diformazan.

When we exposed one eye of a rhesus monkey to the same wavelength of steady blue light (463 nm, 8.8 x 10^{-3} W/steradian over the central 30°) for 120 continuous minutes, quite a different histological picture resulted. The pattern of damage (Fig. 13) was extensive in the retinal pigment epithelium with remarkably little damage to cones. There was no sign of the regularly spaced highly damaged cones found with the intermittent blue light exposure.

We conclude from these studies that the mechanism of the color blinding involves repeated partial bleaching and recovery with differential absorption of light in the blue- or green-sensitive cone photopigments. Since rod vision was found normal in a "blue-blind" eye, we conclude that rods do not serve as blue receptors. The

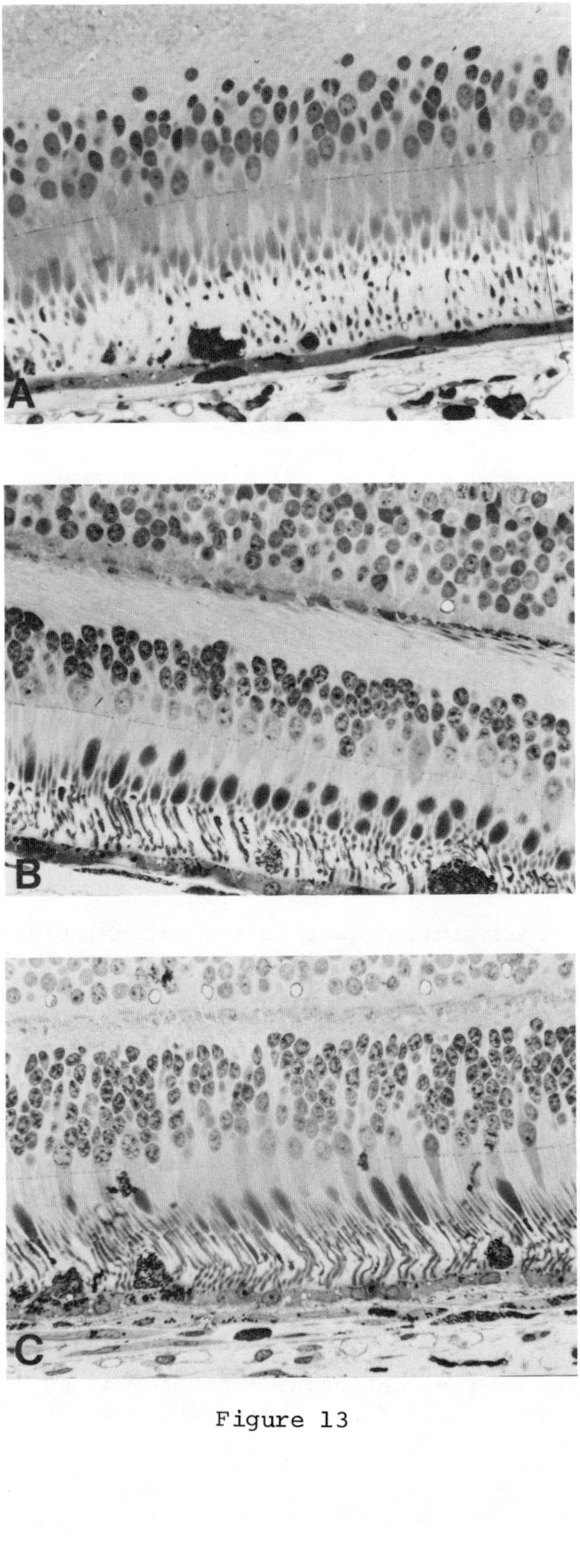

Figure 13

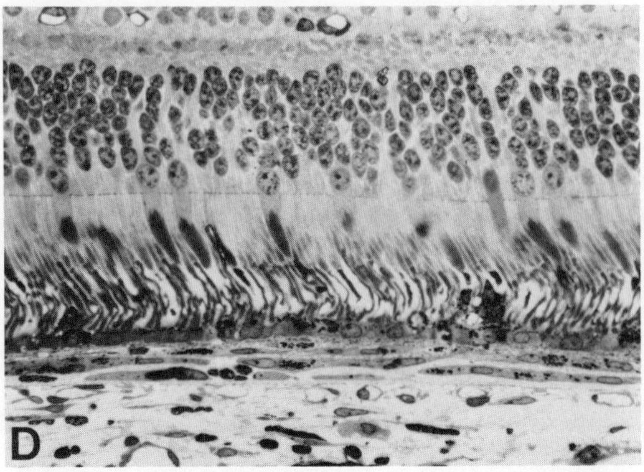

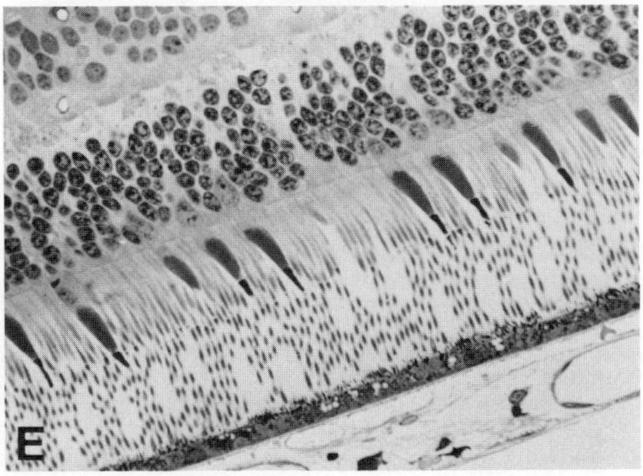

Figure 13. Photochemical damage to the retinal pigment epithelium due to prolonged exposure (central 30°; 463 n, 8.8 x 10° W/steradian, 120 minutes). The damage occurs primarily in the RPE while the cones remain fairly normal. The most extensive damage appears within 9° of the fovea. A: denuded RPE and macrophagic activity at 0.2°; B: similar pathology at 5°; C-D: placque formation well established at 8° and 9° respectively; E: normal RPE and cones at 12° eccentricity.

distribution of red- and green- as well as blue-sensitive receptors in the retinal mosaic of primates has been estimated and eventually will be revealed by further interdisciplinary studies such as those reported here.

In the case of the continuous exposure, it is likely that there was nearly a full bleach of blue-sensitive cone photopigment, resulting in little stress to the photoreceptors, but with extensive damage to the RPE resulting from absorption of energy by the melanin granules. It should be pointed out that the damage seen in Fig. 13, including macrophagic activity, disrupted cells and placque formation, is characteristic of that seen by Ham (14) and others in what he calls the "photochemical" lesion.

We conclude that retinal damage from light in the visible region of the spectrum can be classified into three distinctly different kinds of lesions: (1) The thermal lesion characterized by nonselective absorption of energy in the melanin granules of the RPE and resulting in grossly observable damage. (2) The photochemical lesion described by Ham, which results from continuous exposure to energies of short wavelength light less intense than those which produce gross thermal lesions and primarily damaging the pigment epithelium. (3) The "color blinding" lesion reported here which is wavelength selective, damaging blue-sensitive cones with blue light and green-sensitive cones with green light and depends upon repeated, intermittent exposure.

ACKNOWLEDGEMENT

The author is grateful to the following persons for the various contributions to this effort: Norman A. Sidley, Ronald S. Harwerth, Robert E. Marc, Clement Johnson, Donald Prather, David Garrett and Jack Crawford. This research was supported by National Science Foundation Grant NSF BNS76-07854 and National Institutes of Health Grants EY-00381, EY-01256 and EY-02388.

REFERENCES

1. Crawford, M. L. J. (1976) Vision Res. 16, 117-118.
2. Sidley, N. A. and Sperling, H. G. (1967) J. Opt. Soc. Amer. 57, 816-818.
3. Marks, W. B., Dobelle, W. H. and MacNichol, Jr., E. F. (1964) Science 143, 1181-1183.
4. Sperling, H. G. and Harwerth, R. S. (1971) Science 172, 180-184.
5. Harwerth, R. S. and Sperling, H. G. (1971) Science 174, 520-523.
6. Harwerth, R. S. and Sperling, H. G. (1975) Vision Res. 13, 1193-1204.
7. Crawford, M. L. J. (1976) J. Exp. Anal. Behavior 25, 113-121.
8. Wyszecki, G. and Stiles, W. S. (1967) in Color Sciences, Concepts and Methods, Quantitative Data and Formulas, John Wiley and Sons, Inc., N. Y.
9. Willmer, E. N. (1949) Docum. Ophthal. 3, 194-213.
10. Wald, G. (1966) Proc. Nat'l. Acad. Sci. U. S. 55, 1347-1363.
11. Marc, R. E. and Sperling, H. G. (1976) Science 191, 487-489.
12. Marc, R. E. and Sperling, H. G. (1976) Vision Res. 16, 1211-1224.

13. Marc, R. E. and Sperling, H. G. (1977) Science 196, 454-456.
14. Ham, W. T., Jr., Mueller, H. A., Ruffolo, J. J., Jr. and Clarke, A. M. (1979) Photochem. and Photobiol. 29, 735-743.

Invertebrates

PHOTORECEPTOR MEMBRANE TURNOVER IN ARTHROPODS: COMPARATIVE STUDIES OF BREAKDOWN PROCESSES AND THEIR IMPLICATIONS

A. David Blest

Department of Neurobiology
Research School of Biological Sciences
Australian National University
Canberra City, A.C.T., 2601, Australia

Starting some fifteen years ago, White (1-4) showed that the rhabdoms of larval mosquitoes diminish in volume when illuminated, and re-grow in the dark. At the time, these pioneer observations were regarded as implausible, and their protagonists had to struggle for their acceptance.

We now suppose that all neuronal plasma membrane probably turns over (5) and that in certainly one other special case of re-cycling, retrieval of synaptic membrane must proceed at an exceptionally high rate.

Turnover in arthropods has mostly been scrutinized in terms of the mechanisms proposed by White and his co-workers, and all that need be said is that these have stood the test of time and been shown to be substantially correct (6, 7); most turnover probably proceeds in the manner described.

The discovery that vertebrate rod outer segments shed packets of membrane at their tips (8, 9) and that cones do so as well (10-13) came later. Implicit in these various observations on invertebrate and vertebrate eyes are two questions: Do all photoreceptors turn over their transductive membranes? And, if so, why?

This contribution discusses general strategies of membrane shedding in arthropods. Drawing upon our qualitative work in Canberra over the last two years, it may properly be included in the present Symposium because it is becoming evident that some, at least, of the effects of both continuous light and continuous darkness upon the transductive components of photoreceptors will probably prove to be explicable in terms of imbalances which these conditions create in the ordinary processes of turnover. One of the tasks we have set ourselves is to try to evaluate the behavior of the subcellular organelles which mediate the synthesis of new membrane and its catabolism after it has been shed. We have also shown that there are more routes of membrane disposal than the solitary arthropod pathway which has been described. An essential test of the hypothesis that photoreceptor membrane turnover is necessary for the maintenance of its transductive function is the examination of "hard cases": those in which the known anatomy of an eye and the complement of organelles which it contains fail to reveal symptoms typical of the classical route of membrane turnover outlined by White. Two such cases will be described, and it will be shown that the shedding phase of turnover indeed takes place through two novel strategies.

All our observations have been made in such a way as to obviate the possibility that we have provided our animals with too much light or too little at the wrong times. We store our animals in an outside building, well-illuminated with diffused natural daylight, and maintained at around 25°C during summer. This is less than mid-day summer shade temperatures in Canberra, which commonly reach 30°C and may sometimes be much more. Unless otherwise stated, all sampling of animals is performed under the normal daily cycle of illumination.

PHOTORECEPTOR MEMBRANE TURNOVER IN THE NET-CASTING SPIDER DINOPIS

(i) Shedding and Degradation of Membrane

Dinopis (Fig. 1) exemplifies the "classical" strategy of turnover in arthropods. These tropical and sub-tropical net-casting spiders were originally studied because of the self-evidently unusual optics of the posterior median ocelli with which they locate prey at night (14). During the day, photoreceptor membrane is confined to a thin rind at the hexagonal margin of each receptive segment, and at night, the receptive segment not only doubles its length, but fills up with transductive membrane. The burst of synthesis occupies roughly 2 h after dusk. At dawn, the new membrane is rapidly broken down.

This temporal localization of the two phases of turnover is not unique to Dinopis, for it is found in the receptors of many wholly nocturnal spiders. It has the advantage for analysis that the processes of renewal and shedding are so far separated in time that they can be examined as separate issues.

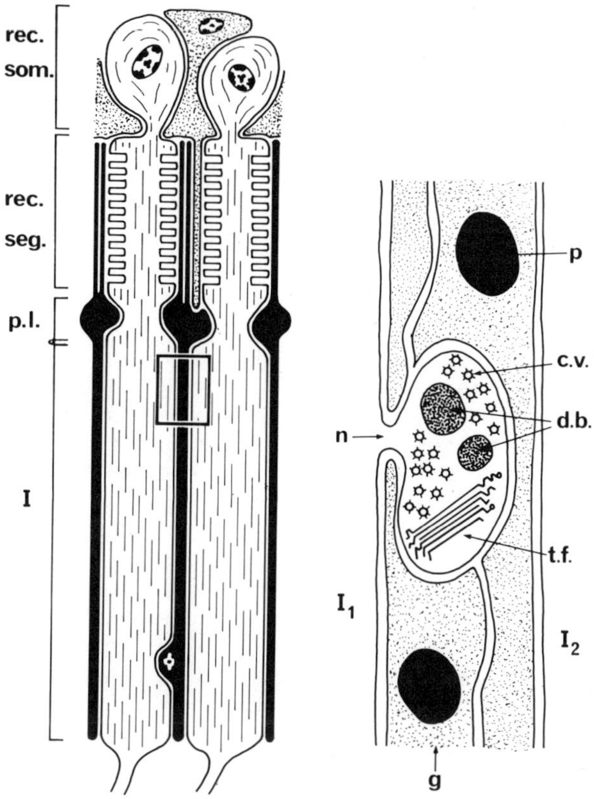

Figure 1. Retinal anatomy of Dinopis. Left: receptors in longitudinal section. rec. som., receptor somata; rec. seg., receptive segments; p. l., dense pigment layer; I, intermediate segments of the receptor axons. Stippled areas represent non-pigmented glial processes, black areas processes of the pigmented glia. Right: diagram to show the components of the region marked by a rectangle. A glial follicle is continuous with an intermediate segment, I_1 by way of a narrow neck, n. It is surrounded by glial processes, g, which contain pigment granules, p. The follicle contains three kinds of organelles, dense bodies, (d.b.), coated vesicles (c.v.) and tubular filaments (t.f.). Each receptor is ca. 22 μm in diameter.

Membrane shedding in Dinopis follows the sequence originally determined for the mosquito (2, 3): Membrane is pinched off from the bases of the rhabdomeral microvilli as coated pinocytotic vesicles. These either adhere to each other to form multivesicular bodies which are not initially membrane-bound, or, they are incorporated into dilated segments of smooth endoplasmic reticulum (15) to form the familiar spherical, membrane-bound multivesicular bodies seen in numerous invertebrate photoreceptors. There is good evidence from freeze-etch preparations of crayfish retinas that individual vesicles enter such multivesicular bodies by secondary endocytosis (7), and we have circumstantial evidence that RER cisternae are transformed to provide the bounding membrane (15). White (3) showed by the use of ferritin as a tracer that some of the coated vesicles of mosquito ocelli are genuinely pinocytotic in origin, and enclose material derived from extracellular space. We have confirmed that the mass production of coated vesicles and multivesicular bodies is associated with the breakdown phase of the daily cycle and not the synthesis (6, 15).

Once formed, multivesicular bodies are rapidly degraded, first to multilamellar bodies, and then to dense residual bodies which are eventually lysed (15). During this process, surviving secondary lysosomes are moved to the intermediate segments, remote from the rhabdoms. In fact, it is seen that any one of the earlier stages *can* be lysed before and without transformation to the next, and *this* presumably relates to the points at which hydrolases are introduced into them. In crayfish, acid phosphatases can first be detected in some multivesicular and multilamellar bodies (7). We have examined the dawn break-down sequence in *Dinopis* with a recent modification (16) of the zinc iodide-osmium tetroxide technique (15, 17) and more recently with acid phosphatase cytochemistry, using either sodium β-glycerophosphate or *p*-nitrophenyl phosphate substrates in tris-maleate buffer at pH 5.0 (18).

Together, these techniques suggest that a substantial part of the endoplasmic reticular system of the big intermediate segments is devoted to autolytic processes. Associated with the degradation of membrane-derived secondary lysosomes there is a massive differentiation of pleiomorphic structures which we consider to be atypical GERL from rough endoplasmic reticulum (17). They enter into intimate association with any one of the various kinds of secondary lysosomes, or may generate the clouds of coated vesicles which in *Limulus* have been termed "*Nebenkerne*" (19). *Nebenkerne* adhere to and are incorporated within residual bodies (18). In their great variety of form, these structures resemble GERL from various tissues of *beige* mice which mimic Chédiak-Higashi syndrome (20, 21), where there are disturbances of lipid metabolism. The parallel with an invertebrate receptor disposing daily of massive quantities of material which, in addition to phospholipids and intrinsic membrane proteins is deduced to contain large quantities of cholesterol (22, 23) is intriguing. As in some *beige* mouse tissues, the GERL-like structures are independent of Golgi stacks.

GERL of this kind is only erratically positive for acid phosphatases, and the reactions, although definite, are not strong (18). Most responses are obtained in barely-transformed endoplasmic reticulum which not only can be shown to give rise to conventional autophagic vacuoles, but also to become incorporated into residual bodies which start to show strong reactions as they lyse.

It is unfortunate that so little histochemical work has been done on non-pathological involutional systems. The closest parallel to the breakdown strategy in Dinopis appears to be the natural involution of guinea-pig corpus luteum at the end of pregnancy (24, 25).

A matter of some importance is the extent of degradation involved in this cycle. Common sense might urge that it would be economical for the spider to conserve its membrane constituents as little decomposed as possible. This would be true were the only object of the exercise to allow adjustment in total rhabdomeral volume of the kind shown so dramatically by Dinopis, and in some crabs (26). Since all secondary lysosomes eventually lyse, there is little doubt that they cannot be mere temporary stores of virtually unmodified chemical components. An interesting problem, then, concerns the fate of retinol, for it or an allied product will presumably be freed by the degradation of either metarhodopsin or rhodopsin; it is insoluble in water, vulnerable to oxidation, and for a predatory animal which can only obtain it from victims which it may or may not catch, a resource wich must surely be conserved near the sites at which it will be required. Indeed, vertebrates are well-known to exchange retinol between receptor outer segment and pigment epithelium during light and dark adaptation (27). On a broader physiological front, studies on the synthesis and release of retinol-binding protein by the liver (28-31) imply that it is a material whose transport presents special problems.

We now know that small follicles in the retinal glia of Dinopis are saccular outgrowths connected to their parent intermediate segments by narrow necks (Figs. 1, 2). These compartments alone contain more-or-less spherical bodies, which are usually electron-dense, assemblies of tubular filaments, often disposed in orderly "crystalline" arrays, and a matrix of large coated vesicles and short, less ordered filaments which fill the glial follicle completely. The tubular arrays are positive to ZlO and such staining reveals that the tubules vesiculate and fragment at their ends and are incorporated into the spherical bodies. Their origin is undetermined, but the follicle walls continuously invaginate small tubules (Fig. 2), and it may be suspected that the arrays are derived from them. There is some qualitative evidence that it is easiest to find glial follicles in sections of day-time retinas. The follicles release their dense bodies into the main volume of intermediate segment cytoplasm.

Initially, we attributed changes in the appearance of the glial sheaths of the intermediate segments to pigment movements (6). It seems certain, now, that the rather scanty pigment granules cannot move, and that the appearances noted are the result of alterations

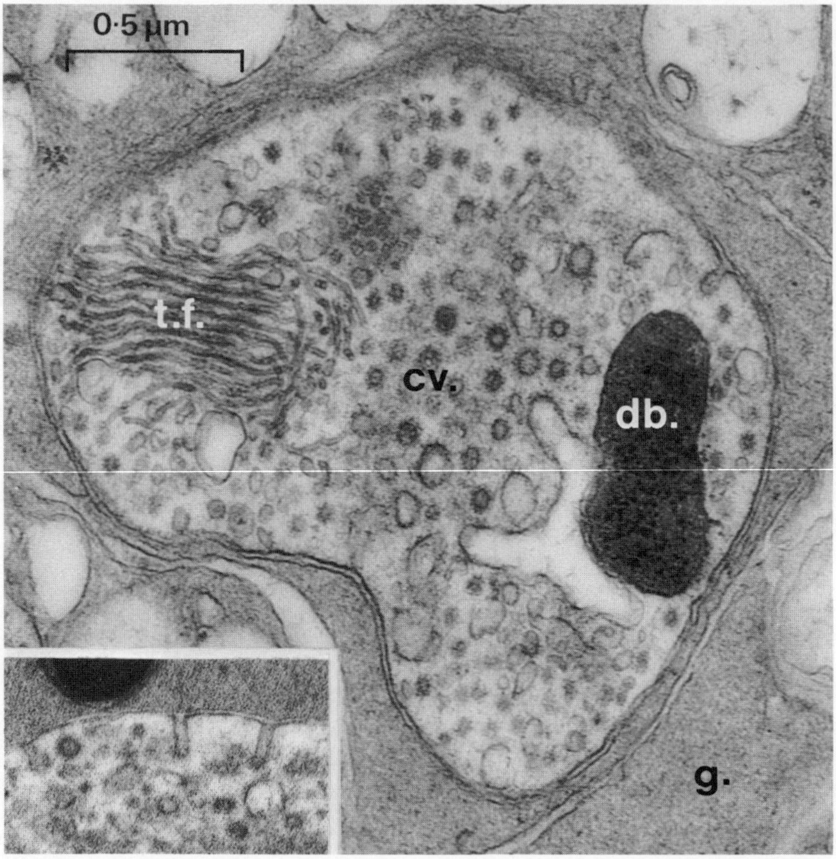

Figure 2. A glial follicle of Dinopis. The section is from a preparation incubated as a control for acid phosphatase localization in medium at pH 5.0 but without substrate. The inset shows tubules invaginating from a follicle wall. t.f., tubular filaments; c.v., coated vesicles; d.b., dense body; g., pigmented glial process.

in the number and contents of glial follicles. Changes in the orange appearance of the distal portion of the intermediate segments and their sheaths are not wholly consistent with respect to time of day. Nevertheless, it is attractive to suggest as a working hypothesis that these very active follicles are concerned with the sequestration and storage of retinol.

(ii) "Normal" and "Abnormal" Patterns of Turnover

Studies of invertebrate photoreceptor membrane turnover were confused for many years by uncertainties as to the normality of the phenomena. Retinas were studied in the laboratory, and since the objective of most researchers was to induce the breakdown of membranes which they believed to be basically stable, long periods of intense illumination were deployed against eyes whose experiences in Nature of environmental luminous fluxes were not measured.

Dinopis has provided the first formal demonstration that these dramatic turnover patterns are quite normal, and doubtless many more will follow. Nassel and Waterman (26) have already provided a similar example in a crab. Nevertheless, distinctions between pathological and normal rhabdomere membrane breakdown have not been clearly made in the invertebrate literature. When *Nephrops*, the Norway lobster, is raised from 30-180 m ocean depth and exposed to ordinary sunlight for 2.5 h, its rhabdomeres suffer irreversible degeneration. The lobster becomes blind, and functional rhabdomeres never re-grow. Degenerating membrane becomes re-organized in whorls, or so-called "myelin figures" (32). Micrographs of rhabdomeres from many different arthropods often show whorls; although it has been established that these may be generated as artifacts by direct fixation in osmium, modern techniques of glutaraldehyde fixation in solutions whose osmolarity is adjusted with sucrose preclude the dismissal of most of these profiles as artifacts. In the spider, *Dolomedes*, single photoreceptors whose rhabdomeres are totally disordered may be found next to wholly normal cells (33).

Superficially, these appearances strikingly resemble vertebrate receptor outer segments which are degenerating following solar retinitis, either naturally induced, or mimicked in the laboratory with lasers (34-37). The early feeling that photoreceptors in the breakdown phase are in some way being "damaged" by light has somewhat obscured the likelihood that regulated breakdown is a quite different phenomenon from the effects of excessive illumination. The highly ordered terminal shedding of vertebrate outer segment membrane, the manner in which it is initiated by light in some species (38) and by a circadian rhythm in others (39), and the rapidity of the process when it occurs, all imply that shedding as a *direct* consequence of damage to the transductive membrane could only be hypothesized in terms of some variant of catastrophe theory. What is the situation in *Dinopis*? First, it is helpful to outline the ecology of the spider, and to consider what demands are imposed upon its eyes by its environment and habits.

The spiders live in scrub habitats which offer little shelter, and which present them daily with extremes of brightness extending over some 6-7 log-units. The eye possesses no pupil mechanism, has an *f*-number of 0.58 and the receptive segments are not equipped with screening pigment. A spider, when ready to catch prey at night, hangs head downwards over dead leaves and detritus, hopefully to detect cockroaches or ants of similar color and reflectance which will run across this unpromising background (14, 40, 41). We have made both ERG-and intracellular recordings from the receptors (42). They are all of one spectral type, with a peak response at 515-520 nm close to that of human rods (peak spectral efficiency at 503 nm). There is a marked cut-off in the UV which is generated by absorption in that region by the lens. After dark-adaptation it is possible to obtain the large, slow (10 mv and above) quantum bumps typical of sensitive receptors of high gain. ERGs reveal that the sensitivity of the whole eye may climb through as much as 1.5 log-units in the first hour of dark adaptation and a further 1.5 log-units in the second, despite the absence of pupillary mechanisms, or of screening pigment distal to the receptive segments.

The breakdown and synthesis of rhabdomere membrane is under the influence of a circadian rhythm which is manifested only indirectly. Eyes kept in the dark over the dawn period do not break down their membrane; kept in the light beyond sunset, they fail to synthesize it. Nevertheless, darkness at noon is not a sufficient stimulus to provoke normal synthesis, which becomes more effective the nearer the "natural" time of synthesis is approached (6).

Clearly, the rind of rhabdomere membranes which remains in each receptor throughout the day (Fig. 3) is not affected by the large luminous fluxes which it must receive. The membrane synthesized each night does not receive light in such quantities while it is being used, and orderly dissolution is provoked by illumination at dawn. What happens if a substantial pulse of light is delivered at midnight, when the retina, under normal circumstances would never receive such an event? The effects of single 30-120 min light pulse well within the range of daytime environmental brightness delivered 3 h after synthesis and followed by periods of darkness up to 12 h are dramatic. Initially, the rhabdomeres swell and become disordered. There is a rapid production of GERL tubules in the intermediate segments (17). Islands of detached rhabdomere membrane are formed. Most of the rhabdomere membrane is broken down into ill-defined, multivesicular masses in association with smooth endoplasmic reticular tubules and saccules of GERL type. Hours after the light-pulse, the majority are found to be further disorganized, but not lysed (Fig. 3). The mass of detritus is devoid of acid phosphatase, other than in local concentrations in large, severely-degraded membrane-whorls. These experiments need extending, but the first implication to be derived from them is that dawn breakdown involves much more than the mere effect of light upon the rhabdomere membrane. The collapse and vesiculation of rhabdomere membrane *en masse* over a period of hours is quite distinct from pinocytosis at the bases of the microvilli. One process is disorderly, while the other is highly regulated.

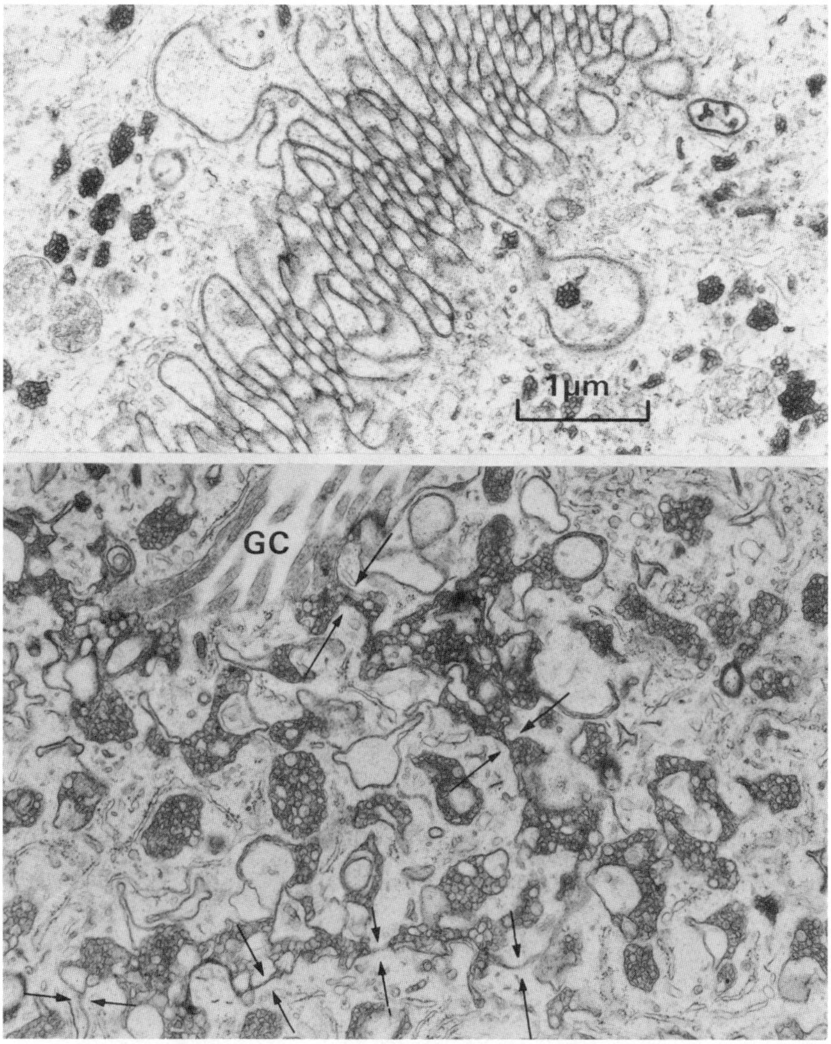

Figure 3. Normal and pathological photoreceptor membrane breakdown in *Dinopis*. Top: Normal boundary between two receptive segments 2-3 h after dawn. Microvilli from the two segments are tightly interdigitated. Bottom: A similar boundary from a pair of receptive segments modestly illuminated for 1 h at midnight, after completion of synthesis, and stored in darkness for 12 h. Breakdown is disorderly; the boundary between receptive segments is picked out with arrows, and its position can only be determined with reasonable certainty by reference to the column of glial strands (GC).

EXTRACELLULAR MEMBRANE SHEDDING IN ARTHROPODS

A survey of ultrastructural papers on arthropod eyes shows that most receptor cytoplasm contains multivesicular bodies or other secondary lysosomes. Some do not: these include jumping-spider receptors, and the receptors of flies with open rhabdomes whose microvilli project into a large extracellular space. We have examined both cases. We were prompted to do so by the observation that in a lengthily dark-adapted pisaurid spider, *Dolomedes*, material appears on exposure to light to be lost to extracellular space as a structureless or minutely micellar detritus (7). In this retina, the glia is divided into innumerable interdigitating processes, and the detritus appears to be "pumped down" into the channels between them until it reaches the level of the intermediate segments, into which it is injected *via* a series of longitudinal endocytotic pleats. In retinas from spiders which have been left in darkness until their rhabdomeres have grown to fill the receptive segments, glial channels are empty; during a few hours of subsequent exposure to light, the rhabdomes are reduced, and the channels fill with a homogeneous detritus which spreads gradually from the extracellular spaces adjacent to the rhabdom to those between the glial sheets which ensheath the intermediate segments. At the same time, the large amount of material which is lost from the rhabdom is not paralleled by as dense a field of intracellular secondary lysosomes as one might expect.

We have examined the anterior lateral eyes of the Australian jumping spider *Plexippus* throughout the daily cycle (43). The basic anatomy of the receptors and a summary of their shedding mechanisms are given in Figs. 4 and 5. Shedding is continuous, although we see signs that there is a modest increase at dawn. It is extracellular, and takes place from the tips of the microvilli. Two kinds of events are seen: Terminal regions of adjacent microvilli may disintegrate, and give rise to a micellar detritus which fills the extracellular space between the microvilli and the glial arms which ensheath them. Or, individual microvilli lengthen at their tips and poke into coated endocytotic pits which develop concurrently in the glial membrane to receive them. The growth of a pit appears to be slow, but pinching-off must be rapid, because we have only observed it in progress once, unfortunately in a less than perfect preparation (Fig. 6). Coated pits which contain a fragment of microvillus can be seen adrift in the glial cytoplasm as a short-lived state which converts to irregular smooth vesicles.

Both kinds of break-down product are dumped in the same place. The detritus produced by the disintegration of microvilli into extracellular space travels in the gap between receptor and glial arm for some distance before being introduced into the glial arms by way of endocytotic clefts. They later branch and ramify to such an extent that it is not certain whether they are true invaginations of the glial plasma membranes or part of a tubulo-cisternal system.

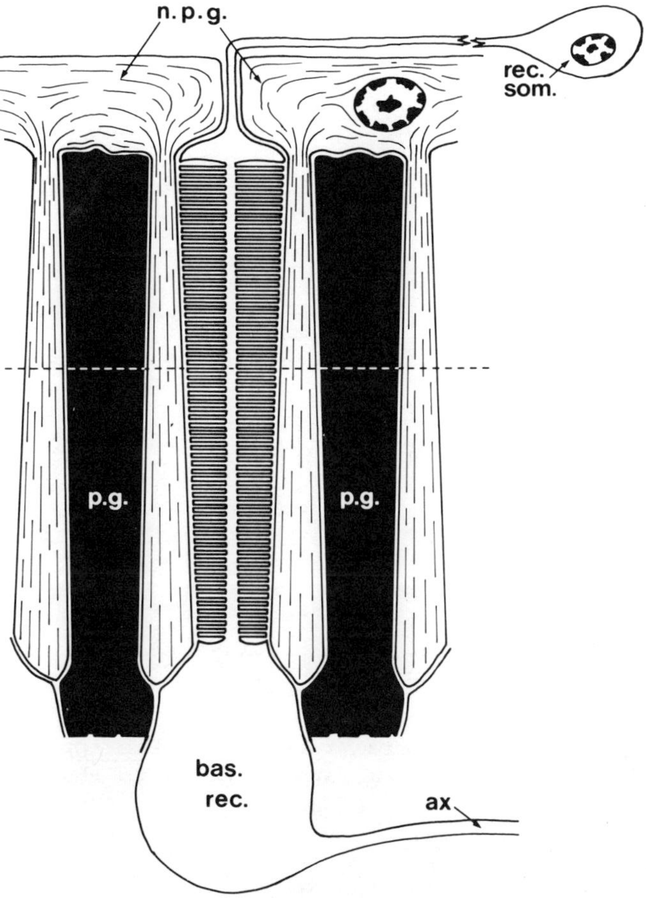

Figure 4. Anatomy of the anterior lateral retina of *Plexippus*, to show relationships between the receptor and pigmented (p.g.) and non-pigmented (n.p.g.) glia. Proportions are arbitrary, and the rhabdom is both longer than shown and also has a twist which is not represented. Note that the rhabdom is flanked by non-pigmented glial arms. Rec. som., receptor soma; bas. rec., basal receptor; ax, axon.

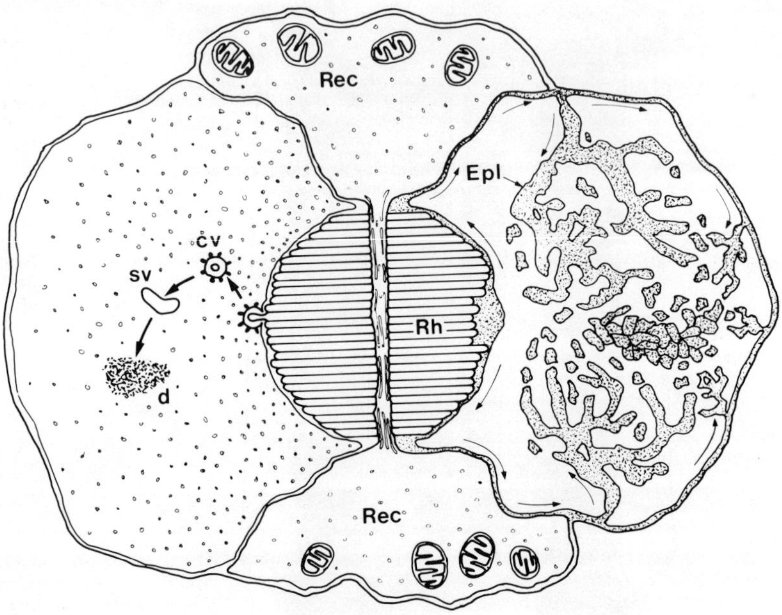

Figure 5. An anterior lateral receptor of *Plexippus*, to show the two routes of microvillus membrane disposal to the non-pigmented glial arms. At left, coated endocytotic pits (cv) pinch off microvillus tips into the glial cytoplasm, convert to smooth vesicles (sv) and appear to give rise to fine detritus (d). At right, amorphous detritus forms at the eroding microvillus tips and is passed along extracellular channels (arrows) to endocytotic pleats (Epl). Tips of pleats seem to break off and accumulate as detritus in the glial cytoplasm. Rec. receptor; Rh, rhabdom.

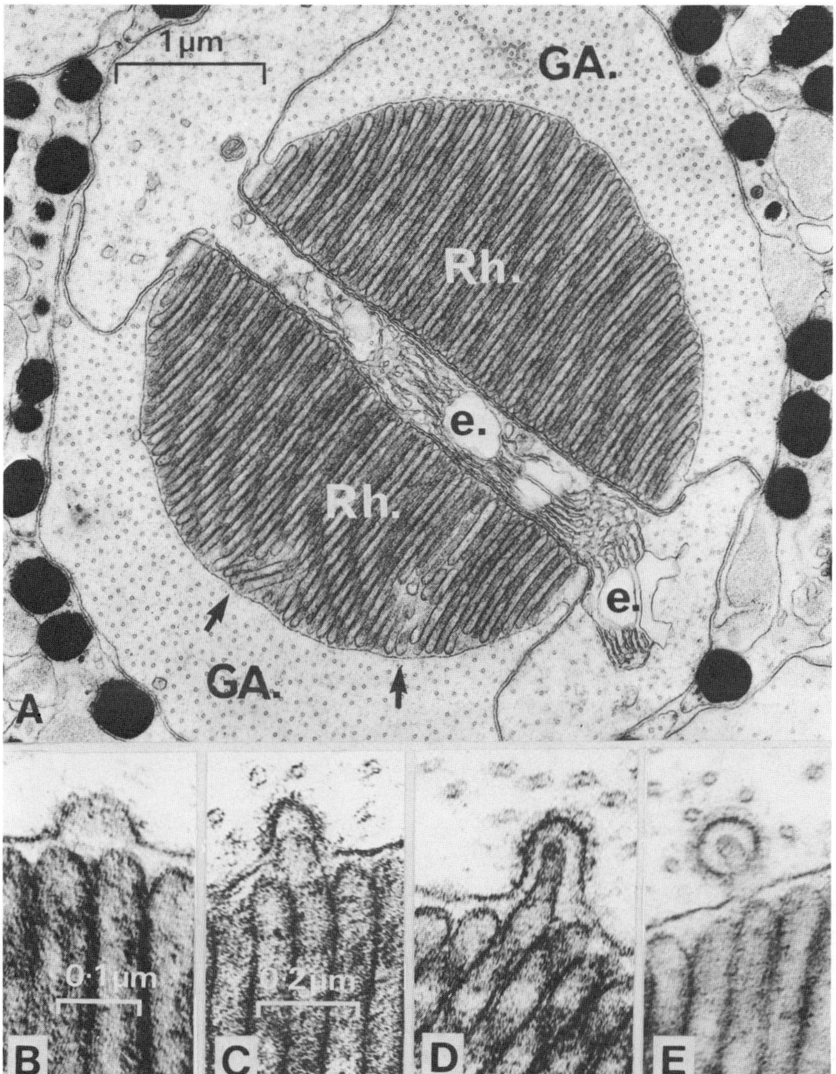

Figure 6. Routes of microvillus membrane shedding in *Plexippus*. A. Transverse section of an AL receptor. The arrows indicate regions where the distal parts of the rhabdomeral microvilli are disintegrating. Rh., rhabdomere; GA, glial arm; e., specialized, smooth endoplasmic reticulum lying between the twin rhabdomeres of the receptor. B, C, D, E. Stages in the lengthening and pinching-off of individual microvilli within coated endocytotic pits in the adjacent glial membrane (see text). D, E: scale as for C.

In correspondence, tubular smooth endoplasmic reticulum resembling GERL is found in the glial arms, especially at the distal end of the retina. Accumulations of membrane detritus are found in the proximal regions - the tips - of the glial arms, but there is little to suggest that they are degraded at that site. The length of glial arm between these tips and the distal region where GERL tubules are in evidence is occupied solely by highly ordered microtubules and scattered material derived from the reorganization of detached endocytotic pits and their contents. It would be attractive to suggest a model in which coated pits are moved after detachment to the middle of the glial arms, transported by cytoplasmic flow to the tips, and then, at some appropriate time, hauled up by microtubules to the distal regions of the glia for degradation. Unfortunately, there is nothing to support so neat and theatrical a picture, and the amounts of detritus found at the tips of the glial arms suggest that it may accumulate over periods of days without special arrangements being made for its disposal.

The receptor cytoplasm of these eyes is remarkably free from organelles, save for mitochondria and an unusual arrangement of endoplasmic reticulum between the two rhabdomeres of each cell. This grows after nightfall, to form massive whorls or networks before dawn which disappear after first light, just as shedding seems to reach its peak. It is probably concerned with the synthesis of microvillus membrane. Interestingly, in view of the increasing evidence for the participation of Golgi bodies in the glycolysation of opsin in vertebrates (44, 45), Golgi saccules are only found sparingly, at the proximal end of the receptor below the rhabdomeres.

An obvious problem concerns the nature of the transformation of the microvillus which causes it to elongate. Examination of many micrographs of shedding suggests that a quite short distal region of each microvillus undergoes alteration, and that it is not the whole microvillus which elongates, but only a small region of it. This conclusion is supported by the work on extracellular shedding in flies described in the next section. Another problem is posed by the development of the endocytotic pit. It appears in its earliest recognizable form before any substantial lengthening of the underlying microvillus has taken place, so that, just as in the case of vertebrate outer segment shedding, one must suppose that the associated microvillus tip, which is often seen just to touch the outer surface of the glial membrane is providing a signal. It has similarly been postulated (45) that completion of the oligosaccharide core of rhodopsin by the addition of fucose and galactose _after_ its insertion in the outer segment membrane may represent the addition of a surface cue which will in due course initiate phagocytosis by the pigment epithelium.

EXTRACELLULAR SHEDDING IN DEPTERA WITH OPEN RHABDOMS

The compound eyes of mosquitoes possess ommatidia of rather primitive design in which there are no large extracellular spaces between the contiguous receptors; their rhabdomeres shed membrane intracellularly as pinocytotic vesicles just as in most arthropods (46). Advanced Diptera in families such as the Calliphoridae and Muscidae have within each ommatidium eight receptors whose rhabdomeres project into a large, common, extracellular space (47-49). This arrangement has been termed an "open rhabdom". Published micrographs of these retinas show few organelles in the receptors other than abundant mitochondria; multivesicular bodies are notably absent. At the tips of the microvilli, there are always clouds of electron-dense material. We argued (43) that these observations suggest that flies with open rhabdoms may use an extracellular route for turnover.

Williams has looked at the receptors of two such flies: the blowfly *Lucilia*, and a large tipulid fly, *Ptilogyna*. The first observations on *Lucilia* (50) seemed to imply extracellular shedding as a major event in turnover, but we now believe their interpretation to be equivocal. They did, however, stimulate the much more informative study on *Ptilogyna* which is still in progress.

The events which are seen in sections after glutaraldehyde or glutaraldehyde/paraformaldehyde fixation are summarized in Fig. 7. Sparse multivesicular and residual bodies are found, it proves, often surrounded by small primary lysosomes. They can be deduced to stem from two kinds of membrane internalization: pinocytosis at the bases of the microvilli, and the formation of endocytotic pits in the retinula cell plasma membrane. The latter can be seen to contain electron-dense, structureless material of the kind lying at the microvillus tips, and must transport it to the interior of the cells. The infrequent nature of these events implies that turnover must be slow. A more dramatic type of shedding, however, involves the collapse of microvilli at the lateral margins of the rhabdomeres to form whorls. That this must be at least an occasional, normal event is indicated by the fact that it is possible to find whorls at various stages of internalization and envelopment by membranes derived from smooth endoplasmic reticulum. Such profiles are far less frequent than those of whorls in the extracellular space; we doubt the validity of these results and are currently attempting to determine the extent to which they can be explained in terms of artifacts of fixation. Here, it must be noted that the uptake of whorls by the retinula cells apparently proceeds by pseudopodial engulfment, and that this mechanism can be seen at work far more convincingly in *Ptilogyna*. One may speculate that extreme environmental conditions may provoke whorl formation with subsequent uptake, and that *Ptilogyna* has exploited, in an evolutionary sense, an existing quasi-pathological pathway.

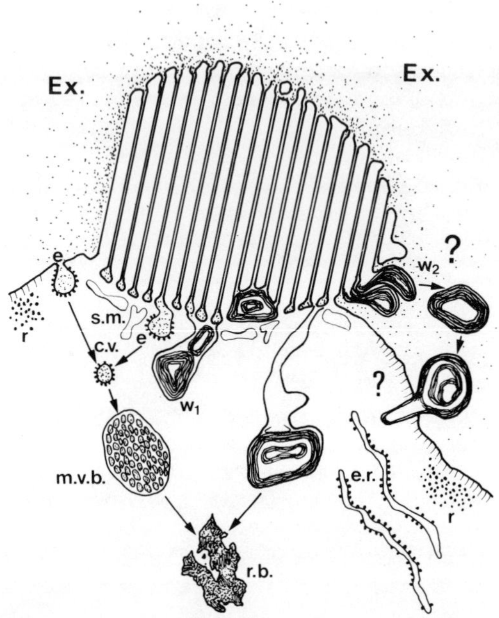

Figure 7. Possible routes of photoreceptor membrane shedding and retrieval in *Lucilia*. Left: Membrane at the microvillus bases is internalized as pinocytotic vesicles and non-rhabdomeral retinula cell plasma membrane as coated endocytotic pits (e). The latter enclose amorphous material lying in the extracellular space (Ex). Coated vesicles are assembled into multivesicular bodies (m.v.b.) and yield residual bodies (r.b.). Material in extracellular space may be derived from shedding at the microvillus tips. Right: Portions of microvilli are shed as whorls (w_1, w_2) which are internalized by endocytosis, enveloped by smooth membranes (s.m.) derived from rough endoplasmic reticulum (e.r.) and degraded. Currently, it seems likely that the majority of whorls seen in sections may be artifacts which nevertheless correspond to rare natural events; this interpretation is suggested by the fact that stages in the degradation of internalized whorls can be observed.

Anatomically, the retinas of Lucilia and Drosophila are quite close. Mutant Drosophila in which the retinas degenerate after eclosion have been prepared by Hotta and Benzer (51). In some mutants, degeneration is accelerated by exposure to light and retarded if the flies are kept in darkness. Recently, a sophisticated model of transduction has been built around the capacity of Norp A alleles (which to a variety of degrees impair transduction) to suppress degeneration as well as receptor potentials when combined with these mutants (52). Our interpretation of turnover processes, and especially our tentative evidence for the induction of local lytic activity within receptors by the presence of its own targets lead to a suspicion that it may be possible to explain these genetic interactions in terms of defects in turnover, as Ostroy (53) has recently suggested. More specifically, one might hypothesize that they are symptoms of failures in the regulation of autophagic systems that are switched on normally either by the presence of their substrates, by light, or by both. Unfortunately, there is no published information about the detailed ultrastructural cytology of retinal degeneration mutants, or of their various combinants with those creating transduction defects.

EXTRACELLULAR SHEDDING IN THE TIPULID FLY PTILOGYNA

One way of solving the problem of whether the coated vesicles seen within the receptor cytoplasm of so many arthropods represent the shedding of membrane into the cell, or the transport of membrane precursors to the microvilli was to find an animal, Dinopis, in which massive shedding is localized at a different point in the daily cycle to synthesis. We felt that our doubts about the reality of extracellular shedding in Lucilia might well be shared even more acutely by those who had not seen the actual material under the electron microscope. The conventional solution, obviously, would be to embark upon high-resolution EM autoradiography. The quick one is to find a nocturnal or crepuscular fly with an open rhabdom, and see what happens at dawn.

We chose to look at the huge Australian crane-fly, Ptilogyna, as a result of Professor G. A. Horridge's comment that he had failed to observe a pseudopupil in the eye, or to transilluminate the ommatidia antidromically with a light-probe. The eye is anatomically very unusual; amongst its many strange features are exceptionally short ommatidia with rhabdomeres only some 50 μm long in the diurnal eye, and a distal iris beneath each cone which shuts down during the day and opens widely at night. The rhabdomeres are of enormous cross-sectional area (Figs. 8, 10). Each of the peripheral receptors (R_{1-6}) is so constructed that in its proximal half the extracellular space around the rhabdomere is enclosed in a "pocket" formed by the retinula cell itself (Fig. 9).

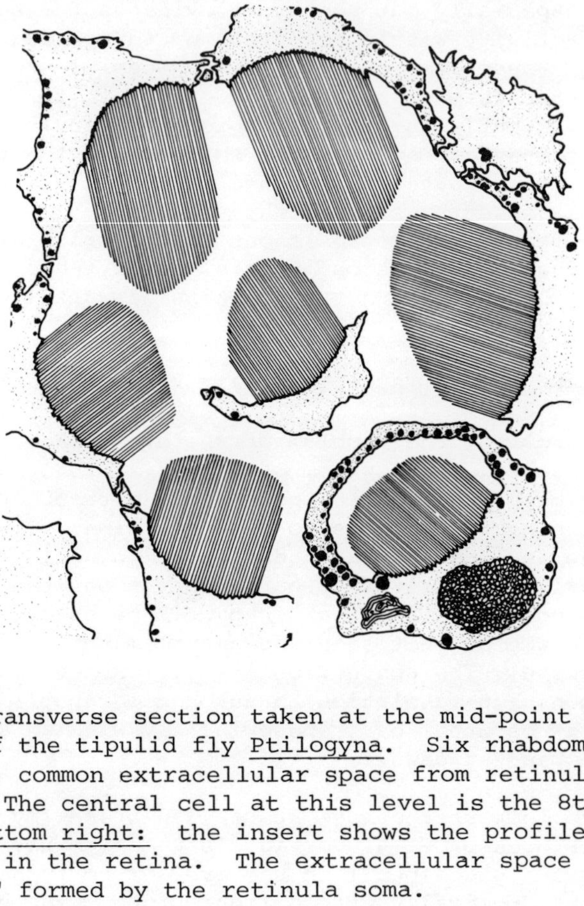

Figure 8. Transverse section taken at the mid-point of the ommatidium of the tipulid fly Ptilogyna. Six rhabdomeres project into a large, common extracellular space from retinula somata which enclose it. The central cell at this level is the 8th retinula cell. At bottom right: the insert shows the profile of a receptor more basally in the retina. The extracellular space is contained in a "pocket" formed by the retinula soma.

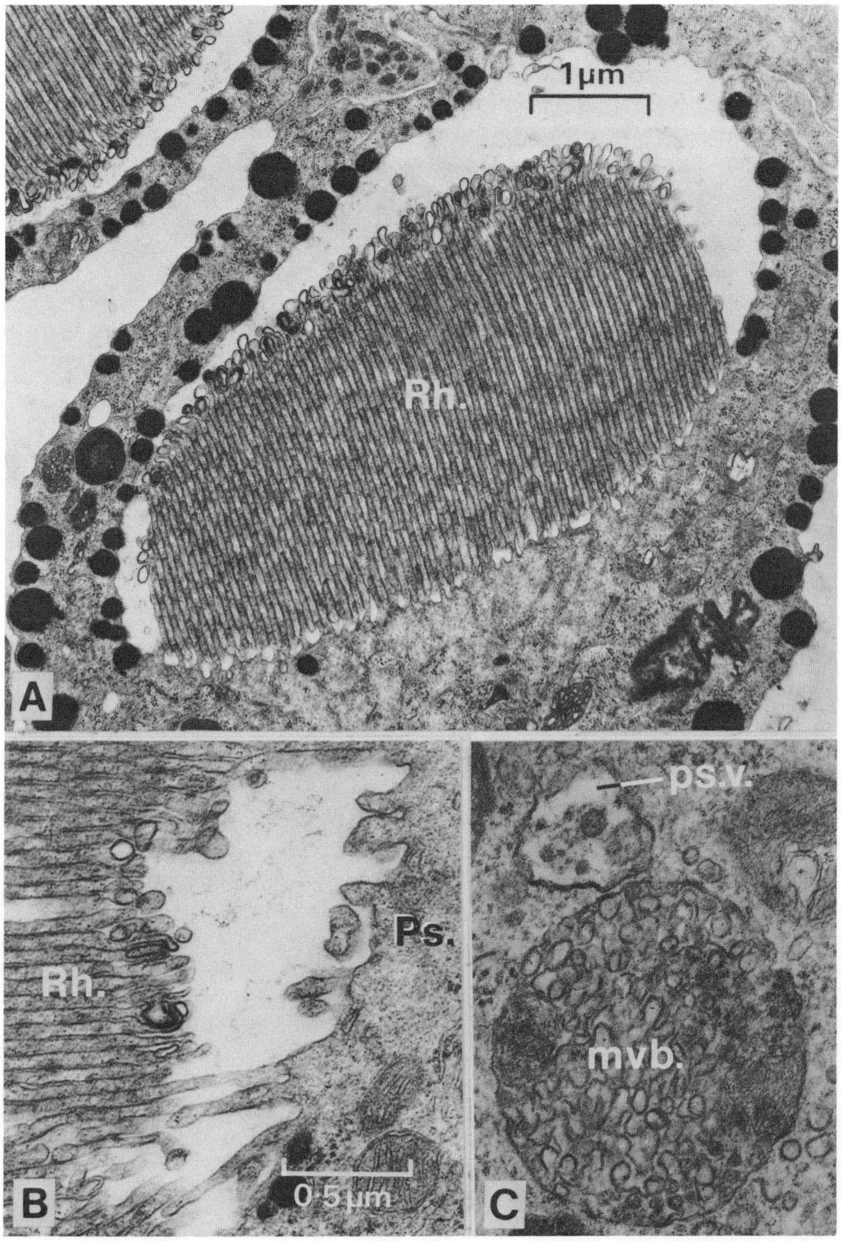

Figure 9. Ptilogyna. Basal region of a receptor to show the disordering of a distal border of the microvilli prior to shedding. B. Pseudopodial processes (Ps.) of a retinula cell close to the tips of its own rhabdomeral microvilli. C. A multivesicular body in the retinula cell cytoplasm. Above it, a vacuole (ps.v.) containing detritus, derived from the pseudopodial engulfment of material in the extracellular space.

Membrane is shed _en masse_ from the tips of the microvilli (Fig. 9A)(54). In our material, shedding was not precisely-timed in relation to the normal diurnal cycle. However, our flies had to be caught some fifty miles or more from the laboratory and brought to Canberra in transparent plastic bags resting on ice while exposed to dim, diffuse daylight, so they may be forgiven some degree of confusion. Receptors in a single ommatidium may be in slightly different shedding states, but, allowing for these variations, it seems clear that shedding reaches a peak at dawn, and that membrane is lost from the microvilli, endocytosed and re-cycled over the next 6-8 hours. Some flies were dissected and fixed immediately after capture in the field at critical points in the breakdown sequences, and their retinas found to be in the same states as those held in the laboratory.

Shedding, nevertheless, shows a different geometry from the patterns seen in _Lucilia_. All microvilli in the transverse profile of a single receptor are always in exactly the same state, and during shedding, they undergo the following sequence:

(i) A distal segment of each microvillus becomes less electron-dense in section. This segment swells, slightly, and may collapse locally to form one or more "micro-whorls" (Fig. 9A).

(ii) Whole blocks of microvillus tips detach and float in the extracellular space. They disintegrate further to a variety of configurations, including vesicles (54).

(iii) The detritus is endocytozed (55). The process must be very rapid, because it is not easy to detect it in progress. If flies are kept in darkness for 36 hours and then provoked to shed membrane at dawn by re-illumination, it is found that extra membrane has been synthesized, and that shedding is massive. Eight hours after such a "delayed dawn" evidence of endocytosis is much more readily seen. Some of the mechanisms of internalization are still problematical, but is is certain that they include the pseudopodial engulfment of masses of detritus (Fig. 9B) in a manner very like that in which colloid droplets of thyroglobulin are endocytozed from the follicular lumina of the thyroid gland (56). The end result is masses of vesicles and membranous fragments contained in vacuoles which can be very large (Figs. 9C, 10): they gradually degrade to tight multilamellar bodies and residual bodies which are ultimately lysed. At late stages in this process, they are surrounded by tubular configurations suggestive of the GERL and smooth tubular systems found in _Dinopis_ (17).

PATTERNS OF SHEDDING IN RELATION TO FUNCTIONAL CONSIDERATIONS

The proposition that all arthropod eyes turn over photoreceptor membrane can probably now be accepted. Exceptions, if there are any, will most probably be found in insects such as mayflies, where the adult lifespan can be as little as twenty-four hours, and

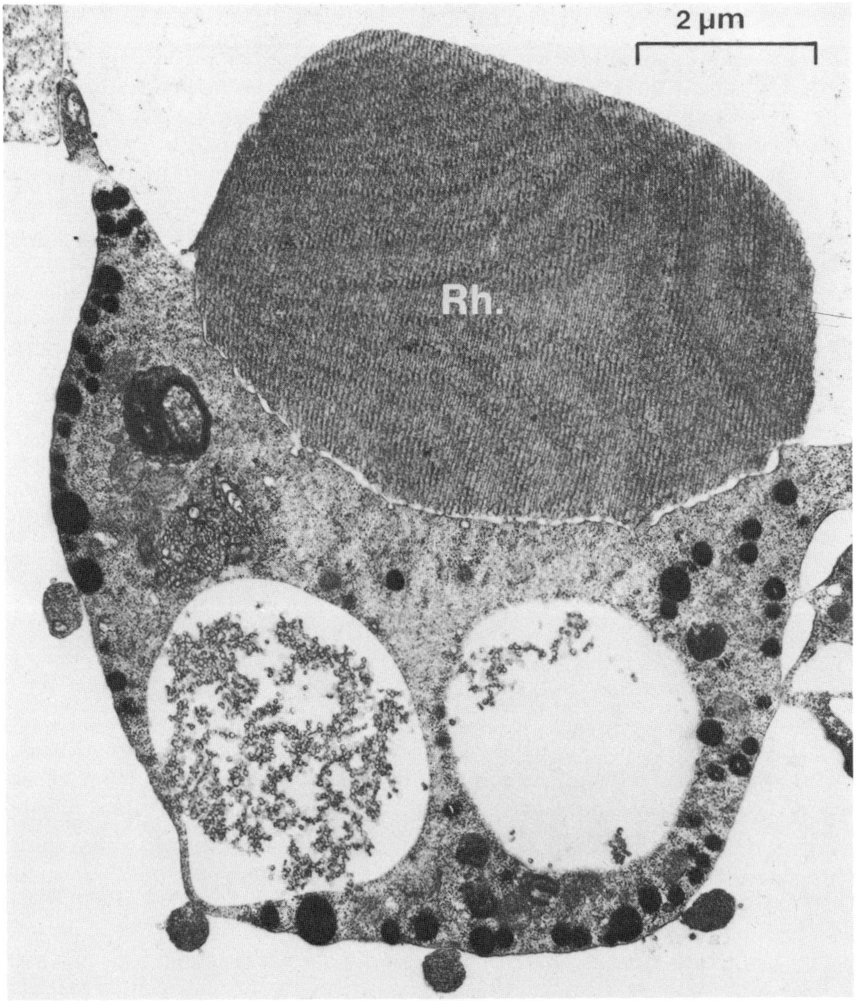

Figure 10. Ptilogyna. A receptor after the completion of shedding. The tips of the microvilli are orderly, the extracellular space is free of detritus, and the receptor cytoplasm contains the transverse profiles of two enormous vacuoles, both derived from pseudopodial uptake of extracellular material.

turnover of membrane might be pointless. This latter prediction rests upon the assumptions one is forced to make about why membrane is turned-over in the first place. There are a number of possible reasons, but they are not all equally plausible (6). Before discussing them briefly, it is worth considering how turnover patterns seem to be distributed. The rate at which turnover proceeds and its geometrical consequences seem to produce two main categories of photoreceptor, which no doubt intergrade:

(i) Receptors with rather few secondary lysosomes in their cytoplasm, with abundant mitochondria, and without any notable increase in rhabdomeral volume at night. This group includes the flies _Lucilia_ and _Calliphora_ (50), and the lateral eyes of the jumping spider, _Plexippus_, studied by us in Canberra (43). These arthropods have receptors designed for use primarily during the day, with low absolute sensitivity.

(ii) Receptors with abundant secondary lysosomes in their cytoplasm, few mitochondria, and gross changes in the amount of rhabdomeral membrane in the course of the daily cycle. This group includes _Dinopis_, the receptors of numerous other nocturnal spiders in several families (unpublished data), mosquitoes, larval and adult (1-4, 46), some crabs (26), and _Limulus_ (Barlow and Chamberlain, this Symposium). These receptors appear to be designed for use at night, or over a very wide range of intensities. The evidence suggests that they have high absolute sensitivities.

It is symptomatic of the paucity of information that so few animals can be listed, for observations from light microscopy alone are not adequate as evidence of turnover. We find that in the dinopid spider, _Menneus_, substantial, spurious "growth" of the rhabdomeres is achieved by increasing the diameters of the microvilli, and this might, in principle, be achieved by mere membrane expansion of the kind which can be induced, _in vitro_, in nuclear membranes (57, 58). The intriguing correlation suggested here needs testing over a broad spectrum of arthropods.

Whatever the nature of the classification which will emerge, two distinct questions are implied: Why does rhabdomere membrane turn over? Why does it do so in such a way as to create two broad classes of receptor, which can be considered to have stable-volume and variable-volume rhabdoms?

Because attention has been drawn to variable-volume rhabdoms late in the history of arthropod physiological optics, it is easy to forget that they were probably, in evolution, the precursors of those with stable volumes. For example, nocturnal spiders preceded diurnal families phylogenetically, and it is only in the very advanced jumping spiders which have optomotor responses of high acuity that we believe there to be true stable-volume rhabdoms (43, 59). Similarly, sophisticated optomotor programs became evident in the free flight of the Odonata and the advanced Diptera (60-62). What selection pressure led to the stable-volume rhabdom?

Two operational areas for natural selection are plausible candidates. Firstly, a recent theoretical analysis (63) has shown that loss of regularity in retinal mosaics degrades the quality of the images which they can abstract. It might be argued that in the case of fused rhabdoms, especially, cyclical degeneration and regeneration of membrane on a large scale might increase the likelihood both of mosaic irregularities and variations in sensitivity from receptor to receptor. These effects are most probable in spider receptors in which the geometry of the cross-sectional profiles is transformed radically at the transitions between one state and another (6). All such spider retinas, however, are nocturnal, and acuity is not therefore at a premium. Grapsid crabs present a more difficult case. Enormous changes in cross-sectional area of their rhabdoms are known to occur (Fig. 11), and these are achieved by the destruction and re-synthesis of rhabdomere membrane in bulk, just as in _Dinopis_ (26, S. J. Stowe, personal communication). Yet crabs have been shown to achieve high levels of optomotor performance (64-66), and this appears to be true of our local Australian _Leptograpsus_ (67, 68) which has been shown by Stowe to adjust its rhabdom volumes.

Other studies from this laboratory (69, 70) show, _a priori_, that for a given optical design, modulation transfer of a moving sinusoidal grating through an eye is optimized with respect to the periodicity of the grating, the speed at which its image moves across the retina, the background luminance, and its intrinsic contrast. This can be expressed in another way: if periodicity, speed, luminance and contrast are held constant, there will be an optimum rhabdom width which will maximize the solution to the modulation transfer function, and thus conserve the maximum amount of information about the stimulus. But the number of photons traversing a receptor and captured by it depends on the rhabdom width as well as its length. Apposition compound eyes or ocelli operating at low light intensities usually have long, wide rhabdoms and sacrifice spatial acuity in order to maximize photon capture. At the other extreme, eyes which operate at high light intensities are designed to maximize spatial acuity and have rhabdoms whose width is limited in the opposite direction by the constraints imposed by their waveguide properties. It seems that such eyes must restrict the amount of light which reaches their receptors or their responses will saturate, and this is usually done by screening pigment movements or pupillary mechanisms. For animals which operate over a wide range of intensities, there may be an advantage in restricting rhabdom width at the brighter end of the range to preserve sharp-peaked angular sensitivity functions and optimal modulation transfer, and in increasing their width at night in order to augment photon catch and improve the signal-to-noise ratio.

Nevertheless, overall sensitivity increase as a consequence of synthesis is likely to be small in relation to that achieved by

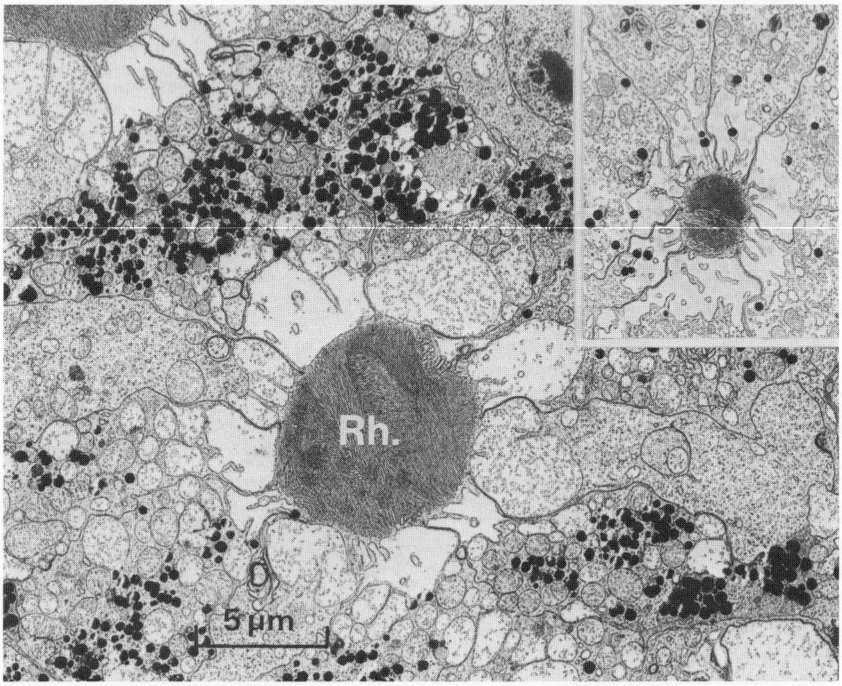

<u>Figure 11</u>. Rhabdoms of the grapsid crab, <u>Leptograpsus</u>, in transverse section at equivalent positions just below the eighth retinula cell. The main figure shows the rhabdom at midnight, after membrane synthesis, the small insert, at the same magnification, the rhabdom at mid-day. Electron micrographs by kind permission of Sally J. Stow.

other devices, and which in arthropod eyes is the product of photomechanical processes and the adjustment of transduction gain. Boosted sensitivity derived from the synthesis of new transductive membrane may prove impossible to isolate from that generated by other such causes, simply because in a real eye there are always alternative, more powerful sources of sensitivity modulation (42). Adaptational performance at retinal level is, in any case, designed with reciprocal reference to the deeper optic neuropils (71), thus adding yet another complicating factor to the situation which we have to unravel.

It has been mentioned that renewal of photoreceptor membrane may be primarily necessary because, in some sense, it ages (6, 43). It is unsatisfactory that an explanation based upon aging is merely forced on us by a process of elimination, and supported by little more than circumstantial evidence. Aging might either be intrinsic (the result, for example, of thermal perturbations of its constituents); work-dependent (for example, the cumulative acquisition of configurational accidents resulting from bleaching and regeneration of rhodopsin, or the opening and closing of sodium channels); or even, in one sort of explanation (52) a consequence of mechanisms generated by a hypothetical transmitter system involved in transduction itself.

We have shown that within the arthropods, the mechanisms of membrane disposal are very varied. The orderly processes of normal turnover can be discriminated from pathological shedding, just as they can in vertebrate photoreceptors. In particular, the newly-disclosed shedding geometries of *Lucilia* and *Ptilogyna* establish unequivocally that while shedding may be initiated by light, it does not happen as a direct result of a damaging effect of light on particular regions of the membrane itself. Indeed, a comparison between shedding in *Lucilia* and *Ptilogyna* is especially illuminating; both have open rhabdoms, and the optics of their ommatidia are not likely to be radically different. In *Lucilia*, shedding, however achieved, is slow. This deduction follows from the low density of organelles in the receptor cytoplasm. Despite the ability of the cells to endocytoze whorls, it is likely that in the healthy fly most membrane is shed by internalization. In *Ptilogyna*, a distal portion of each microvillus is transformed prior to shedding, and the latter event is achieved by such regions dropping off and vesiculating. In a given transverse profile, all microvilli exhibit exactly the same sharply-defined point of transformation with respect to the distance from the microvillus base at which it can be seen to have occurred, yet adjacent rhabdomeres may be in markedly different shedding states. Thus, whatever influence light may have on the process, (primarily, it appears, as a trigger), it may not determine the length of microvillus tip which will be shed; the transformations which reveal incipient shedding can be seen in many cells to have taken place before dawn.

Shedding, then, should be regarded as involving programmed developmental events whose precise nature and geometry must have been adjusted by natural selection to deal with particular exigencies whose nature we do not at present understand. Perhaps the main importance of these observations on flies with open rhabdoms is that they align the ways in which we must look at shedding in vertebrate and invertebrate receptors; they prove to be phenomena of very much the same kind.

Tempting but premature speculations about possible mechanisms which might allow local regions of microvillus to collapse on schedule, such as withdrawal of cholesterol from the membrane or changes in components of the sub-membrane cytoskeleton beneath it should perhaps be restrained until there is direct experimental evidence.

To conclude: what would happen if transductive membrane turnover failed, and the photon-trapping systems were made to become static? As presently evolved, photoreceptors can be expected to have their turnover programs so regulated as to anticipate damage or deterioration, however caused, rather than to accompany or lag behind it. Professor A. W. Snyder, (personal communication) arguing from the behavior of man-made sensors of very high sensitivity and low threshold, has suggested that damage caused by any form of aging is most likely in the first instance to manifest itself as a decreased signal-to-noise ratio in the receptor output rather than as frank decrements in sensitivity *per se*. It is important, now, that a preparation should be found of sufficient simplicity to allow these matters to be tested.

ACKNOWLEDGEMENTS

A large number of people have contributed to our research program. It owes much to enthusiastic support from Professor G. A. Horridge, FRS, and from the former Director of the Research School of Biological Sciences, Professor Sir Rutherford N. Robertson, FRS. I thank my collaborators for their stimulating contributions. In chronological order they are: Dr. M. F. Land (Sussex, England), W. A. Day (Christchurch, New Zealand), A. D. Austin (Sydney, Australia), Dr. S. B. Laughlin, to whom I am indebted for a critical reading of the last section of this paper, Ling Chen Kao, Karen Powell, Joanne Maples, Sally Stowe and Dean Price (Canberra). In particular, results on *Diptera* with open rhabdoms are the energetic work of a post-graduate student, David S. Williams to whom I am indebted for permission to cite his results. Early stages of the studies received financial support from the U. K. Science Research Council (to M. F. L.), from the Erskine Foundation of the University of Canterbury, and the University of Canterbury Research Fund (to A. D. B.) and from the New Zealand State Lotteries Scientific

Distribution Committee (to the Otago Museum). Research on *Dinopis* would be impossible without the continuing access to field facilities generously provided by Professor D. T. Anderson, FRS, and the School of Biological Sciences, Sydney University. Line drawings were prepared by Chris Snoek, to whom I am grateful for his speed and skill. Figure 6, modified from (43), is reproduced by kind permission of the Council of the Royal Society.

REFERENCES

1. White, R. H. (1964) *Am. Zool.* 4, 433.
2. White, R. H. (1967) *J. Exp. Zool.* 166, 405-425.
3. White, R. H. (1968) *J. Exp. Zool.* 169, 261-278.
4. White, R. H. and Lord, E. (1975) *J. Gen. Physiol.* 65, 583-598.
5. Holtzman, E., Schacher, S., Evans, J. and Teichberg, S. (1977) *Cell Surface Reviews* 6, 165-246.
6. Blest, A. D. (1978) *Proc. R. Soc. Lond.* B, 200, 463-483.
7. Eguchi, E. and Waterman, T. H. (1976) *Cell Tiss. Res.* 169, 419-434.
8. Young, R. W. (1967) *J. Cell Biol.* 33, 61-72.
9. Young, R. W. and Droz, B. (1968) *J. Cell Biol.* 39, 169-184.
10. Anderson, D. H., Fisher, S. K. (1976) *J. Ultrastruct. Res.* 55, 119-141.
11. Anderson, D. H., Fisher, S. K. and Steinberg, R. H. (1978) *Invest. Ophthalmol.* 17, 117-133.
12. Hogan, M. J., Wood, I. and Steinberg, R. H. (1974) *Nature, Lond.* 252, 305-307.
13. Young, R. W. (1978) *Invest. Ophthalm.* 17, 105-116.
14. Blest, A. D. and Land, M. F. (1977) *Proc. R. Soc. Lond.* B, 196, 197-222.
15. Blest, A. D., Kao, L. and Powell, K. (1978) *Cell Tiss. Res.* 195, 425-444.
16. Duce, I. R. and Keen, P. (1977) *Cell Tiss. Res.* 185, 263-279.
17. Blest, A. D., Powell, K. and Kao, L. (1978) *Cell Tiss. Res.* 195, 277-297.
18. Blest, A. D., Price, D. G. and Maples, J. *In preparation.*
19. Fahrenbach, W. H. (1970) *Z. Zellforsch.* 105, 303-316.
20. Essner, E. and Haimes, H. (1977) *J. Cell Biol.* 75, 381-387.
21. Essner, E. and Oliver, C. (1974) *Lab. Invest.* 30, 596-607.
22. Abrahamson, E. W., Fager, R. S. and Mason, W. T. (1974) *Exp. Eye Res.* 18, 51-67.
23. Hamdorf, K. and Schwemer, J. (1975) *Photoreceptor Optics,* (A. W. Snyder and R. Menzel, eds.), pp. 264-289, Springer-Verlag, Berlin-Hedelberg-New York.
24. Paavola, L. G. (1978) *J. Cell Biol.* 79, 45-58.
25. Paavola, L. G. (1978) *J. Cell Biol.* 79, 59-73.
26. Nässel, R. and Waterman, T. H. (1979) *J. Comp. Physiol.* (in press).

27. Dowling, J. E. (1960) Nature, Lond. 188, 114-118.
28. Muto, Y., Goodman, DeW. S. (1972) J. Biol. Chem. 247, 2533-2541.
29. Poole, A. R., Dingle, J. T., Mallia, A. K. and Goodman, DeW. S. (1975) J. Cell Sci. 19, 379-394.
30. Smith, J. E., Borek, C. and Goodman, DeW. S. (1978) Cell 15, 865-873.
31. Smith, J. E., Muto, Y. and Goodman, DeW. S. (1975) J. Lipid Res. 16, 318-323.
32. Loew, E. R. (1976) Proc. R. Soc. Lond. B, 193, 31-44.
33. Blest, A. D. and Day, W. A. (1978) Phil. Trans. R. Soc. Lond. B, 283, 1-23.
34. Djamgoz, M. B. A. and Ruddock, K. H. (1978) Neuroscience Letters 7, 251-256.
35. O'Steen, W. K. and Lytle, R. B. (1971) Am. J. Anat. 130, 227-234.
36. Tso, M. O. M. (1973) Invest. Ophthalmol. 12, 17-34.
37. Tso, M. O. M., Wallow, I. H. L. and Powell, J. O. (1973) Arch. Ophthalmol. 89, 228-234.
38. Basinger, S., Hoffman, R. and Matthes, E. S. (1976) Science, N. Y. 194, 1074-1076.
39. LaVail, M. M. (1976) Science, N. Y. 194, 1071-1074.
40. Austin, A. D. and Blest, A. D. (1979) J. Zool. Lond. (in press).
41. Robinson, M. H. and Robinson, R. (1971) Am. Mid. Nat. 85, 85-96.
42. Laughlin, S. B., Stowe, S. J. and Blest, A. D. In preparation.
43. Blest, A. D. and Maples, J. (1979) Proc. R. Soc. Lond. B. 204, 105-112.
44. Bok. D., Basinger, S. F. and Hall, M. O. (1974) Exp. Eye Res. 18, 225-240.
45. O'Brien, P. J. (1976) Exp. Eye Res. 23, 127-137.
46. Brammer, J. D. and Clarin, B. (1976) J. Exp. Zool. 195, 33-40.
47. Melamed, J. and Trujillo-Cenóz, O. (1968) J. Ultrastruct. Res. 313-334.
48. Boschek, B. C. (1971) Z. Zellforsch. 118, 369-409.
49. Trujillo-Cenóz, O. and Melamed, J. (1966) Functional Organizatio of the Compound Eye, (C. G. Bernhard, ed.) pp. 339-361, Oxford.
50. Williams, D. W. In preparation.
51. Hotta, Y. and Benzer, S. (1970) Proc. Natl. Acad. Sci. U.S.A. 67, 1156-1163.
52. Harris, W. A. and Stark, W. S. (1977) J. Gen. Physiol. 69, 261-291.
53. Ostroy, S. E. (1978) J. Gen. Physiol. 72, 717-732.
54. Williams, D. S. and Blest, A. D. In preparation.
55. Blest, A. D. and Williams, D. S. In preparation.
56. Wetzel, B. K., Spicer, S. S. and Wollman, S. H. (1965) J. Cell Biol. 25, 593-618.
57. Wunderlich, F., Günter, G. and Bucherer, C. (1978) J. Cell Biol. 79, 470-490.

58. Wunderlich, F., Wallach, D. F. H., Speth, V. and Fischer, H. (1974) Biochem. Biophys. Acta 373, 34-43.
59. Eakin, R. M. and Brandenburger, J. L.(1971) J. Ultrastruct. Res. 37, 616-663.
60. Collett, T. S. and Land, M. F. (1975) J. Comp. Physiol. 99, 1-66.
61. Collett, T. S. and Land, M. F. (1978) J. Comp. Physiol. 125, 191-204.
62. Land, M. F. and Collett, T. S. (1974) J. Comp. Physiol. 89, 331-357.
63. French, A. S., Snyder, A. W. and Stavenga, D. G. (1977) Biol. Cybernetics, 27, 229-233.
64. Horridge, G. A. (1966) J. Exp. Biol. 44, 233-245.
65. Horridge, G. A. (1966) J. Exp. Biol. 44, 247-254.
66. Horridge, G. A. (1966) J. Exp. Biol. 44, 263-274.
67. Erber, J. and Sandeman, D. C. (1976) J. Comp. Physiol. 112, 189-197.
68. Sandeman, D. C. and Erber, J. (1976) J. Comp. Physiol. 112, 181-188.
69. Horridge, G. A. (1978) Phil. Trans. R. Soc. Lond. 285, 1-59.
70. Snyder, A. W., Stavenga, D. G. and Laughlin, S. B. (1977) J. Comp. Physiol. 116, 183-207.
71. Laughlin, S. B. and Hardie, R. C. (1978) J. Comp. Physiol. 128, 319-340.

LIGHT AND A CIRCADIAN CLOCK MODULATE STRUCTURE AND FUNCTION IN *LIMULUS* PHOTORECEPTORS

Robert B. Barlow, Jr. and Steven C. Chamberlain

Institute for Sensory Research
Syracuse University
Syracuse, New York 13210

Efferent nerve signals transmitted from the brain modulate the response characteristics of sensory organs in a number of animals (1) and thereby may play an important role in processing the sensory information the brain receives. Extensive studies of the *Limulus* visual system have contributed significantly to our understanding of the retinal mechanisms that process the information transmitted to the brain (2, 3). Recent work in our laboratory now reveals that the response characteristics of the retina of the *Limulus* lateral eye are strongly influenced by efferent nerve signals transmitted from a circadian clock located in the brain (4).

In this paper we review the evidence for circadian rhythms in visual sensitivity of the *Limulus* eye and describe the effects of both light and the circadian clock on the morphology of *Limulus* ommatidia. We also discuss the clock's control of membrane turnover in *Limulus* photoreceptors as well as present preliminary results on the effects of constant light and constant dark on the structure of *Limulus* photoreceptors. Thus far our studies suggest that the circadian efferent input to the retina adapts the eye for optimal function at very low light levels and controls, in part, the metabolic processes that maintain photoreceptor function.

A Circadian Clock Increases Visual Sensitivity

Fig. 1A shows the circadian rhythm in the amplitude of the ERG recorded from the lateral eye. In this experiment an animal was clamped to a rigid platform in a seawater aquarium located in a lightproof, shielded cage. A corneal electrode was positioned on one of the lateral eyes, a fiber-optic bundle was aligned in front

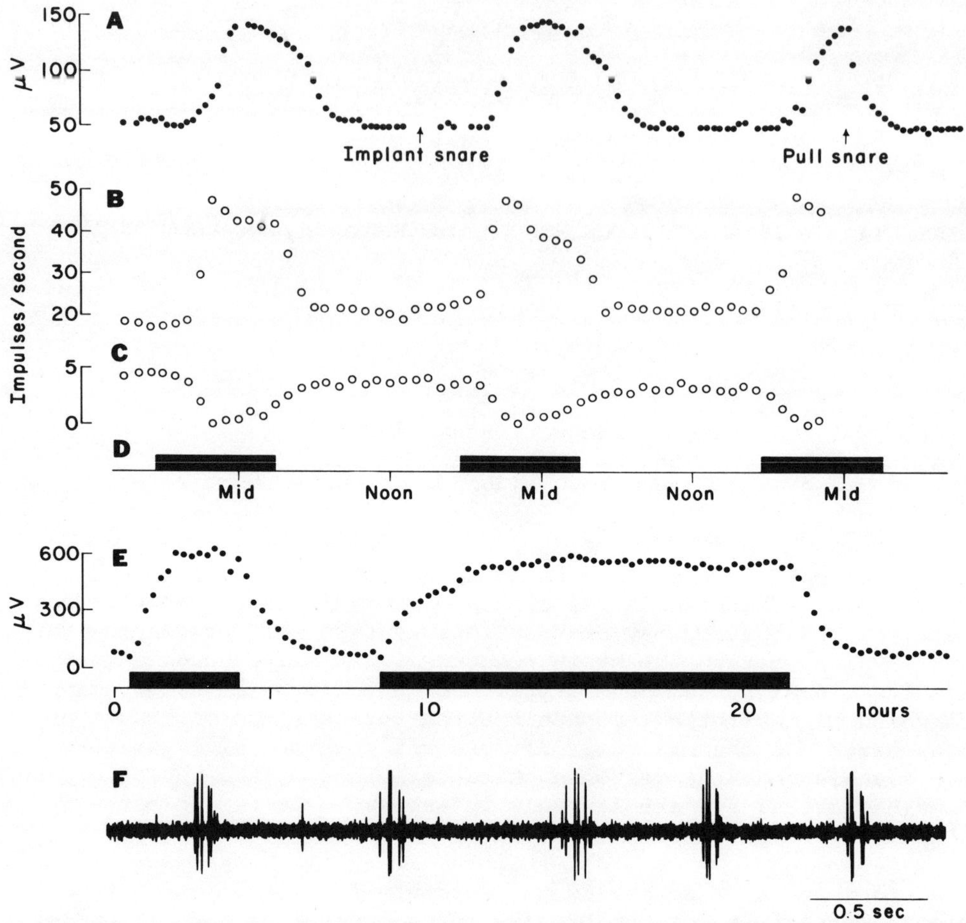

Figure 1. (A) Circadian rhythm in the amplitude of the ERG of the *Limulus* lateral eye. Cutting the optic nerve ("pull snare") abolished the freerunning rhythm. (B) Circadian rhythm in the optic nerve response of a single, dark-adapted ommatidium. Test flash intensity corresponds to log I = -7 in Fig. 2. (C) Circadian rhythm in the spontaneous optic nerve activity of a dark-adapted ommatidium. Data in (B) and (C) were recorded from the same optic nerve fiber. (D) Circadian rhythm of the endogenous efferent optic nerve activity recorded from the proximal stump of the cut optic nerve. Dark bars give the periods of general efferent activity. (E) Effects on the ERG amplitude caused by shocking the optic nerve *in situ* (dark bars). (F) Endogenous efferent spike activity recorded from the optic nerve trunk. The 3.5-sec recording was made at 7:30 p.m. when efferent activity is about maximum. Taken from (4).

of the eye, the cage was closed, and ERG responses to 50-msec test flashes were recorded every 30 minutes while the animal remained in darkness. Note that the ERG responses were higher at night than during the day. Also, cutting the optic nerve ("pull snare") at midnight of the third day caused a rapid decline of the ERG amplitude to the low daytime level. These results suggest that, at night, tonic efferent activity in the optic nerve increases the ERG amplitude.

Efferent activity can indeed be recorded from the optic nerve at night, but not during the day. Fig. 1F shows that regular bursts of impulses were discharged by several efferent fibers of the cut optic nerve. The frequency of bursting was maximum during the early evening hours. Little or no efferent activity was recorded during the day. The efferent optic nerve activity (dark bars in Fig. 1D) exhibited a circadian rhythm with approximately the same periodicity as that of the ERG responses (Fig. 1A).

Mimicking the endogenous efferent activity by shocking the optic nerve *in situ* produces the same effects as the circadian clock. Fig. 1E shows that pulses delivered at the rate of 2 per second to the distal end of the cut optic nerve increased the ERG amplitude. The shock rate of 2 per second is about equal to the burst frequency of the efferent nerve fibers during the early evening hours (4). When the current pulses were turned off, the ERG amplitude decayed to the normal daytime level with a time constant of about 1 hour. When the current pulses were turned on again several hours later, the ERG amplitude again increased and remained elevated throughout the 13-hour shocking period (indicated by the black bar).

Optic nerve responses exhibit a circadian rhythm which follows the periodicity of the endogenous efferent activity. An animal was mounted in an aquarium as in the ERG experiments, and the spike discharge of a single fiber *in situ* was recorded without interrupting the efferent input to the retina. Details of the surgery and recording technique are presented elsewhere (5-6). A fiber-optic light pipe was aligned in contact with the cornea to optically isolate the recorded ommatidium. Fig. 1B gives the steady-state rate of discharge in response to a 6-second flash delivered to the ommatidium every hour for 56 hours while the animal remained in the dark. Fig. 1C gives the rate of spontaneous activity recorded from the same nerve fiber. The circadian rhythm in both the optic nerve and ERG responses indicates that the action of the circadian efferent input is to increase retinal sensitivity at night.

The circadian rhythm in the spontaneous optic nerve discharge (Fig. 1C) revealed an unexpected result: retinal "noise" is reduced at night. It appears that the circadian clock in the *Limulus* brain increases the sensitivity of the lateral eyes at night by increasing the amplitude of the retinal "signal" and decreasing the amplitude of retinal "noise". In short, the clock increases retinal sensitivity by increasing the signal-to-noise ratio for optic nerve responses.

The efferent input to the retina also influences the intensity-coding characteristics of the ommatidia. Following the procedure given above for the optic nerve recordings, we measured the steady state response of the recorded ommatidium over a wide range of test intensities during the day and at night. Fig. 2 reveals that during the day the ommatidium exhibited a threshold of about log I = -9, a graded response over at least a 9 log-unit range of light intensity, and a distinct plateau in response at moderate intensities. At night, the same ommatidium exhibited a threshold of less than log I = -11, a graded response over at least a 9 log-unit range, and no

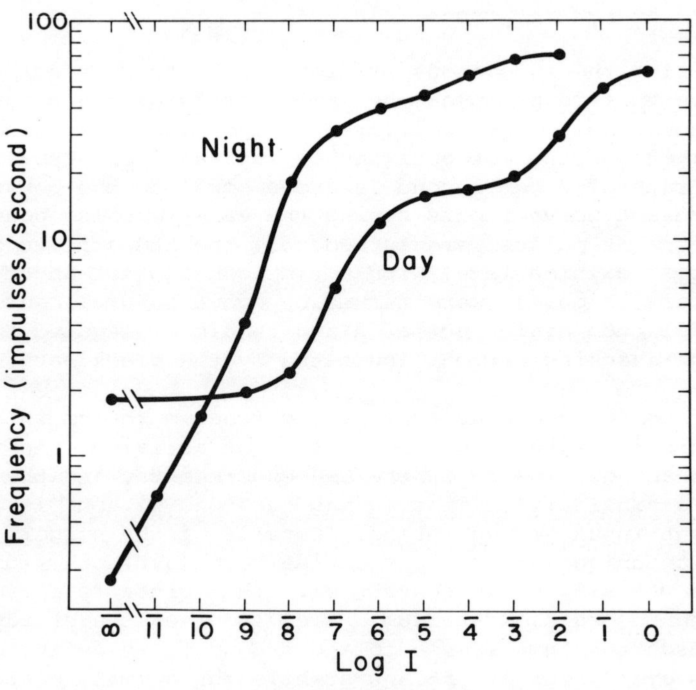

Figure 2. Intensity characteristics for daytime and nighttime responses from a single, dark-adapted ommatidium *in situ* with the optic nerve trunk uncut. The mean firing rate of the single optic nerve fiber during the last 3 sec of a 6-sec flash is plotted on the ordinate as a function of the logarithm of the relative light intensity plotted in the abscissa (log I = 0 corresponds to 10^{12} photon/sec at the cornea from 400 to 700 nm). "Day" responses were recorded from 3 to 4 p.m. and "night" from 9 to 10 p.m. The animal remained in the dark throughout the experiment. Taken from (4).

clear plateau. Although the threshold for steady-state responses decreased about 2 log units at night, the sensitivity of the retina for suprathreshold stimuli at night can exceed daytime levels by as much as 10^5. For example, Fig. 2 shows that a steady response of 20 impulses/sec requires 5 log units less light intensity at night than during the day.

Changes in the position and shape of the intensity-response functions in Fig. 2 indicate changes in both the quantum catch and gain of the photoreceptors (3). As described below, circadian changes in the morphology of the photoreceptor and surrounding pigment cells can provide a ready explanation for changes in quantum catch (number of photons absorbed by the retinular cells at each light intensity) but not for changes in gain or in spontaneous activity. We do not yet understand the cellular mechanisms underlying the circadian changes in gain and spontaneous activity.

Intracellular records from retinular cells indicate that the circadian efferent input to the retina acts directly on the photoreceptor cells. Preliminary experiments carried out with Dr. Ehud Kaplan (The Rockefeller University) show that the clock enhances visual sensitivity by elevating photoreceptor response (receptor potential) and lowering photoreceptor noise (spontaneously generated fluctuations in membrane potential) (7). The properties of the receptor potential and of potential fluctuations in the unexcised eye are reported elsewhere (8). Our experiment followed the techniques described above for the ERG recordings with the exception that we removed a small section of cornea in order to impale single photoreceptor cells with a glass micropipette (8). Stable recordings of single cells were maintained over a sufficiently long period of time (6 hrs.) to observe the physiological effects of the circadian clock. Recordings covering the period of onset of efferent activity (5 to 7 p.m.) show a progressive increase in the receptor potential elicited by a fixed intensity test flash and a progressive decrease in the spontaneously generated potential fluctuations.

The increase in photoreceptor sensitivity and decrease in photoreceptor noise were not accompanied by significant changes in either cell resting potential or membrane resistance. The observed reduction in the frequency of the spontaneous quantum bumps with the onset of the efferent activity appears to be the origin of the observed decrease in the spontaneous optic nerve discharge (Fig. 1C). Reduction of photoreceptor noise indicates that the efferent input to the retina acts at a very early stage in the process of visual transduction.

The efferent origin of the circadian rhythms in visual sensitivity is consistent with the existing anatomical evidence for efferent fibers in the optic nerve trunk (9) and efferent terminals on retinular and pigment cells of the ommatidium (10). Fig. 3A shows an efferent terminal with a synaptoid specialization adjacent to a pigment cell (arrow). Efferent processes are shown invaginating an eccentric-cell dendrite in Fig. 3B and a retinular cell in Fig. 3C. The efferent terminals contain granules (Fig. 3D-F) which we presume to be neurosecretory. Note that the dense granules appear cylindrical with an invagination at one end (Fig. 3E, arrows). Although we have not yet determined the central origin of the efferent nerve fibers, Fahrenbach (11) found that some terminal processes of nerve cells in the corpora pedunculata of the brain contain granules similar in appearance to those in Fig. 3D-F. The precise identification of the efferent cell bodies in the brain should provide a better understanding of the cellular basis of the circadian clock.

A Circadian Clock and Light Modulate Photoreceptor Structure

Circadian changes in the morphology of the ommatidia of the *Limulus* eye parallel the circadian changes in visual sensitivity described above (12). Fig. 4 shows reconstructions of dark-adapted ommatidia fixed during the day and at night. At noon, the retinular cells are separated from the base of the corneal lens by about 30 μm. Surrounding pigment cells form a small aperture (≃17 μm dia) between the lens and the retinular cells. Screening pigment in the retinular cells has moved centrally to the edge of the photosensitive rhabdom but remains outside the ends of the rhabdomeral rays (see Fig. 5). The rhabdom of the central ring lies next to and extends the entire length of the dendrite of the eccentric cell. At midnight, near the period of peak efferent activity, the retinular cells have shifted position and now lie within 4 μm of the corneal lens. In addition, the pigment-cell processes have moved radially away from the ommatidial axis, increasing the aperture diameter to more than 40 μm. The retinular cells appear to be compressed against the base of the lens; the rhabdom is shortened by 40% and widened by 50%. The individual rays become bent and folded and appear as loops in the longitudinal reconstruction. The numerous bends in the distal portion of the eccentric-cell dendrite apparently result from the deep folds in the adjacent rhabdom structure.

Mimicking the endogenous efferent input to the retina during the day by shocking the cut optic nerve during the day changes the morphology of the ommatidium from the noon to the midnight state (13). We therefore conclude that it is the efferent input to the retina from the circadian clock that changes the internal structure of the ommatidium.

Fig. 4 (top) shows that the circadian changes in morphology correspond with changes in the field of view of a single

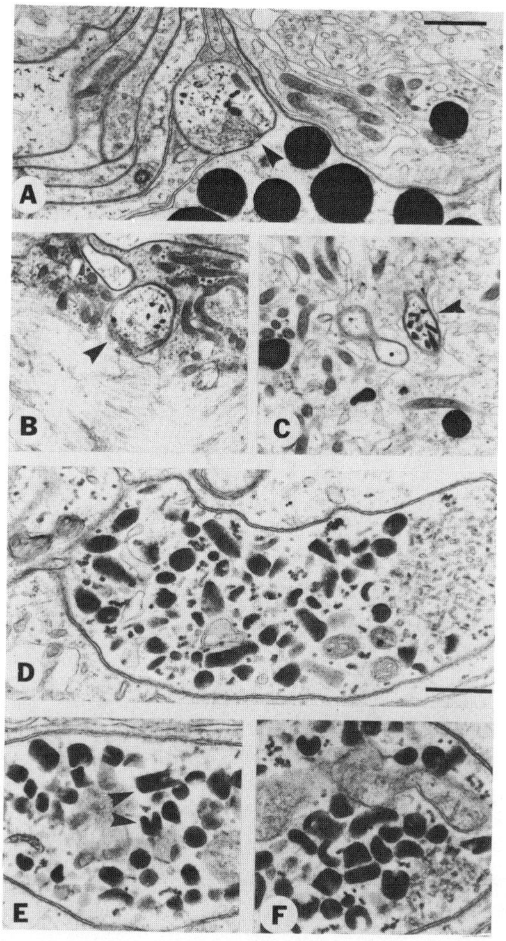

Figure 3. Electron micrographs of efferent terminals in *Limulus* ommatidia. (A) Arrow indicates a synaptoid specialization in an efferent terminal adjacent to a pigment cell. A retinular cell lies on the upper right. (B) Arrow shows an efferent process invaginating the base of an eccentric-cell dendrite. (C) Arrow points to an efferent terminal in a retinular cell. Bar in (A) represents 1 μm for the upper three micrographs. (D-F) Dense granules in efferent terminals. Granules are generally cylindrical with an invagination in one end (arrows in E). Bar in (D) represents 0.4 μm for the lower three micrographs.

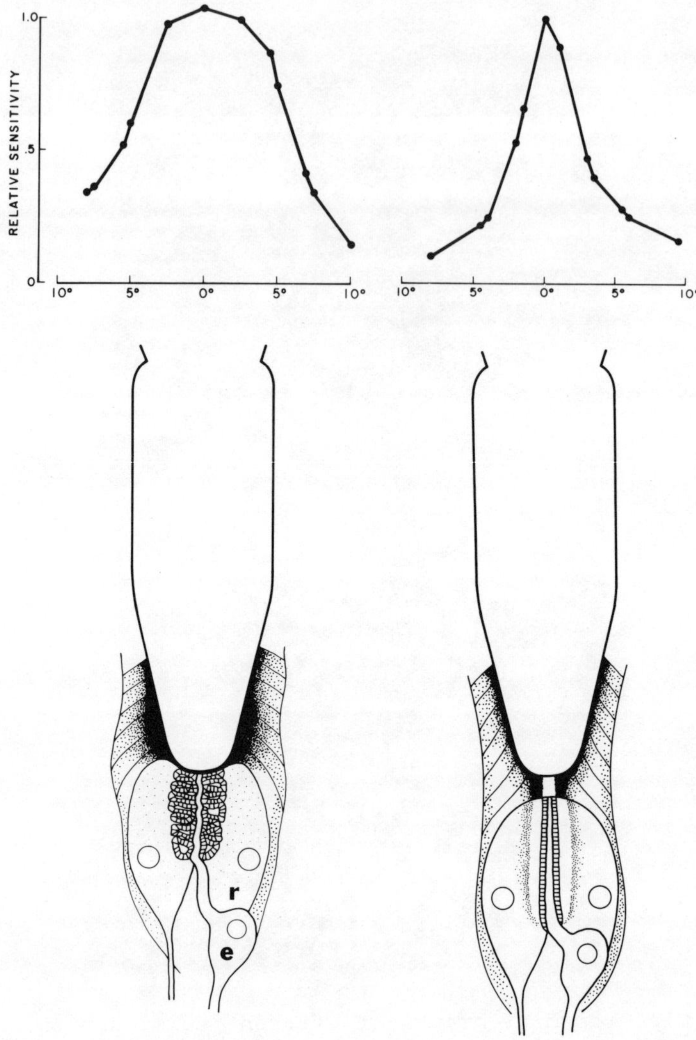

Figure 4. Circadian changes in the field of view and the morphology of an ommatidium. Top: Relative sensitivity of the optic nerve discharge from a single, dark-adapted ommatidium *in situ* as a function of the angle of incidence of test flashes delivered from a point light source. The acceptance angle increased from 6° during the day (right) to 13° at night (left). Bottom: Longitudinal reconstructions of dark-adapted ommatidia fixed during the day (right) and at night (left). Outlined are the corneal lenses, retinular (r) and eccentric cells (e), and the surrounding pigment cells. The width of the microvillus structure of the rhabdom has been increased for diagrammatic purposes. Taken from (13).

ommatidium (13). The acceptance angle (width at half maximum) of an ommatidium increases from 6° during the day to 13° at night. Cutting the lateral optic nerve abolishes the circadian changes in acceptance angle, leaving it in the narrow daytime state. Shocking the distal end of the cut optic nerve during the day increases the acceptance angle to the nighttime state. Efferent optic nerve activity therefore mediates the circadian changes in acceptance angle.

This circadian rhythm in the spatial summation of an ommatidium can be accounted for by the combination of the fixed optics of the corneal lens and the changing morphology of the retinular and surrounding pigment cells. During the day, separation of the rhabdom from the base of the lens and formation of a small aperture in front of the rhabdom by surrounding pigment cells restricts the incident rays to only those lying near the optic axis of the ommatidium. At night, enlargement of the aperture, distal movement of the rhabdom, and its distribution across the base of the lens allows more divergent rays to reach the photosensitive membrane. Our analysis of a simplified model of the corneal lens indicates that it is the nonimage-forming, reflected rays that determine the acceptance angle at night. Apparently, the efferent input from the circadian clock moves the photosensitive membrane of the retinular cells into a position that makes optimal use of the light-collecting characteristics of the corneal lenses. In sum, the modulation of ommatidial morphology by the circadian clock allows the *Limulus* eye to trade daytime acuity for nighttime sensitivity.

Serotonin mimics the effects of the endogenous efferent input on both the structure and function of *Limulus* ommatidia (7). Injection of 20 μL of serotonin (0.5 mM) under the cornea of the eye *in situ* during the day elevated the ERG amplitude by 200% and lowered the spontaneous optic nerve activity for about 3 hours. The same dose also increased the acceptance angle from 6° to 13° and changed the morphology of the ommatidium from the daytime state (Fig. 4, right) to the nighttime state (Fig. 4, left). In short, serotonin reproduced all of the effects we have thus far been able to attribute to the activity of the circadian clock. No other transmitter we tested was equally effective; however, we have not yet shown that serotonin is the transmitter released by the efferent terminals.

Exposing the eye to light introduces additional changes in ommatidial morphology. The light micrographs in Fig. 5 show cross-sectional views of ommatidia in four different states: (A) dark-adapted midnight, (B) dark-adapted noon, (C) light-adapted at noon for 15 min, and (D) light-adapted at noon for 2 hours. The first two states correspond to the cases in Fig. 4. In both dark-adapted states, large pigment granules are found at the ends of the rhabdomeral rays. The large granules are located in proximal pigment-

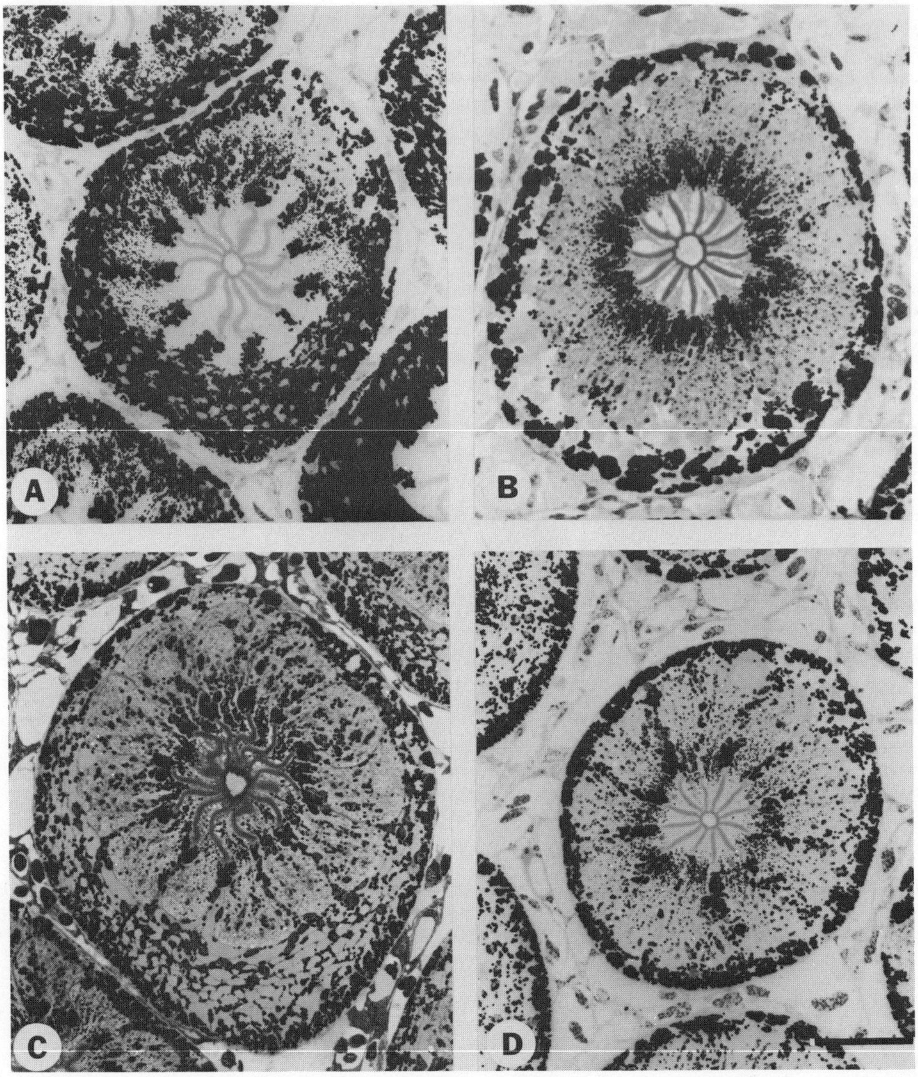

Figure 5. Light micrographs of ommatidia in four different states: (A) dark-adapted midnight, (B) dark-adapted noon, (C) light-adapted at noon for 15 min, and (D) light-adapted at noon for 2 hours. The plane of the transverse sections is about midway up the eccentric-cell dendrite for (B-D) and near the distal end of the dendrite in (A). Bar represents 50 μm.

cell processes that run between the retinular cells. At midnight, smaller pigment granules are dispersed in the periphery of the retinular cells; however, at noon the granules have moved centrally to the edge of the rhabdom just outside the ends of the rays. Fifteen minutes of light exposure moved the retinular-cell pigment inside the rays near the central ring of the rhabdom with no detectable change in the pigment-cell processes (Fig. 5C). After 2 hours of light exposure, the retinular-cell pigment has moved outside the rhabdom region and the pigment-cell processes remained unchanged (Fig. 5D). The small pigment granules of the retinular cells appear to move rapidly after light onset presumably to screen the rhabdom from incident light. Although extended light exposure does not significantly influence the processes of the proximal pigment cells, it may further constrict the aperture (see above) which is formed from the processes of the distal pigment cells.

A Circadian Clock and Light Control Photoreceptor Membrane Turnover

The onset of light triggers a rapid, synchronous turnover of membrane in all of the photoreceptor cells in the lateral eye (14). This turnover of *Limulus* photoreceptor membrane appears analogous in several respects to the process of shedding and renewal of vertebrate photoreceptors (15-17). One important difference, however, is that an essential requirement for membrane turnover in the *Limulus* eye is the efferent input to the retina from the circadian clock in the brain (14).

Fig. 6 shows the sequence of light-induced changes in the rhabdomeres of the retinular cells. In each electron micrograph the plane of the section is perpendicular to the optic axis of the ommatidium, and the view is near the central region of the rhabdom. Fig. 6A shows the rhabdom from the eye of an animal that was maintained in darkness throughout the night and was fixed in darkness on the following day. Included in the electron micrograph are portions of the well-defined microvillar arrays from the rhabdomeres of three neighboring retinular cells and a portion of the central ring surrounding the dendrite of the eccentric cell. Fig. 6B shows that a 15-minute exposure to sunlight triggered an extensive breakdown of the light-sensitive rhabdomeres. The procedure for this experiment was the same as that for Fig. 6A except that the animal was exposed to light for 15 minutes before fixative was injected in the eye. Note that the microvilli were almost completely disassembled by the brief light exposure, and whorls of microvillar membrane have formed lamellar bodies in the nearby palisade region. A third animal was exposed to 30 minutes of sunlight before the eye was fixed. Fig. 6C shows that by this time the rhabdom was nearly restored to its original state. No lamellar bodies were present and numerous small multivesicular bodies (MVB) lined the rhabdom rays (Fig. 6D). After 1 hour of exposure the small MVB

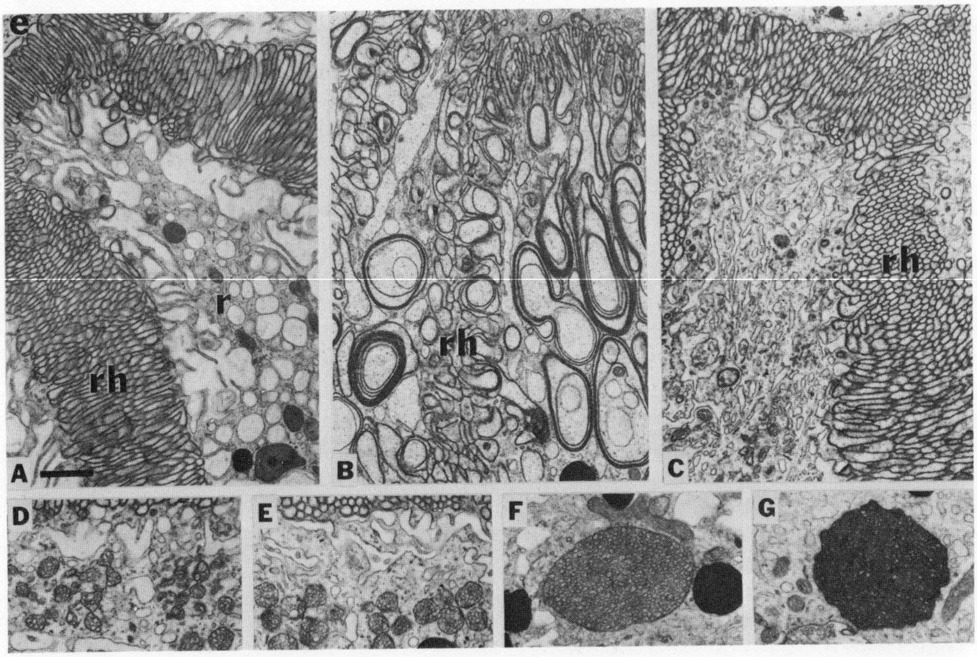

Figure 6. Electron micrographs showing the temporal sequence of membrane turnover in *Limulus* photoreceptors. (A) Part of the rhabdom from an eye that was not exposed to light. (B) Breakdown of the rhabdom after 15 min of continuous exposure to sunlight. (C) Renewal of the rhabdom after 30 min of continuous exposure to sunlight. (D) Multivesicular bodies (MVB) lining a rhabdom after 30 min of continuous exposure. (E) MVB have coalesced and moved away from the rhabdomere after 1 hour. (F) Large MVB have moved to the periphery of the rhabdom after 4 hours. (G) Large dense inclusion bodies appear in the periphery of retinular cells after 8 hours. Bar represents 1 µm. Taken from (14).

began to coalesce, forming larger MVB (Fig. 6E) which moved away from the rhabdom. After 4 hours, several large MVB formed along the peripheral margin of the rhabdom (Fig. 6F), and after 8 hours, large dense inclusion bodies were present in the peripheral region of the photoreceptor cells (Fig. 6G). The appearance of MVB near the rays of the rhabdome is a useful indicator of membrane turnover, but we have no direct evidence that the MVB contain all the membrane of the disassembled microvilli. The exact fate of membrane from microvillar disassembly and the origin of membrane for assembly are not yet known.

The first light onset each day triggers membrane turnover; a second exposure on the same day does not. Thus the time of the day during which the animal is in darkness appears to influence the turnover process. Since both the structure and sensitivity of the photoreceptors are also linked to the time of the day, we investigated the possible influence of the circadian clock on membrane turnover by cutting one lateral optic nerve. In each of five animals the nerve transection was performed shortly before dark and the animals were stored in the dark overnight. The following day both lateral eyes of each animal were fixed after 15 minutes of light exposure. In each case the eyes with intact optic nerves showed evidence of turnover (as in Fig. 6B); the eyes with cut optic nerves did not (as in Fig. 6B). Blocking the efferent input from the circadian clock to the photoreceptors blocks membrane turnover.

Mimicking the efferent input to the retina restores membrane turnover. The procedure of the preceding experiment was repeated with the exception that, when the animals were placed in the dark, current pulses were delivered to the distal end of the cut optic nerve to mimic the endogenous efferent input to the retina. At dawn the following morning the current pulses were turned off and the animals were kept in the dark. At noon, the animals were exposed to light, and 15 minutes later the lateral eyes were fixed. Electron micrographs show that membrane turnover occurred in the retinular cells of both eyes. We therefore conclude that turnover of photoreceptor membrane requires both the onset of light and the efferent input to the retina from a circadian clock.

In view of the extensive degree of rhabdom disassembly in Fig. 6B, we considered the possibility that the turnover process may influence visual sensitivity. Preliminary experiments were carried out with Dr. John Z. Levinson (University of Maryland) to test whether a reduction in sensitivity could be detected during the turnover process. In these experiments animals were maintained in the dark overnight and the ERG responses of one lateral eye was monitored to determine the phase of the circadian clock. After the "nighttime" period of elevated sensitivity was complete, the recorded eye was exposed for 5 minutes to a light intensity sufficient to initiate membrane turnover. Immediately following the exposure, we tracked the dark adaptation of the ERG response. Four hours later after the eye had completely dark adapted, the 5-min light exposure was repeated and the time course of dark adaptation of the ERG response was again measured. The results were not conclusive; the rate of dark adaptation for the initial exposure was only slightly slower than for the second exposure. Since membrane turnover occurred only after the first exposure, we expected the resulting breakdown and buildup of microvillar membrane to slow the course of dark adaptation with respect to that measured for the second exposure; however, it did not. One possible interpretation is that dark adaptation and the turnover process have similar time courses. Indeed, both processes are nearly complete within 30 minutes of the onset of light. It would appear that during the early stages of turnover and dark adaptation in these experiments sufficient rhabdomeral membrane remained functional to elicit criterion ERG responses to relatively bright test flashes.

To summarize, the nighttime period of efferent activity which normally precedes membrane turnover appears to serve an important priming function. Without the efferent input, turnover is blocked. The circadian clock thus not only increases visual sensitivity at night but also controls cellular processes that build up and break down photosensitive membranes.

Why do photoreceptors periodically break down and renew their photosensitive membranes? Perhaps some aspect of visual transduction is irreversible or the membrane which supports transduction lacks long-term stability. In this regard it is interesting to note that membrane turnover may not be restricted to photoreceptors. For example, cochlear hair cells contain numerous MVB as well as an abundant efferent innervation. A better understanding of membrane turnover may shed light on the metabolic processes that maintain receptor function.

Constant Light and Constant Darkness Change Photoreceptor Structure

In view of the striking effects of both light and the circadian clock on photoreceptor structure and function, we decided to investigate the relative influences of these retinal inputs in greater detail. We were particularly interested in whether or not the circadian modulation of photoreceptor structure would continue under the conditions of constant light and constant darkness. We also wished to determine the effects of constant lighting conditions on the rhabdom structure both with and without the priming and trigger mechanisms for photoreceptor membrane turnover.

To investigate these and other possible influences of prolonged light and dark conditions, we prepared eight Limuli in the following manner. The animals were maintained under a natural light-dark cycle for several days, and then in the late afternoon on a given day one lateral optic nerve of each animal was transected. Several hours later, at sundown, four of the animals were placed in an aquarium in constant light (3×10^{-4} watts/cm^2 between 400 and 700 nm at the cornea) and four were placed in an aquarium in constant darkness. At noon, seven days later, two animals were removed from each of the aquaria and their lateral eyes were fixed by injecting 3 cc of a buffered aldehyde fixative (10) beneath the corneas. The eyes were then excised, embedded in plastic, and sectioned. At midnight on the same day the eyes of the four animals remaining in the aquaria were also prepared for sectioning. In each case the animals were maintained in the constant light or dark condition until the eyes were thoroughly fixed. Although we had no direct measure of the phase of the circadian rhythm in any of the animals, we recorded for the seven-day period the ERG responses of one animal that had been kept with the experimental animals before the optic nerve transections were performed. On the seventh day the ERG responses of the one animal were maximum at midnight and minimum at noon. We assumed that the circadian rhythms of the experimental animals had a similar phase and sacrificed the animals accordingly.

Our experimental procedure thus manipulated the three following parameters: light, time of day, and efferent input to the retina. All of the retinas were investigated with both light and electron microscopy with particular emphasis on the ultrastructure of the rhabdom and the general morphology of the ommatidium. The results are shown in Figs. 7 and 8 and summarized in Table I. In both figures, the micrographs in a vertical column are taken from separate eyes of the same animal. Each electron micrograph shows a rhabdomeral ray separating two retinular cells, and each light micrograph shows an ommatidium in cross section with the plane of section lying on or very close to the ommatidial axis. A segment of the eccentric-cell dendrite appears in every light micrograph; however, the entire length of the dendrite is not visible in each case because the ommatidia are often bent and no plane of symmetry could be located. As a consequence, our interpretation of the light micrographs is based on the inspection of many sections lying adjacent to those shown in Figs. 7 and 8 as well as a study of the results from the eyes of the other four animals kept under duplicate conditions.

Comparing the results in Fig. 7 (A and B) with those in Fig. 4 indicates that even after seven days of darkness the efferent input to the retina continues to mediate circadian changes in ommatidial morphology. At night, the aperture between the lens and the retinular cells widens and the retinular cells move distally and lie next to the corneal lens. These effects reverse during daytime. Prolonged darkness, however, reduces the volume of retinular-cell

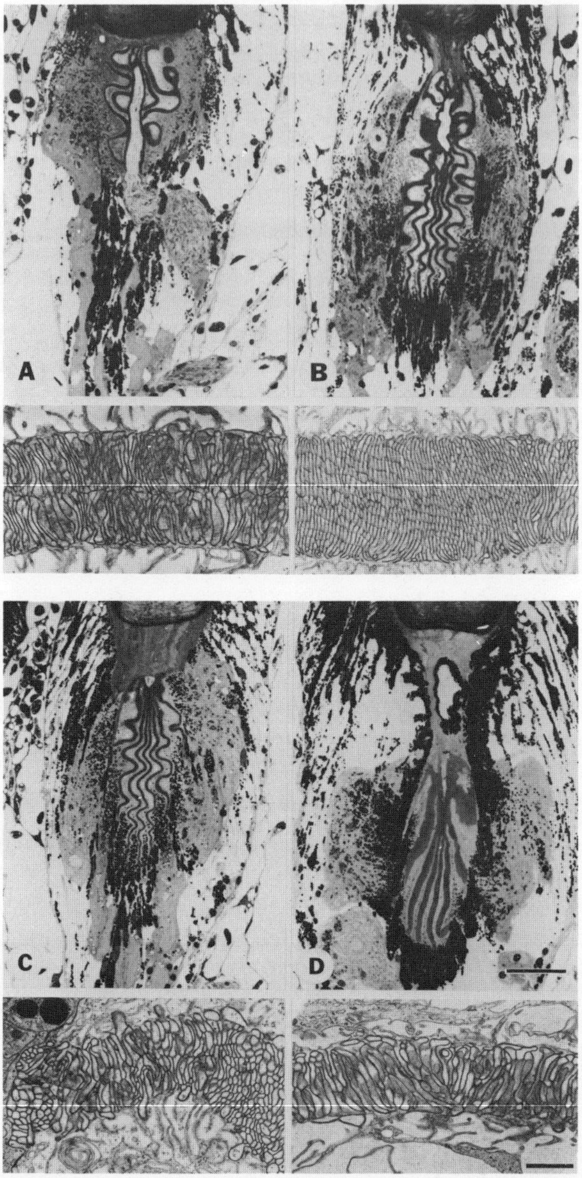

Figure 7. Ommatidial morphology and rhabdom ultrastructure after one week of constant darkness. (A) Eye fixed at night with optic nerve intact. (B) Eye fixed during the day with optic nerve intact. (C) Eye fixed at night with optic nerve cut. (D) Eye fixed during the day with optic nerve cut. Micrographs in a vertical column are taken from separate eyes of the same animal. Bars in (D) represent 50 μm for light micrographs and 1 μm for electron micrographs.

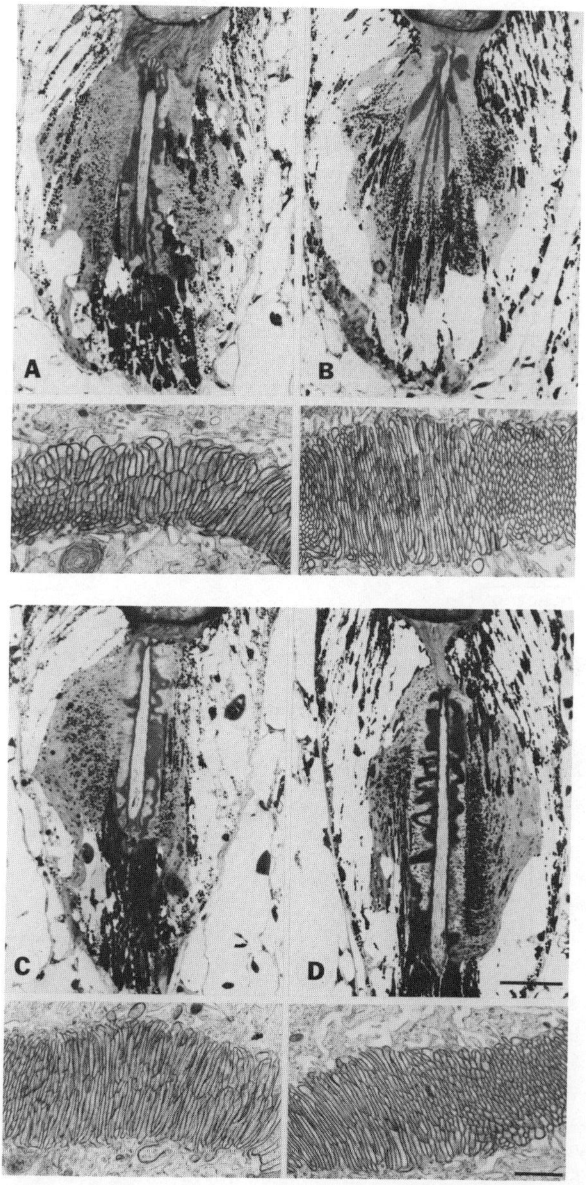

Figure 8. Ommatidial morphology and rhabdom ultrastructure after one week of constant light. (A) Eye fixed at night with optic nerve intact. (B) Eye fixed during the day with optic nerve intact. (C) Eye fixed at night with optic nerve cut. (D) Eye fixed during the day with optic nerve cut. Micrographs in a vertical column are taken from separate eyes of the same animal. Bars in (D) represent 50 μm for light micrographs and 1 μm for electron micrographs.

TABLE I

Summary of the Effects on Photoreceptor Structure of Lighting Condition, Time of Day, and Efferent Input to the Retina

Rhabdom Ultrastructure:

Well Organized

Lighting	Time	Efferent Input
Dark	Day	Intact ON
Light	Day	Intact ON
Light	Night	Cut ON
Light	Day	Cut ON

Moderately Organized

Lighting	Time	Efferent Input
Dark	Night	Intact ON
Light	Night	Intact ON

Disorganized

Lighting	Time	Efferent Input
Dark	Night	Cut ON
Dark	Day	Cut ON

Aperture Width:

Wide

Lighting	Time	Efferent Input
Dark	Night	Intact ON
Light	Night	Intact ON

Medium

Lighting	Time	Efferent Input
*Light	Day	Intact ON
Dark	Night	Cut ON
Light	Night	Cut ON

Narrow

Lighting	Time	Efferent Input
Dark	Day	Intact ON
Dark	Day	Cut ON
Light	Day	Cut ON

Pigment Granules in Retinular Cells:

Dispersed

Lighting	Time	Efferent Input
Dark	Night	Intact ON
Dark	Day	Intact ON
Light	Night	Intact ON
Light	Day	Intact ON
*Light	Night	Cut ON

Gathered

Lighting	Time	Efferent Input
Dark	Night	Cut ON
Dark	Day	Cut ON
Light	Day	Cut ON

Vacuoles in Retinular Cells:

Peripheral

Lighting	Time	Efferent Input
Light	Night	Intact ON
Light	Day	Intact ON
Light	Night	Cut ON
Light	Day	Cut ON

Palisades

Lighting	Time	Efferent Input
Dark	Night	Intact ON
Dark	Day	Intact ON
Dark	Night	Cut ON
Dark	Day	Cut ON

*Results that appear inconsistent with our present understanding of the factors controlling structural changes. Boxes emphasize correlated results.

cytoplasm, reduces the total amount of screening pigment (especially in the distal pigment cells), and also appears to reduce the area of photosensitive membrane in the rhabdome. The electron micrograph in Fig. 7A shows that after prolonged darkness the microvilli become thick and only moderately organized at night; whereas, after just one day in darkness the microvilli are thin and well organized both during the day and at night (results not shown). Thus, continuous darkness appears to reduce the total amount of photosensitive membrane available at night. The result that circadian changes in morphology are detectable after seven days of constant darkness is consistent with our finding that circadian rhythms in the ERG responses persist in constant darkness.

The circadian clock continues to influence ommatidial morphology regardless of whether the animal is kept in constant darkness or in constant light. The summary of results in Table I indicates that the clock partially influences at least three morphological features: rhabdom ultrastructure, aperturn width, and location of retinular cell pigment. When the efferent input is present (Intact ON), the rhabdom is well organized during daytime (Fig. 7B and 8B) and moderately organized at night (Figs. 7A and 8A) regardless of the lighting condition. When the efferent input is absent (Cut ON), the rhabdom remains large and well organized in constant light (Figs. 8C and D) but becomes attenuated and disorganized in constant darkness (Figs. 7C and D). Thus, without efferent input, constant light maintains the rhabdom and prolonged darkness disrupts it. These results may not be consistent with the findings of Behrens and Krebs (18) who reported progressive reduction in the rhabdom with progressive light adaptation up to 12 hours in duration. Our experimental conditions yielded reduced rhabdoms only during continuous darkness with cut optic nerves, or when an animal with an intact optic nerve was maintained in constant light and sacrificed during its subjective night when the efferent fibers were presumably active.

The circadian clock continues to modulate the size of the aperture located below the lens which controls in part the amount of light reaching the retinular cells. At night the aperture is wide regardless of the lighting conditions (Fig. 7A and 8A); however, during the day the aperture is narrow only in constant darkness (Fig. 7B) and not in constant light (Fig. 8B). The partial opening of the aperture during the day under constant light is an unexpected, but consistent, result in both test animals. When the optic nerve is cut, the slight changes in the aperture from day (Figs. 7D and 8D) to night (Fig. 7C and 8C) regardless of lighting conditions may reflect an endogenous circadian rhythm of the retina or circadian changes of a substance in the circulatory system released either by the other lateral eye which has an intact optic nerve or by another organ in the animal.

The position of the retinular cells within the ommatidium was not included in Table I because the results do not provide a clear picture. In prolonged darkness the retinular cells change position with time of day as expected when the optic nerves remain intact (Figs. 7A and B, compare with Fig. 4), but small circadian changes in position occur after the optic nerve has been cut regardless of lighting conditions (Figs. 7C and D, 7C and D). These residual changes may be correlated with the small changes in aperture occurring under the same conditions. However, the proximal movement of the retinular cells at night under constant light in the presence of efferent activity (Fig. 8A) and the distal movement during the day under otherwise similar conditions (Fig. 8B) are not consistent with our other results and remain unexplained.

The location of pigment granules within retinular cells is strongly influenced by the efferent input to the retina. When the optic nerves remain intact, the pigment granules are dispersed regardless of the time of day or prolonged lighting condition. With one exception the granules are gathered near the edge of the rhabdom structure when the optic nerves are cut regardless of the other experimental conditions. The one exception occurred during the night in constant light (Fig. 8C) in which case the granules are dispersed. Here again the result was unexpected but was observed in both test animals.

All of the characteristics considered thus far are influenced in part by the circadian clock. Effects attributed solely to the conditions of constant light and darkness appear to be restricted to abnormal changes in the cytoplasm and volume of the photoreceptor cells. For example, both lighting conditions reduce the volume of cytoplasm in the periphery of retinular cells but not in the rhabdom region. The light micrographs in Fig. 7 and 8 show shrunken retinular cells. As indicated in Table I, the prolonged exposure to light leads to the formation of vacuoles in the peripheral cytoplasm of retinular cells; whereas, prolonged darkness usually causes coalescence of the palisade structure near the rhabdomeral rays. Vacuoles in the peripheral cytoplasm are visible in the light micrographs of Fig. 8A, B, and D. Although the electron micrographs show only limited regions of the rhabdoms, vacuoles in the palisade structure are apparent in Figs. 7A and D.

Pigment-cell pigment is the last characteristic we considered. In only two cases (Figs. 7B and D) did the concentration of large pigment granules in pigment cell appear about normal. In all other cases the concentration of pigment-cell pigment was reduced. Figs. 9A and B show normal pigment granules for the two cases in Figs. 7B and D. Figs. 9C and D show several normal granules plus ones that are either empty or partially empty or pigment. These electron micrographs correspond to the cases in Figs. 8B and C. One possible explanation for empty granules is that constant light disrupts the

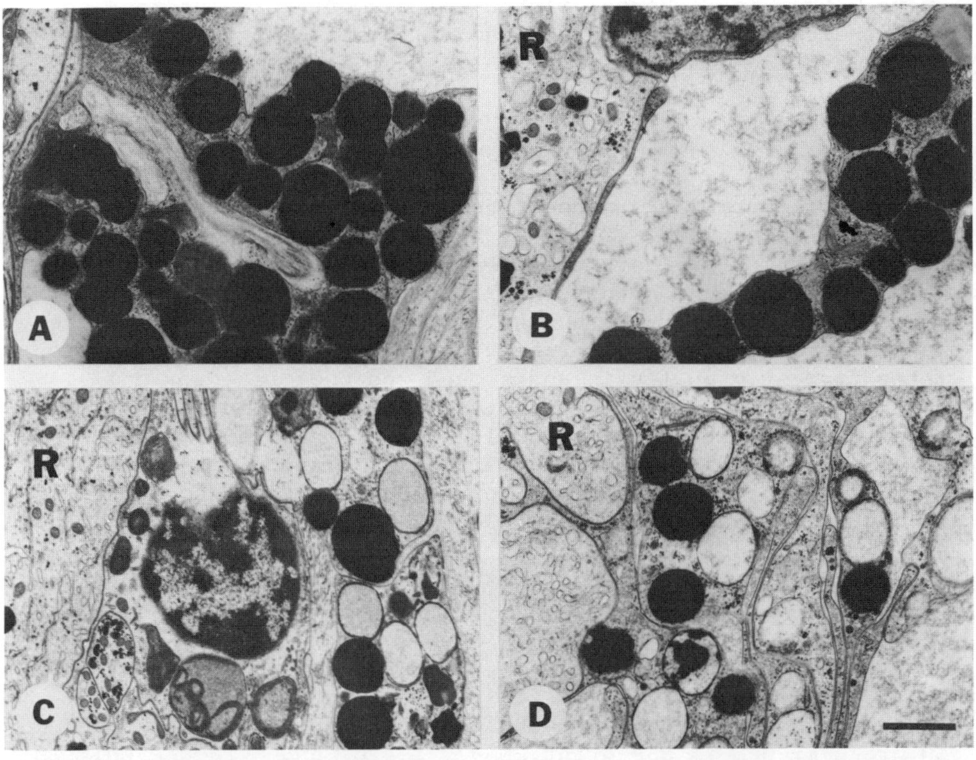

Figure 9. Pigment granules in processes of pigment cells. Normal pigment granules were observed only after prolonged darkness when the eye was fixed during the day whether the optic nerve was (A) intact or (B) cut. In all the other cases in Figs. 7 and 8 granules were observed that were normal, empty, and partially empty. (C) Conditions of constant light, intact optic nerve, and fixed during the day. (D) Conditions of constant light, cut optic nerve, and fixed at night. The metabolism of screening pigment appears to be influenced by the lighting condition. Retinular cells are indicated by "R". Bar represents 1 μm.

normal cycling of pigment granules, and as a consequence more pigment is destroyed than is synthesized. The results in Figs. 9A and B suggest that prolonged darkness disrupts the cycling of pigment granules at night but not during the day regardless of the efferent input. These results are not readily understandable in terms of our present knowledge of the circadian effects in the *Limulus* eye; however, the clear changes within the pigment granules in Figs. 9C and D indicate that the metabolism of screening pigment is influenced by the lighting condition.

SUMMARY

Photoreceptors are indeed the dynamic components of the *Limulus* visual system. Each day they break down and renew a substantial portion of their photosensitive membranes and shift the position of these membranes with respect to the incident light. In addition, the aperture controlling the light incident on the photoreceptors changes with the time of day as does the location of the pigment granules in and around the photoreceptor cells. These circadian changes in structure are controlled primarily by the efferent input from a clock located in the brain; however, the structural changes can be modified by disrupting the clock's input or subjecting the animal to the conditions of constant light and dark. The efferent input to the retina appears to adapt the eye for optimal function at very low light levels and to control in part the metabolic processes that maintain photoreceptor function.

ACKNOWLEDGMENTS

We thank R. Fladd for technical assistance. Supported by NIH grant EY-00667, NSF grant BNS-77 19436, and Syracuse University Senate Research Fund.

REFERENCES

1. Vision: Engbretson, G. A. and Lent, C. M. (1976) *Proc. Nat. Acad. Sci. USA* 73, 654-657; Miles, F. A. (1970) *Science* 170, 992-995; Ogden, T. E. (1968) in *Structure and Function of Inhibitory Neuronal Mechanisms* (von N. Y.; Pearlman, A. L. and Hughes, C. P. (1976) *J. Comp. Physiol.* 166, 123-131. Audition: Fex, J. (1962) *Acta Physiol. Scand.* 55, Suppl. 189, 1-68; Galambos, R. (1956) *J. Neurophysiol.* 19, 424-437; Wiederhold, M. L. and Kiang, N. Y. S. (1970) *J. Acoust. Soc. Am.* 48, 950-965. Lateral-line system: Flock, A. and Russell, I. J. (1973) *Nature* 243, 89-91; Russell, I. J. (1968) *Nature* 219, 177-178.

2. Ratliff, F. and Hartline, H. K. (1974) Studies on Excitation and Inhibition in the Retina, The Rockefeller University Press, New York.
3. Barlow, R. B., Jr. and Quarles, D. A., Jr. (1975) J. Gen. Physiol. 65, 709-730.
4. Barlow, R. B., Jr., Bolanowski, S. J., Jr. and Brachman, M. L. (1977) Science 197, 86-89.
5. Barlow, R. B., Jr. and Kaplan, E. (1971) Science 174, 1027-1029.
6. Kaplan, E. and Barlow, R. B., Jr. (1975) J. Gen. Physiol. 66, 303-326.
7. Barlow, R. B., Jr., Chamberlain, S. C. and Kaplan, E. (1977) Biol. Bull. 153, 414.
8. Barlow, R. B., Jr. and Kaplan, E. (1977) J. Gen. Physiol. 69, 203-220.
9. Fahrenbach, W. H. (1971) Z. Zellforsch. Mikrosk. Anat. 114, 532-545.
10. Fahrenbach, W. H. (1969) Z. Zellforsch. Mikrosk. Anat. 93, 451-483.
11. Fahrenbach, W. H. (1978) Neuroscience Abstr. 4, 589.
12. Chamberlain, S. C. and Barlow, R. B., Jr. (1977) Biol. Bull. 153, 418-419.
13. Barlow, R. B., Jr., Chamberlain, S. C. and Levinson, J. Z. (1979) Science (submitted).
14. Chamberlain, S. C. and Barlow, R. B., Jr. (1979) Science (submitted).
15. LaVail, M. M. (1976) Science 194, 1071-1074.
16. Basinger, S., Hoffman, R. and Matthes, M. (1976) Science 194, 1074-1076.
17. Besharse, J. C., Hollyfield, J. G. and Rayborn, M. E. (1977) J. Cell Biol. 75, 507-527.
18. Behrens, M. and Krebs, W. (1976) J. Comp. Physiol. 107, 77-96.

TURNOVER OF PHOTORECEPTOR MEMBRANE IN THE LARVAL MOSQUITO OCELLUS: RHABDOMERIC COATED VESICLES AND ORGANELLES OF THE VACUOLAR SYSTEM

Richard H. White, Dian Gifford and Norman A. Michaud

Department of Biology
University of Massachusetts/Boston
Boston, Massachusetts 02125

We have presented evidence in previous papers showing that the degradation and renewal of photoreceptor membrane in larval mosquito ocelli is regulated by light. Mosquito rhabdomeres grow larger in darkness and smaller in light as the amount of membrane increases and decreases. The kinetics of rhabdomere growth and diminution indicate that these changes arise from variations in the rate of membrane turnover (1).

The present paper deals with the ultrastructural mechanism of the degradative side of rhabdomere turnover. The cells of the mosquito ocellus contain (as do arthropod photoreceptors generally) a prominent assemblage of organelles, including multivesicular, lamellar and dense bodies. In other kinds of cells such organelles - designated vacuolar organelles - have been shown to compose a system whose function is endocytosis coupled with lysosomal digestion (2). The role played by these organelles in the uptake and digestion of extracellular material is well established. Plasma membrane is necessarily taken into a cell during endocytosis, and in some situations turnover of redundant membrane rather than phagocytosis has been seen as the primary function of the vacuolar system. For example, endocytosis is coupled with exocytosis in secretory cells and neurons in order to retrieve membrane added to the cell surface by the release of secretory or synaptic vesicles (2-5).

We and others have argued from several lines of evidence that the vacuolar systems of arthropod photoreceptors have a similar retrieval function in the turnover of photoreceptor membrane (6-20). A relationship between rhabdomere microvilli and the vacuolar organelles is suggested by a similarity of membrane structure. The membranes of both were seen in freeze-fracture preparations of crab retinulae to be densely studded with particles believed to be rhodopsin molecules (9). Moreover, experiments using ferritin to label the pathway of endocytosis showed that the vacuolar system collects vesicles cast off by the mosquito rhabdomere (11). Furthermore, the vacuolar organelles of the mosquito ocellus proliferate when rhabdomeres grow smaller as they lose membrane during the first hours of light adaptation (11). Eguchi and Waterman (9) confirmed that the vacuolar system is a lysosomal digestive system (Fig. 5) by demonstrating acid phosphatase activity in lamellar and mixed lamellar-vesicular bodies.

One expects to find membrane renewal accompanying loss. Our data suggest that the degradation and renewal of mosquito rhabdomeres are balanced in full light adaptation. Under continuous illumination, the rhabdomere eventually settles to a constant size (1), but the vacuolar system remains active (11). Hence the rhabdomere must then be in equilibrium between opposed processes of loss and renewal. If such a light adapted rhabdomere is transferred to darkness, the amount of photoreceptor membrane increases threefold before leveling off again after 4-5 hrs. A striking feature of rhabdomere growth is that it begins immediately, with no lag period, at the onset of darkness. A possible explanation is that membrane loss from the rhabdomere immediately ceases with darkness while membrane renewal continues. If that is what happens, then the generation of vacuolar organelles should stop immediately with darkness. The various classes of organelles should disappear from the ocellar cells in the order of their appearance during light adaptation, the time required for each sort of organelle to disappear depending upon the rate at which it changes into the next member of the developmental sequence. We have found that the vacuolar system is reduced, following the expected sequence, beginning at the onset of dark adaptation.

METHODS

Eggs of the yellow fever mosquito _Aedes aegypti_, kindly provided by Professor George B. Craig, Jr., Biology Department, Notre Dame University, were hatched and reared as before (1). Since ocelli lacking accessory pigment are essential for many of our observations and experiments (1, 21), we examined both wild type (Rockefeller strain) and white eye mutants. As we found no differences in the vacuolar systems of wild type and white eye ocelli, the data were pooled.

Larvae were exposed to white fluorescent light of approximately $1mW/cm^2$ (1) for 48 hr to assure morphological equilibrium of the rhabdomere and vacuolar system. Fourth instar larvae were fixed in 0.12M phosphate-buffered glutaraldehyde or in the 0.1M cacodylate-buffered formaldehyde-glutaraldehyde mixture of Karnovsky (22),

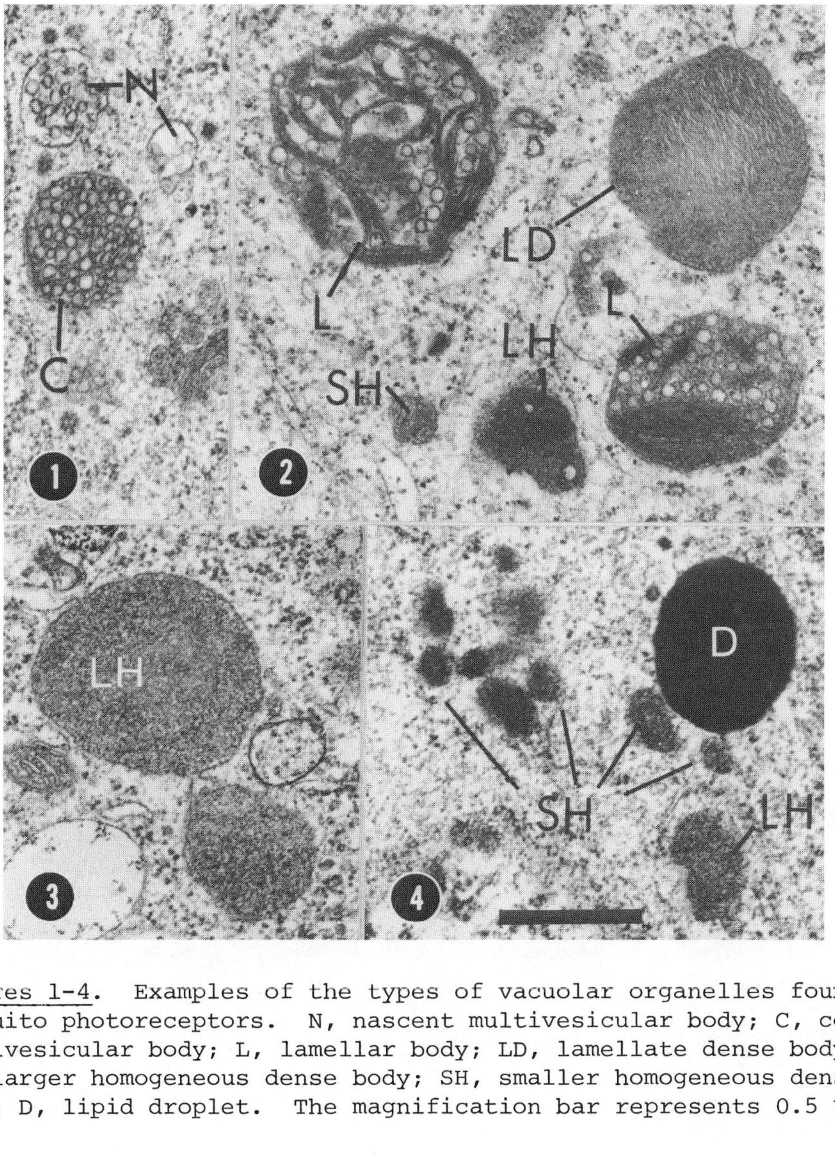

Figures 1-4. Examples of the types of vacuolar organelles found in mosquito photoreceptors. N, nascent multivesicular body; C, compact multivesicular body; L, lamellar body; LD, lamellate dense body; LH, larger homogeneous dense body; SH, smaller homogeneous dense body; D, lipid droplet. The magnification bar represents 0.5 μm.

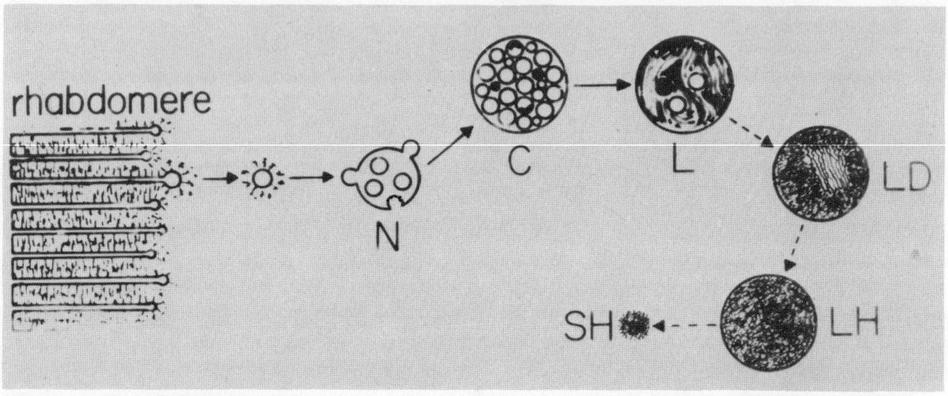

Figure 5. Diagrammatic summary of the origin and developmental relationship of vacuolar organelles in mosquito photoreceptors. Designations are those used in Figs. 1-4. Solid arrows indicate relationships established by experiments using ferritin as a label for endocytosis. The broken arrows indicate relationships inferred from the progressive changes in the vacuolar system that occur during light and dark adaptation. The whole scheme is consistent with the organization of vacuolar systems found in many other kinds of cells. Coated vesicles bud off the inner ends of the rhabdomere microvilli. These endosomes lose their coats and coalesce to form nascent multi-vesicular bodies (N). These become compact multivesicular (C) and lamellar (L) bodies (secondary lysosomes). They then proceed through the remaining developmental sequence to dense (residual) bodies (LH and SH).

post-fixed in buffered 1% osmium tetroxide, stained before embedment with uranyl acetate, and stained after sectioning with lead citrate. Light-adapted larvae were dissected and fixed in white light. Dark-adapting larvae were fixed after period ranging from 15 min to 7 days of continuous darkness. They were dissected in deep red light but fixed in darkness.

Embedded heads were oriented so that the plane of section passed vertically through the ocelli grouped at the side of the head (21). Sections so oriented include profiles of two or three of the four ocelli cut parallel to the axes of their photoreceptor cells. Since cellular organelles are distributed assymetrically along the cellular axis, the only sections used for morphometric analysis were those including both rhabdomere and nuclei, marking respectively the distal and proximal ends of the cells.

For the enumeration of vacuolar organelles, composite low magnification electron micrographs were made of single complete sections of any one group of ocelli. These included profiles of about 25 receptor cells. Each data point on the graph (Fig. 6-8) represents 4 to 5 such sections, or about 100 cells, for the earlier more critical measurements (0 to 7.5 hr), and 2 sections at 1 to 7 days (except at 2 days, represented by a single section). Cell, rhabdomere, and nuclear boundaries were traced on transparent overlays and all organelles of interest traced and categorized. Organelles were counted and expressed as (number)/(μm^2 cytoplasmic area). [Note: nuclear area was subtracted in order to obtain only the cytoplasmic area.] The distribution of organelles in the immediate vicinity of the rhabdomere was determined by counting organelles falling within a 2 μm cytoplasmic perimeter of the rhabdomere.

The scheme by which we classified the vacuolar organelles for enumeration is presented below.

(a) Nascent multivesicular bodies (Fig. 1): Small organelles only loosely filled with vesicles. They contain neither dense vesicles nor lamellae. We estimate their average diameter at about 0.5 μm.

(b) Compact multivesicular bodies (Fig. 1): Those filled with vesicles. They generally contain some dense vesicles and membrane lamellae as well. They range to larger sizes than the nascent multivesicular bodies of category a, averaging around 1 μm diameter in light adapted cells.

(c) Lamellar bodies (Fig. 2): Those made up predominantly of membrane lamellae and dense material. Most also possess some vesicles. Lamellar bodies and compact multivesicular bodies cover the same size range.

(d) Lamellate dense bodies (Fig. 2): Bodies with loose stacks of lamellae surrounded by flocculent material. Their average diameter is about 1 μm in light-adapted cells.

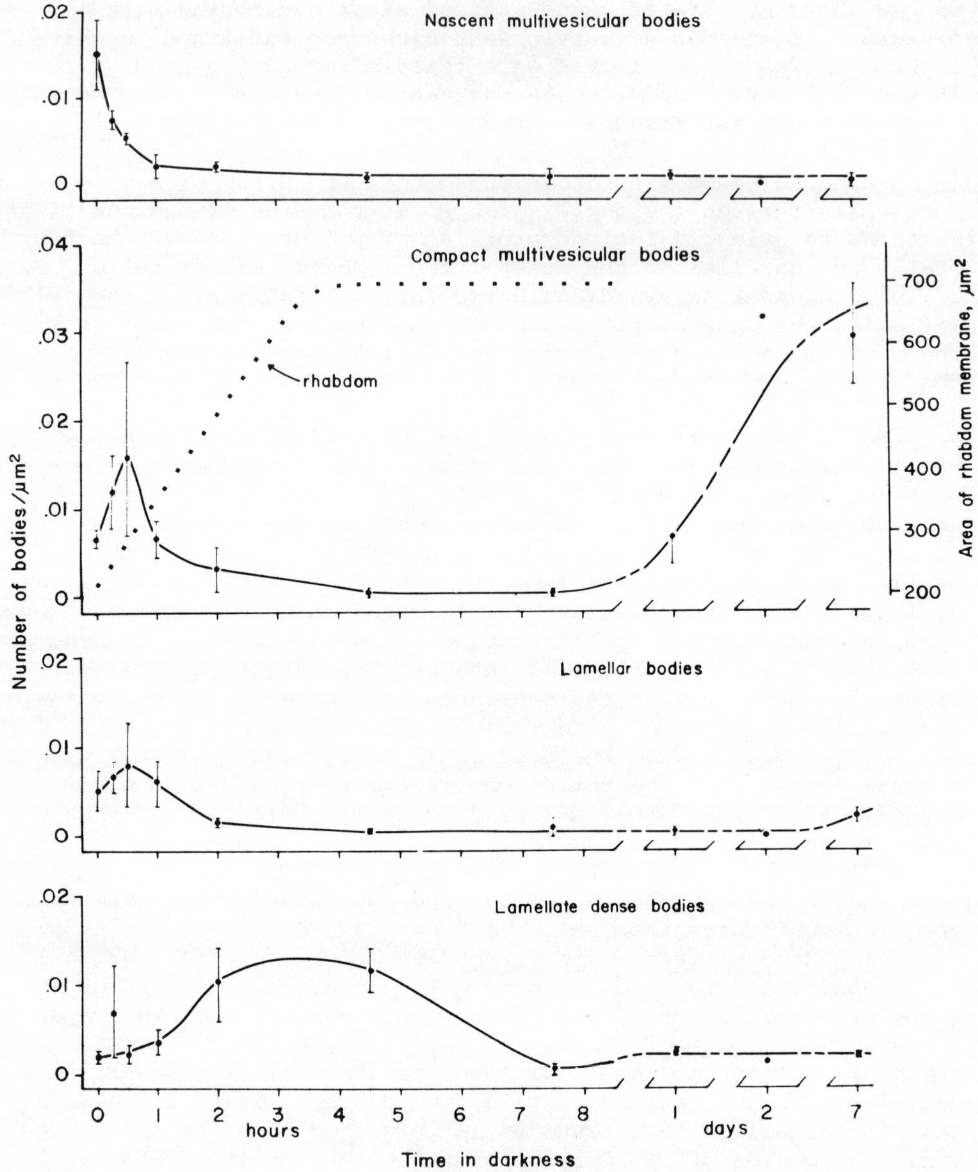

Figure 6. Changes in the abundance of the various vacuolar organelles during the course of dark adaptation. The increase of photoreceptor membrane (calculated from Fig. 7, reference 1) during the same period is also shown. The error bars represent standard error of the mean.

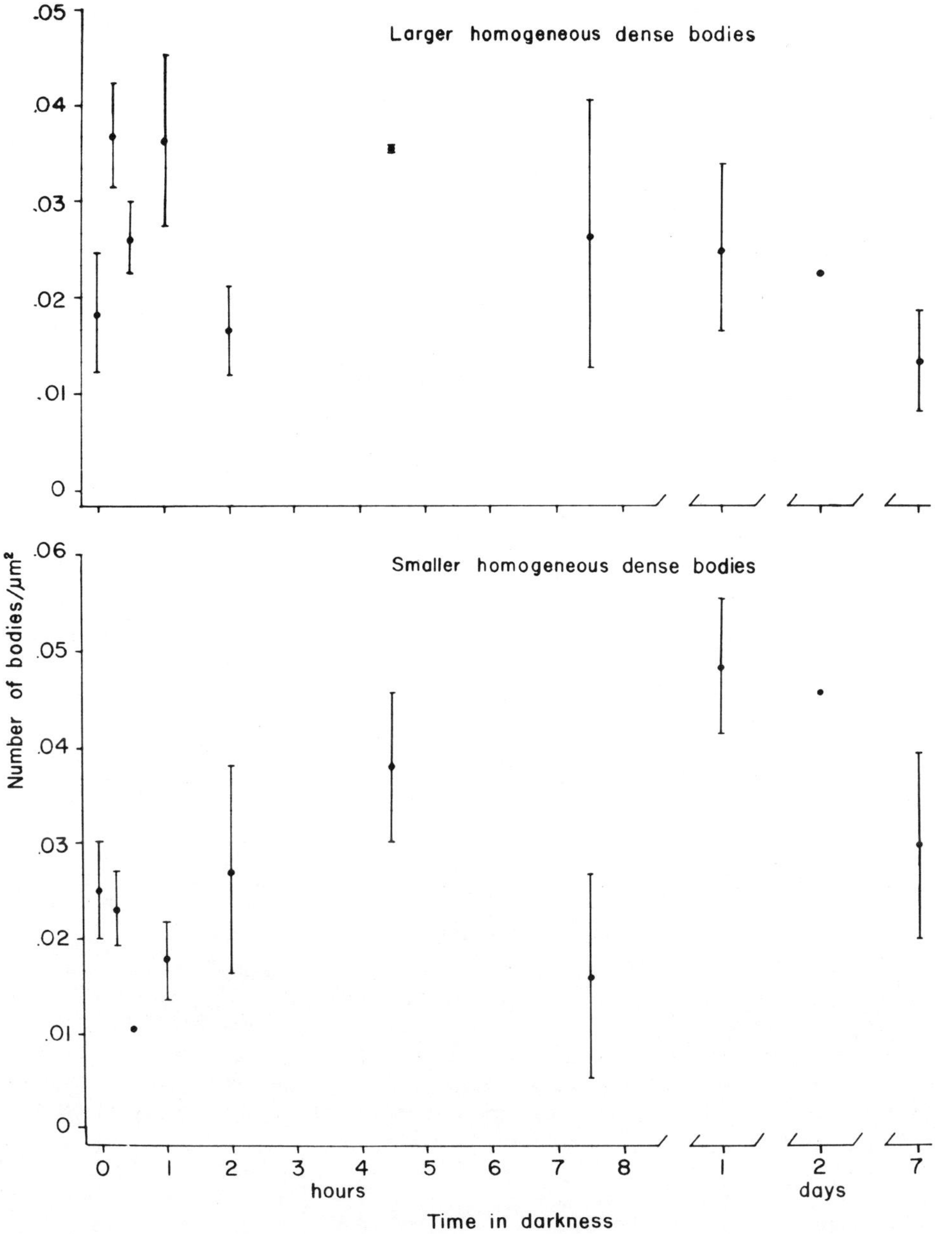

Figure 7. Changes in the abundance of the various vacuolar organelles during the course of dark adaptation. The increase of photoreceptor membrane (calculated from Fig. 7, reference 1) during the same period is also shown. The error bars represent standard error of the mean.

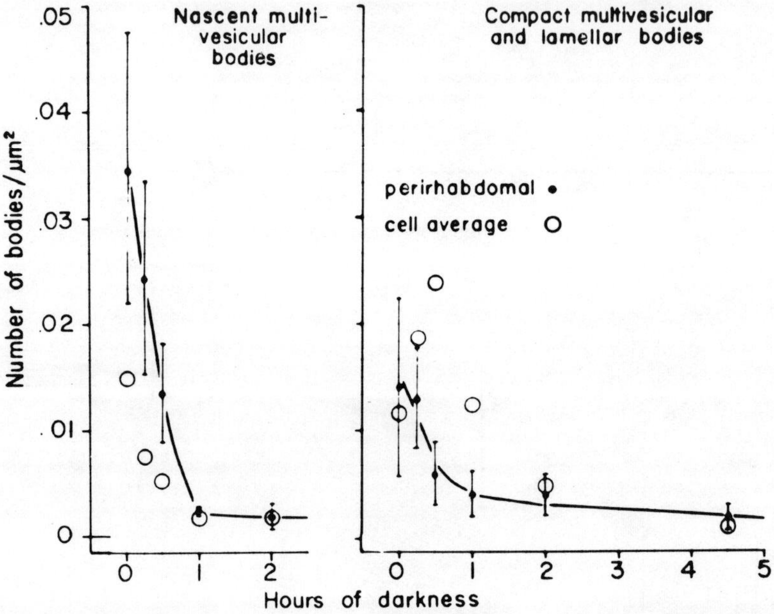

Figure 8. The abundance of vacuolar organelles within a 2 μm perimeter of the rhabdomere compared with cell averages from Fig. 6. Numbers of nascent multivesicular bodies are plotted to the left. To the right, compact multivesicular and lamellar bodies have been combined, since they behave as though members of a single class.

(*e*) Larger homogeneous dense bodies (Fig. 2): Bodies similar to those of category *d*, but lacking lamellae.

(*f*) Smaller homogeneous dense bodies (Fig. 4): Arbitrarily those smaller than 0.2 μm. This distinction is mainly one of convenience since such small bodies cannot be clearly distinguished in some of our micrographs. Nevertheless, some of the smaller dense bodies may in fact fall into a morphologically distinct category, since many appear to lack the external membranes that bound all the other organelles. However, these are uncertain distinctions since grazing sections of larger bodies would have a similar appearance.

We wish to be clear about the origin and significance of this taxonomic scheme, and how it was used in collecting data. Its basis is purely morphological. That is, the various categories of organelles were distinguished for the purpose of data collection because they can be conveniently picked out in electron micrographs. They were established independently of the inferred developmental relationships shown in Fig. 5. Only "nascent multivesicular bodies" were given a name implying a developmental-functional role based on earlier observations (11). Although size differences are noted above, size was not used as a criterion in assigning organelles to the appropriate bins during data collection: there are methodological difficulties in dealing with size of organelles in thin sections that we wanted to avoid. Finally, it should be emphasized that the various organelles are members of a continuous morphological spectrum with intermediate forms. Care was taken that data collection should be consistent when intermediates had to be arbitrarily assigned to one or another bin.

Coated vesicles along the inner cytoplasmic border of the rhabdom were counted in composite micrographs. Only those within 0.5 μm of the rhabdom were counted; those attached to the rhabdomere were scored separately from those free in the ctyoplasm (Fig. 10). The total length of the cytoplasm-rhabdomere interface was measured, and the data expressed as (coated vesicles)/(μm interface). To assess patterns of distribution of coated vesicles along the interface, observed distributions were compared with Poisson series describing random distributions. Variance from the null hypothesis was tested using the statistical procedure of Svedberg described in Kershaw (23).

In order to trace the pathway of endocytosis, ocelli were exposed to ferritin (Sigma, Cd content less than 0.1%). Injecting ferritin into the larval head capsule gives variable results because of leakage (11). Therefore the hemocoel was opened by cutting off the anterior end of the head at the level of the antennae. With the hemocoel breached, a larva was placed in a pool of ferritin solution made up from ferritin stock (88 mg/ml in 0.5M Na Cl) diluted 1:1 with insect Ringer. In this way the ocelli, which

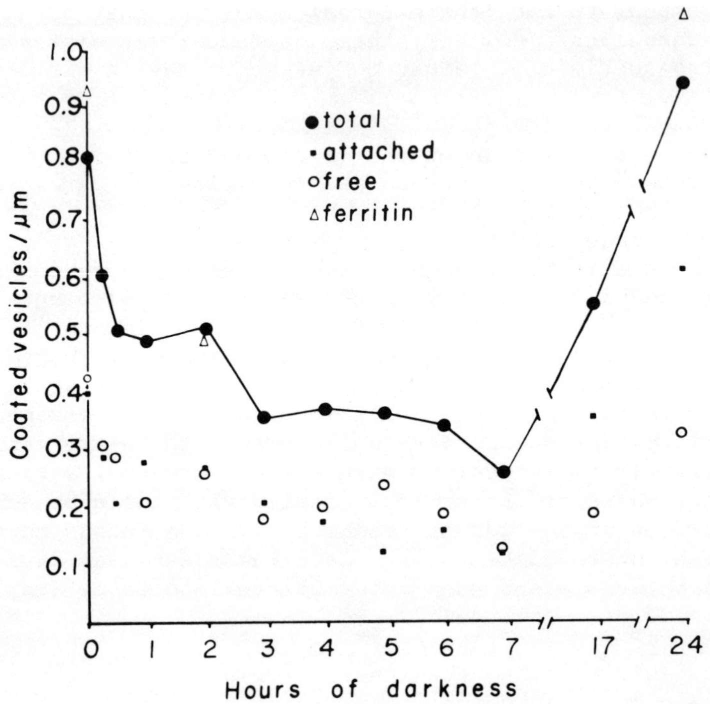

Figure 9. The frequency of coated vesicles along rhabdomere-cytoplasm interface. Each point represents data averaged from 2 to 4 cells. The distinction between free and attached vesicles is illustrated in Fig. 12. The points labeled "ferritin" (triangles) represent the total number of coated vesicles in ocelli exposed to ferritin for 30 min. The proportions of free and attached vesicles in such ocelli were similar to those in unlabeled eyes.

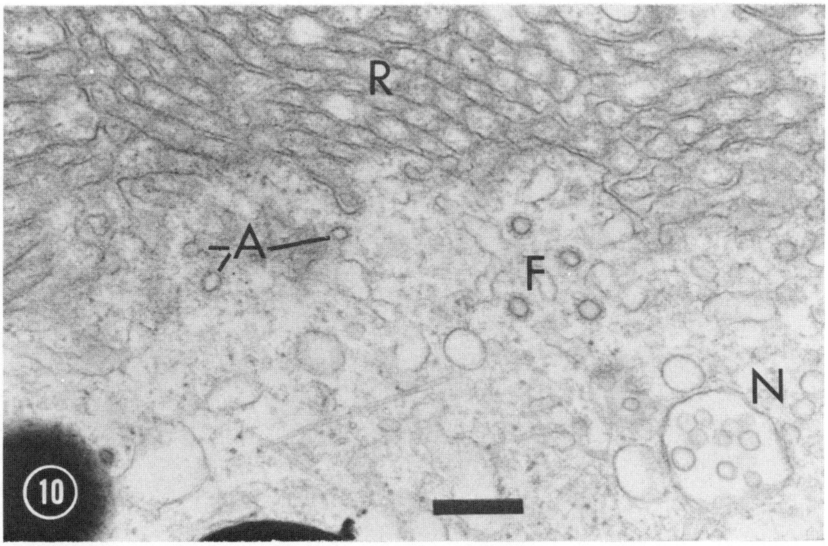

Figure 10. The rhabdomere-cytoplasm interface in a light-adapted cell. R, rhabdomere; F, free coated vesicle; A, attached coated vesicle; N, nascent multivesicular body. Note the abundant smooth ER adjacent to the rhabdomere. The bar represents 0.25 μm.

hang in the hemocoel, were exposed directly to the ferritin for 30 min, and then fixed. Ferritin uptake was compared in light adapted for 2 or 24 hr. The latter were manipulated under deep red light, but treated and fixed in darkness.

RESULTS

Vacuolar Organelles

At the equilibrium of full morphological light adaptation, the relative abundance of vacuolar organelles was found to be 10 nascent multivesicular bodies: 4 compact multivesicular bodies: 3 lamellar bodies: 1 lamellate dense body: 20 large homogeneous dense bodies: 20 small homogeneous dense bodies. Figs. 6 and 7 show the changes that occurred in this population of organelles during the course of morphological dark adaptation. Nascent multivesicular bodies declined immediately, dropping to 10% of their light-adapted frequency (from $0.015/\mu m^2$ to $0.002/\mu m^2$) by 1 hour of dark adaptation, remaining at that level thereafter. Compact multivesicular and lamellar bodies behaved as if they were members of the same class during early dark adaptation. That is, they followed the same temporal pattern of increase and decline, with their relative numbers remaining constant. They doubled in number during the first half-hour of dark adaptation, and then declined to low frequency by 2 hr of darkness.

Lamellate dense bodies, relatively infrequent in light adapted photoreceptors, increased sixfold during the first 5 hr of dark adaptation before declining again to below the light adapted frequency by 7 hr.

The numbers of homogeneous dense bodies, both large and small, were variable at all stages examined. Much of this variation probably arose from the difficulty of distinguishing them with certainty from mitochondria and tangentially sectioned pigment granules in some specimens. Because of the difficulties of data collection we feel that the variability seen in Fig. 7 is not significant. However, the larger dense bodies appear to have gradually declined to about half the light-adapted frequency by 7 days of darkness.

After 1 day of darkness, compact multivesicular bodies began to increase again, reaching at least three times the light-adapted frequency by 2 to 7 days of darkness. They differed from those of light-adapted ocelli in rarely displaying lamellar elements or dense vesicles. Moreover, lamellar bodies remained rare. Hence, the compact multivesicular bodies of prolonged dark adaptation do not grade into lamellar organelles as they do in light adaptation.

Nascent multivesicular bodies tended to lie distally in the light-adapted receptor cells (Fig. 11). Their frequency within a 2 μm perimeter of the rhabdomere was double the cell average. As dark adaptation proceeded they became more evenly distributed across the cell until, by 1 hr of darkness, the perirhabdomal frequency was that of the cell average (Fig. 8). In light adaptation, the perirhabdomal frequency of compact multivesicular and lamellar bodies was the same as the cell average. Fig. 8 shows that their perirhabdomal frequency declined smoothly through the first hours of dark adaptation. That is, their transient increase during the period occurred deeper in the cell, away from the rhabdomere. The perirhabdomal frequency (not plotted) of all three classes of dense body was the same as the cell average in light adaptation and remained at the cell average throughout dark adaptation.

Rhabdomeric Coated Vesicles

The frequency of coated vesicles along the rhabdomere-cytoplasm interface also declined sharply at the beginning of dark adaptation, falling to about 60% of the light-adapted value after 30 min of darkness. Their frequency continued to drop slowly over the next several hr, but by 17 hr had risen again, and after 24 hr of darkness had risen above the light-adapted level. In light and early dark adaptation, the populations of coated vesicles were about evenly divided between those that were continuous with the rhabdomere membrane (attached) and those that were free. After 7 hr of darkness, the relative proportion of attached vesicles began to rise until there were twice as many attached as free vesicles at 24 hr. At all stages, coated vesicles appeared to be abundant along some regions of the rhabdomere-cytoplasm interface and sparse in between. This observation was confirmed statistically. The distribution of vesicles varied significantly (below the 1% level) from a random Poisson series at all stages. These sites of vesicle concentration, however, appeared themselves to be randomly distributed along the interface.

Ultrastructural Features

As described previously (11), the cytoplasm adjacent to the rhabdomere was found in light adaptation to be packed with short elements of rough ER, smooth ER particularly along the margin of the rhabdomere, and vacuolar organelles (Fig. 11). Many small Golgi complexes were also scattered throughout this region. As dark adaptation proceeded, vacuolar organelles disappeared, the rough ER was reorganized to longer elements, and the Golgi complexes became more conspicuous, with dilated cisternae (Fig. 14). After a prolonged period in darkness, the cytoplasm was characterized by the reappearance of the multivesicular bodies noted above, as well as by numerous free vesicles (Fig. 16).

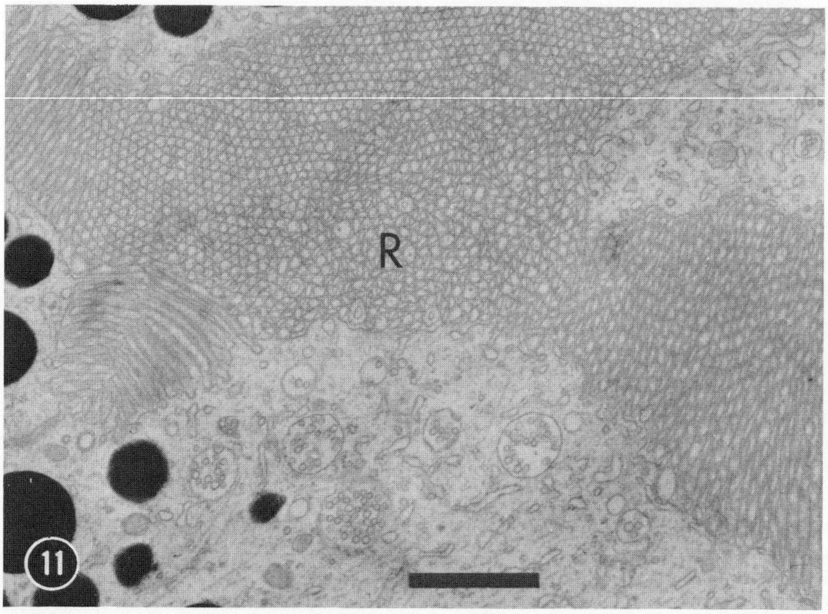

Figure 11. Light-adapted cell. R, rhabdomere; N, nascent multi-vesicular body; R, rhabdomere. The bar represents 1μm.

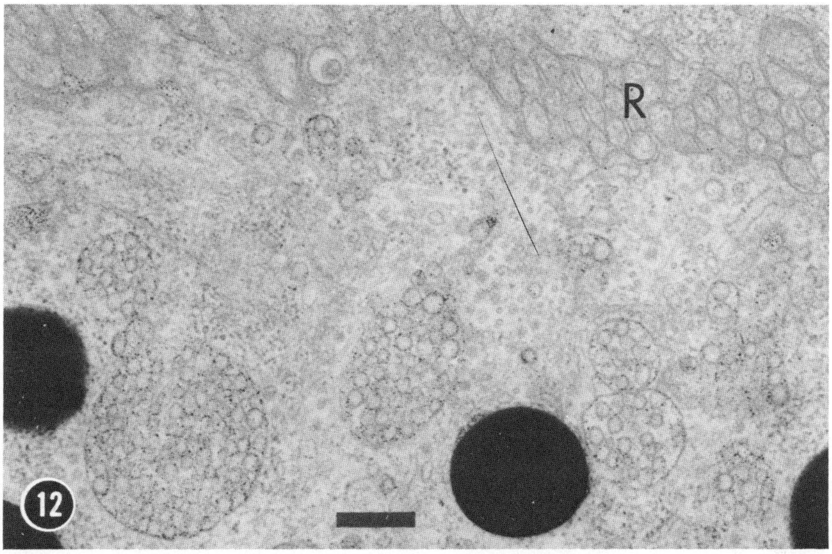

Figure 12. Light-adapted cell exposed to ferritin for 30 min. There are heavy concentrations of ferritin in the intervillous space and multivesicular bodies. R, rhabdomere. The bar represents 0.25 μm.

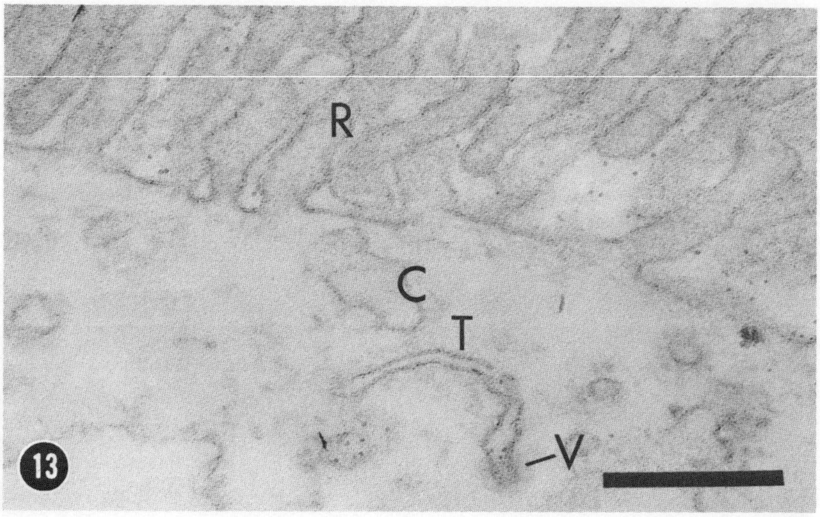

Figure 13. Ferritin-treated, light-adapted cell showing one of the short tubular elements (T) that collect ferritin directly from coated vesicles (V). C, cisternum of smooth ER into which ferritin does not enter. R, rhabdomere. The bar represents 0.25 μm.

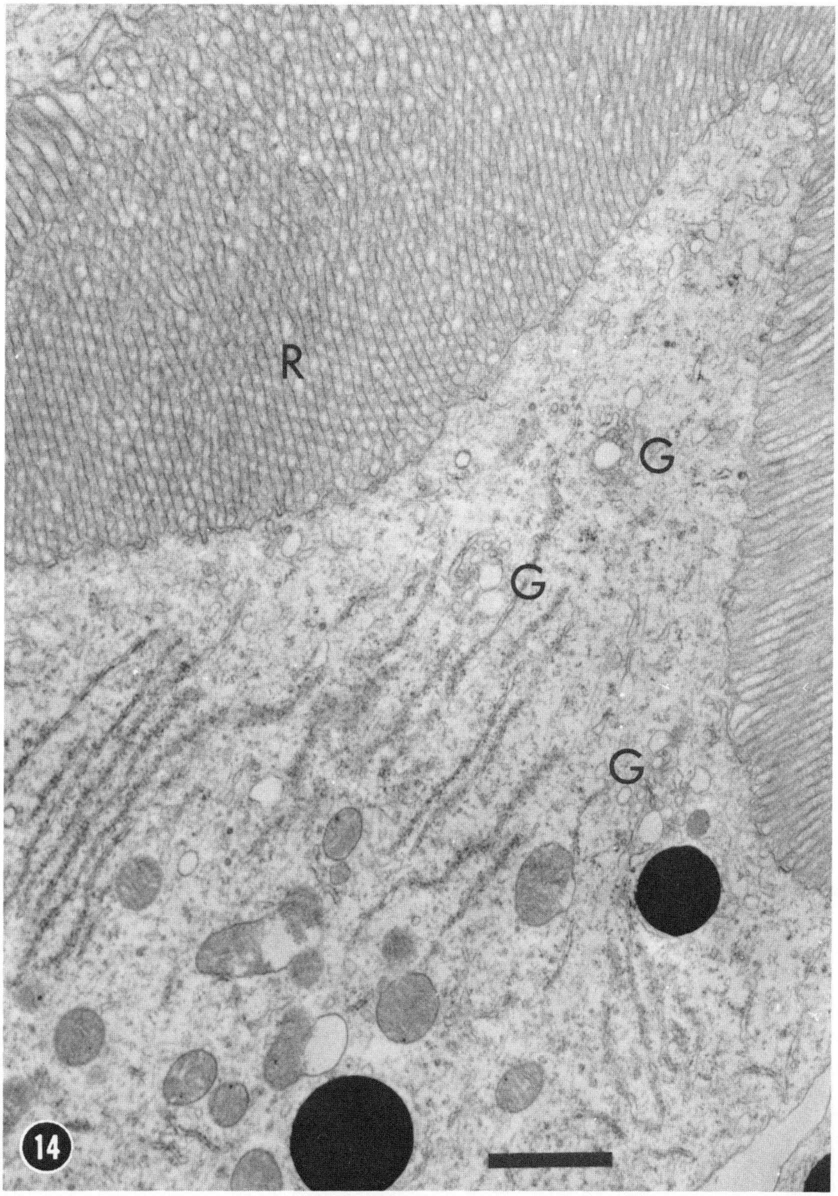

Figure 14. Receptor cell after 4 hr dark adaptation. Note the coated vesicles along the rhabdomere, the absence of vacuolar organelles, the prominence of small Golgi-GERL complexes (G) and long elements of rough ER. R, rhabdomere. The bar represents 1 μm.

Ferritin Endocytosis

Mosquito larvae swim normally for at least 2 hr in a pool of insect Ringer after the anterior tips of their head capsules have been cut off. This is an effective way to deliver various agents to the ocelli, whose cells hang in the cephalic hemocoel separated from it by only a thin basement membrane (11). Relatively large molecules - horseradish peroxidase, dextran and ferritin - diffuse rapidly from the Ringer into the extracellular space between contiguous receptor cells. We report here only on ferritin experiments since peroxidase and dextran did not prove to be as effective for marking the pathway of endocytosis.

As in previous experiments (11), we found that ferritin penetrates the extracellular space between the rhabdomeric microvilli, is concentrated in the pockets between their inner ends, and from there is taken into the cell by coated endocytic vesicles. Occasionally we found ferritin indiscriminately scattered throughout the cytoplasm rather than selectively packaged within organelles. These were assumed to be cells that had been damaged either before or during fixation, and they were discarded. In undamaged cells, ferritin was found localized within coated vesicles, certain coatless vesicles, short tubular elements and vacuolar organelles (Figs. 12, 13, 15, 17). It was notably absent from other elements of the endoplasmic reticulum in the vicinity of the rhabdomere. In particular, ferritin molecules rarely entered the cisternae of smooth ER that lie just beneath the inner ends of the rhabdomeric microvilli.

In early dark adaptation, when the vacuolar system is depressed, loose clusters of ferritin-containing coated vesicles, smooth vesicles and tubules were often seen at particular sites along the margin of the rhabdomere (Fig. 15). Farther into the cytoplasm these ferritin-loaded bodies converged on more closely bunched empty vesicles resembling multivesicular bodies with incomplete outer membranes. In the same cells, the usual sort of membrane-bound multivesicular bodies were rare, but they took up ferritin. In ocelli kept dark for many hours, ferritin accumulated in the compact multivesicular bodies and scattered uncoated vesicles characteristic of that condition. The presence of ferritin did not alter the frequency of rhabdomeric coated vesicles, nor did it change the relative numbers of free and attached vesicles in any of the states of adaptation (Fig. 9).

DISCUSSION

Several lines of evidence, outlined in the Introduction, have lead to the conclusion that the multivesicular bodies of arthropod photoreceptors collect membrane shed from the rhabdomere. Nascent multivesicular bodies have been identified as endosomes

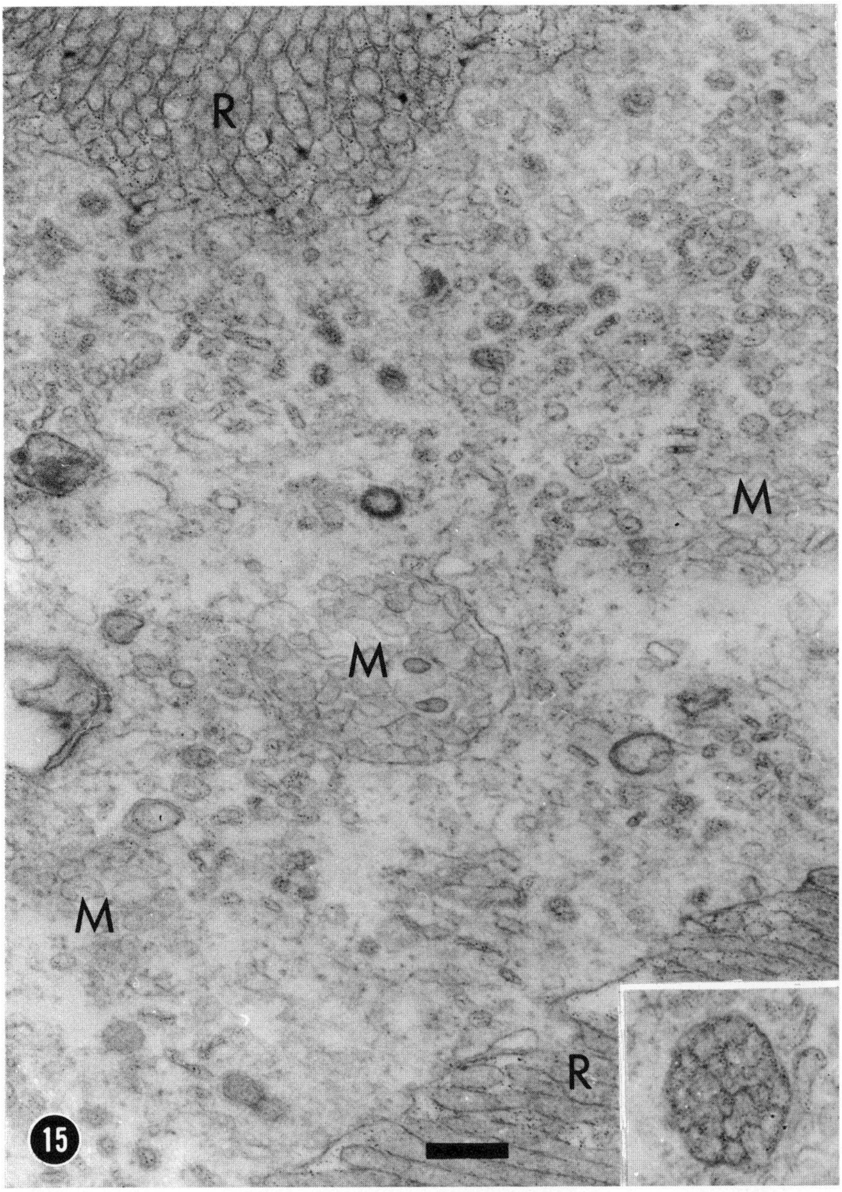

Figure 15. Two hr dark-adapted cell exposed to ferritin for 30 min. Note the clusters of ferritin labeled vesicles and tubules, and the unlabeled vesicles bunched like incomplete multivesicular bodies (M). The inset show that the rare multivesicular bodies in the the same cell accumulated ferritin. R, rhabdomere. The bar represents 0.25 μm.

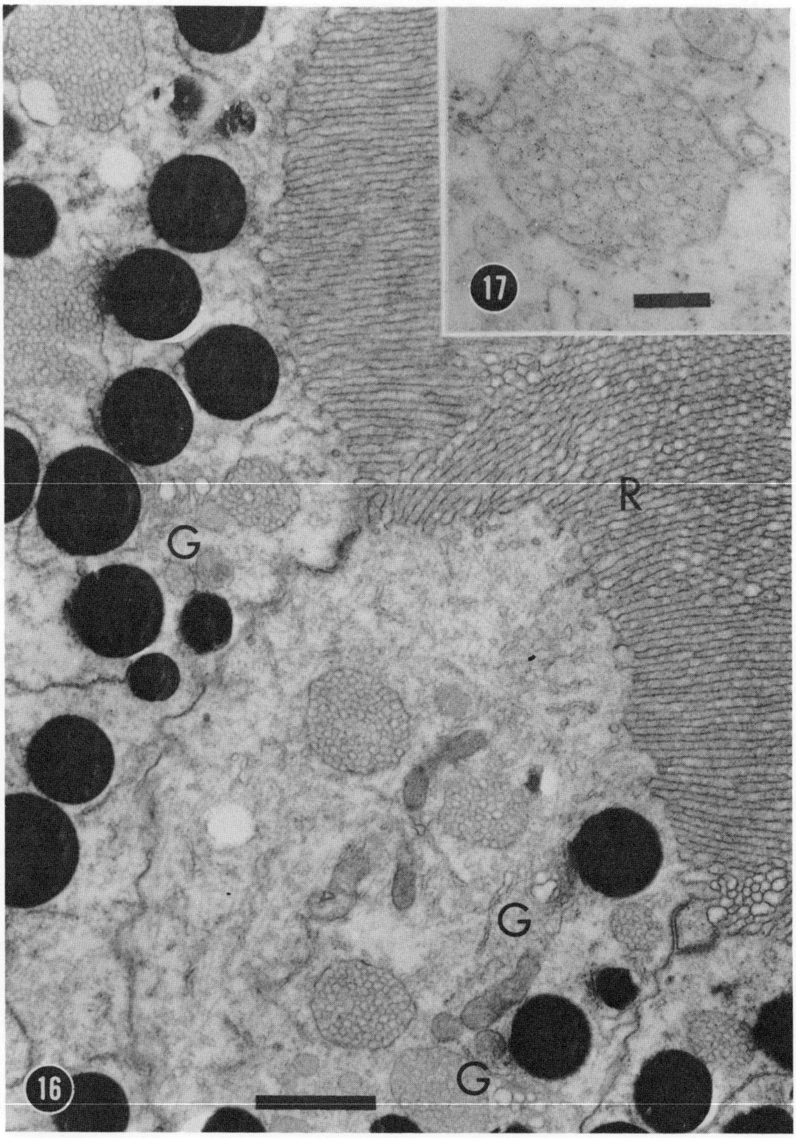

Figure 16. Receptor cell in darkness for 24 hr. Coated vesicles are abundant along the rhabdomere, and compact multivesicular bodies have begun to accumulate. G, Golgi-GERL complexes; R, rhabdomere. The bar represents 1 μm.

Figure 17. Ferritin-labeled multivesicular body from a cell darkened for 24 hr and exposed to ferritin for 30 min. The bar represents 0.25 μm.

formed from vesicles that pinch off the rhabdomere (11). This mode of origin is consistent with our present finding that nascent multivesicular bodies are concentrated near the light adapted rhabdomere (Fig. 8). Their numbers start dropping at the beginning of dark adaptation, and their perirhabdomal frequency declines to the cell average within 30 min. These data indicate that endocytosis is curtailed abruptly with the onset of darkness, while those multivesicular bodies that were last formed continue to disperse and mature during the first minutes of dark adaptation. These observations support the hypothesis that the rhabdomere begins linear growth with the onset of darkness (1) because the loss of membrane is then curtailed. Growth of the rhabdomere would be due to the continuation of ongoing renewal.

The overall pattern of change in the vacuolar system during the first hours of dark adaptation also supports previous inferences. The progressive ultrastructural changes of light adaptation (11) led us to propose that the various vacuolar organelles are the sequential stages of lysosomal degradation shown in Fig. 5. If that scheme is correct, and the formation of multivesicular bodies is cut off sharply at the beginning of dark adaptation, then the other classes of organelles should decline subsequently in the order of their appearance during light adaptation. Fig. 6 shows that they do decline in the expected sequence during the first 8 hr of darkness.

These data also allow us to estimate the lifetimes of each sort of organelle. Nascent multivesicular bodies drop to low frequency by 1 hr of dark adaptation, while compact multivesicular and lamellar bodies are infrequent by 2 hr, indicating that it takes about an hr for the nascent multivesicular bodies to coalesce and another to pass through the lamellar stage. The data from dark adaptation agree with our observations of light adaptation: an hr passes between the first appearance of nascent multivesicular bodies and the first appearance of compact multivesicular and lamellar bodies. Our present data also agree with the rate at which ingested ferritin sequentially appears in multivesicular and lamellar bodies (11). Lamellate dense bodies remain abundant until at least 5 hr of dark adaptation and then decline to low frequency by 7 hr. Therefore they have lifetimes of 4 to 5 hr. Homogeneous dense bodies show no certain pattern of change during 7 days of dark adaptation except that they seem fewer and smaller by then. The lifetimes of the vacuolar organelles of the mosquito ocellus are similar to those estimated for comparable organelles in the mammalian lymphocyte (24).

In its simplest form our hypothesis predicts that the abundance of the various vacuolar organelles in the fully light-adapted ocellus would be determined by their relative lifetimes and the rate of membrane input from the rhabdomere. In subsequent dark adaptation, with the input of membrane suddenly withdrawn, the organelles of a particular class would remain at their light-adapted level while their precursor organelles were being converted. Then they would decline in turn. However, our data show that the situation is more complicated. Before declining, compact multivesicular and lamellar bodies double in number during the first ½ hr of darkness. Although the standard errors of these data are large, the pattern of transient increase is later repeated more convincingly by lamellate dense bodies. Relatively rare in light adaptation, they increase sixfold and become larger between 1 and 6 hr of darkness before declining (Fig. 6). Thus dark adaptation brings a transient bulge in the population of compact multivesicular and lamellar bodies, which then passes through the lamellate dense body stage. Furthermore, the frequency of lamellate dense bodies is only about 1/5 that of the multivesicular and lamellar classes in full light adaptation. This is obviously incompatible with a lifetime of 4-5 hr for the former, and only 1 hr for the latter measured during dark adaptation. A possible explanation is that the rates of conversion of the organelles become slower in darkness. We suggest that lamellate dense bodies are infrequent in the light because they are rapidly converted to homogeneous dense bodies, while in darkness their rate of transformation is slowed to the extent that they accumulate in large numbers.

The gradual build-up of compact multivesicular bodies, but not of other vacuolar organelles, after 24 hr of darkness is another unexpected feature of prolonged dark adaptation. We had found previously (11) that very few vacuolar organelles of any sort are present in ocelli from mosquito larvae hatched and reared in darkness for several days. This difference points up the distinction between light deprivation and dark adaptation after prior illumination; there are a number of ultrastructural differences that distinguish these two states.

Both the transient accumulation of lamellate dense bodies during the first hours of dark adaptation and the amassment of compact multivesicular bodies later suggest that the degradation of the vacuolar organelles proceeds more slowly in darkness, probably because the incorporation of lysosomal hydrolases is curtailed. Since the multivesicular bodies of prolonged dark adaptation accumulate exogenous ferritin they may be presumed to arise from membrane shed by the rhabdomere, just as in light adaptation. Eguchi and Waterman found that acid phosphatase activity appears with the conversion of multivesicular to lamellar bodies (9). Thus the multivesicular bodies of protracted darkness represent accumulated rhabdomere membrane that has been shed but not degraded. These observations indicate that light influences both the formation and destruction of multivesicular bodies, but that these processes are under separate control in the mosquito ocellus, as Blest and coworkers (18, 20) have clearly shown them to be in the spider *Dinopis*.

Overall, the present quantitative assessment of the vacuolar system in dark-adapting Aedes has confirmed our view that the membrane shed by the rhabdomere is conveyed to the vacuolar system for degradation. The data further support our contention that enlargement or diminution of the rhabdomere results when the balance between assembly and degradation of photoreceptor membrane is upset. This interpretation is also supported by recent studies on compound eyes in the mosquito (25, 26), Limulus (17, 27) and shrimp (28, 29), and especially by a thorough analysis of rhabdomere turnover in spiders (18-20). As others have pointed out recently (2, 20) the mechanism for the degradation of rhabdomere membrane appears to be the same as that utilized by other cells, particularly neurons, for the turnover of plasma membrane.

The Question of Renewal

We had hoped that a study of dark adaptation would give us some insight into the process of rhabdomere renewal, since it would then be proceeding without concomitant degradation. Our attention was drawn to the rhabdomeric coated vesicles. Like the multivesicular bodies, they decline sharply with darkness, but only to about half their light-adapted frequency, while the vacuolar organelles are reduced by 90%. This discrepancy lead us to wonder if coated vesicles might not be increments of rhabdomere renewal as well as of endocytic membrane retrieval. In light adaptation, half would be conveying membrane away from the rhabdomere, half would be bringing new membrane to it. In darkness, only the latter would remain. This idea was bolstered by the (unpublished) observation that coated vesicles are prominent features of differentiating rhabdomeres in the developing compound eye of the mosquito. In fact, coated vesicles are among the earliest features of microvillus differentiation.

The ferritin experiments were aimed at testing this hypothesis, with the expectation that the coated vesicles of the dark-adapting rhabdomere would not be labeled. However, when substantial amounts of ferritin are present in the extracellular space of the rhabdomere, very few coated vesicles remain empty in either light or darkness (Figs. 12, 15, 17). The simplest interpretation of these results is that all the coated vesicles along the margin of the rhabdomere have been shed from it. However, the ferritin technique only indicates confluent cellular compartments. It does not trace directly the origin of membrane or the pathway of its flow. Ferritin-labeled coated vesicles could be carrying membrane onto the rhabdomere if they originate from a cellular compartment that also receives endocytic vesicles.

In any event, it is not clear how the coated vesicle and endomembrane configurations seen in early dark adaptation (Fig. 15) are to be interpreted. If they are not part of a renewal or recycling system, these clusters of ferritin-labeled and empty vesicles, the latter resembling abortive multivesicular bodies, would appear to be membrane shed by the rhabdomere. According to this interpretation, labeled elements might be those that formed during the 30 min pulse of ferritin, the unlabeled vesicles, those shed earlier. Hence, shedding does not entirely cease with the onset of darkness.

We now understand the general process of rhabdomere breakdown in arthropods, and have some idea of the factors involved in its control. However, the mechanism of renewal remains obscure, and the possibility that photoreceptor membrane may be recycled without degradation under some circumstances - as in presynaptic plasma membrane in neurons (2) - has not been much considered. The more provocative questions of why the turnover of arthropod photoreceptor membrane should be particularly vigorous (1, 18, 19, 20), and how turnover is related to transduction, remain completely open.

SUMMARY

Aspects of the ultrastructure of dark adaptation related to rhabdomere turnover were studied in the larval ocellus of the mosquito, *Aedes aegypti*. Coated endocytic vesicles and the organelles of the vacuolar system - multivesicular, lamellar and dense bodies that function in the lysosomal degradation of rhabdomere membrane - were counted in light-adapted photoreceptors, and in cells dark adapted for periods of time ranging from 15 min to several days. Nascent multivesicular bodies, in accord with their function as phagosomes collecting membrane cast off the rhabdomere, were concentrated near the rhabdomere in light adaptation, falling to low frequency by 30 min of darkness. During the following hours of darkness, the other vacuolar organelles followed each other in a pattern of transient increase and the decline. The overall pattern of change during the early hours of dark adaptation was consistent with the abrupt cessation of endocytosis at the beginning of dark adaptation, with those phagosomes that were last formed passing through the successive stages of lysosomal degradation. After a day of darkness a particular class of multivesicular bodies began to accumulate again and rose to high frequency, presumably because their degradation was blocked through the absence of hydrolytic enzymes. Rhabdomeric coated vesicles also declined in dark adaptation, but not to the extent of the vacuolar organelles. An hypothesis that coated vesicles might carry membrane to the rhabdomere as well as remove it was not supported by experiments using exogenous ferritin to distinguish endocytosis from exocytosis. Coated vesicles in both dark-and light-adapted cells accumulated ferritin, indicating their prior attachment to the rhabdomere.

ACKNOWLEDGEMENTS

This work was supported by grants EY01012-03 from the National Institutes of Health and BNS-76-18623 from the National Science Foundation. We thank Emily Stone for assistance.

1. White, R. H. and Lord, E. (1975) J. Gen. Physiol. 65, 583-598.
2. Holtzman, E., Schacher, S., Evans, J. and Teichberg, S. (1977) in The Synthesis of Cell Surface Components (Post, G. and Nicholson, G. L., eds.) pp. 165-246, American Elsevier Pub. Co., New York.
3. Heuser, J. G. and Reese, T. S. (1973) J. Cell Biol. 57, 315-344.
4. LaVail, M. M. and LaVail, J. H. (1974) J. Comp. Neur. 157, 303-358.
5. Teichberg, S., Holtzman, E., Crain, S. M. and Peterson, E. R. (1975) J. Cell Biol. 67, 215-230.
6. Melamed., J. and Trujillo-Cenoz, O. (1966) Z. Zellforsch. 74, 12-31.
7. Eguchi, E. and Waterman, T. H. (1967) Z. Zellforsch. 79, 209-229.
8. Eguchi, E. and Waterman, T. H. (1968) Z. Zellforsch. 84, 87-101.
9. Eguchi, E. and Waterman, T. H. (1976) Cell Tiss. Res. 169, 419-434.
10. Eguchi, E., Waterman, T. H. and Akiyama, J. (1973) J. Gen. Physiol. 62, 355-375.
11. White, R. H. (1968) J. Exp. Zool. 169, 261-278.
12. Burton, P. R. and Stockhammer, K. A. (1969) J. Morph. 127, 233-258.
13. Wachmann, E. (1969) Z. Zellforsch. 99, 263-276.
14. Tuurala, O. and Lehtinen, A. (1971) Ann. Acad. Sci. Fenn. A, IV Biologica 176, 1-9.
15. Tuurala, O. and Lehtinen, A. (1971) Ann. Acad. Sci. Fenn A, IV Biologica 177, 1-8.
16. Meyer-Rochow, V. B. (1972) Z. Zellforsch. 133, 59-86.
17. Behrens, M. and Krebs, W. (1976) J. Comp. Physiol. 107, 77-96.
18. Blest, A. D. (1978) Proc. Roy. Soc. Lond. B 200, 463-483.
19. Blest, A. D., Powell, K. and Kao, L. (1978) Cell Tiss. Res. 195, 277-297.
20. Blest, A. D., Kao, L. and Powell, K. (1978) Cell Tiss. Res. 195, 425-444.
21. Brown, P. K. and White, R. H. (1972) J. Gen. Physiol. 59, 401-414.
22. Karnovsky, M. J. (1965) J. Cell Biol. 27, 137a.
23. Kershaw, K. A. (1973) Quantitative and Dynamic Plant Ecology, 2nd ed., pp. 128-144, American Elsevier Pub. Co., New York.
24. Biberfeld, P. (1971) J. Ultrastruct. Res. 37, 41-68.
25. Brammer, J. D. and Clarin, B. (1976) J. Exp. Zool. 195, 33-40.

26. Brammer, J. D., Stein, P. J. and Anderson, R. A. (1978) J. Exp. Zool. 206, 151-156.
27. Chamberlain, S. C. and Barlow, R. B. (1978) Invest. Ophthalmol. Visual Sci., Suppl. (ARVO abstracts) p. 134.
28. Itaya, S. K. (1976) Cell Tiss. Res. 166, 265-273.
29. Itaya, S. K. (1976) Tissue and Cell 8, 583-590.

VISUAL PIGMENT REGENERATION RATE AND SUSCEPTIBILITY TO PHOTIC DAMAGE

Ellis R. Loew

Section of Physiology
Department of Physical Biology
Cornell University
Ithaca, New York 14853

A number of animals are known to undergo photoreceptor degeneration when kept under conditions of constant, low-level illumination, including rats (1), rabbits (2), monkeys (3) and lobsters (4). However, a far greater number show little, if any, harmful effects from long-term exposure even at intensities equal to that of sunlight. For example, flesh flies kept under constant, approximately 82,000 lx white light showed no signs of photoreceptor damage after 30 days of treatment (personal observation), whereas lobsters exposed for less than 5 minutes to approximately 300 lx showed complete photoreceptor degeneration.

Such differences in susceptibility to photic damage force one to consider what the terms "constant" and "low-level" really mean to a given species. It will be shown that for lobsters, at least, inherent visual pigment regeneration rate and the euphotic range of environmental light (i.e., that intensity range normally experienced by the animal in the environment) set boundary conditions for defining the above terms.

THE STRANGE CASE OF THE NORWAY LOBSTER

Ecology and Behavior

The Norway lobster, Nephrops norvegicus (L) (the common European "Scampi"), exists in many distinct populations along the coasts of the North Atlantic Nations (e.g., Scotland, England, etc.). They live in burrows in the coastal shelf sediment at depths varying from somewhat less than 10m to over 100m. Reports of *Nephrops* scavenging in the intertidal zone are also available.

Because they represent an important "cash crop", much work has gone into studying the behavior of *Nephrops*, particularly those factors regulating out-of-burrow activity since it is at these times that *Nephrops* can be potted or trawled. The picture that emerges, based mostly on the extensive work of Chapman et al. (5) on several Scottish populations, is that ambient light is one of the primary factors regulating burrow emergence. For the Loch Torridon population it can be shown that *Nephrops* leave the burrow only when the ambient light intensity is within a preferential range of 10^{-5} to 1 lx (6). In terms of activity, this leads to two peaks in shallow water where this intensity range is encountered around dawn and dusk, while only at high noon is this intensity range reached in deeper waters resulting in a single activity peak. Only rarely are lobsters of the Torridon population found out of their burrows when the light is brighter than 1 lx and only then for short periods. Bright-light avoidance is a definite attribute of the Torridon lobsters.

Other populations of *Nephrops* show the same type of behavior pattern although some also show night-time activity as well as emergence at light intensities above the 1 lx limit reported for the Torridon lobsters (7).

In Vitro Photoreceptor Degeneration

Most of my thoughts on light-induced photoreceptor degeneration evolved as a result of work I performed on the Loch Torridon population of *Nephrops*. When I was first approached to examine the visual cells of *Nephrops*, it was for the purpose of ascertaining the λ_{max} of the visual pigment so that an "actinic" photometer could be constructed for ecological/behavioral measurements. A previous attempt by Prof. H. J. A. Dartnall (personal communication) to extract the visual pigment from *Nephrops* was unsuccessful, so microspectrophotometry was the obvious technique to try. Arrangements were made to catch and ship a number of live *Nephrops* from the Loch Torridon (Scotland) population to the laboratory at the University of Sussex. The animals were to be caught and raised to the surface at night and placed immediately into light-tight containers for shipment. Upon arrival the lobsters were unpacked and checked under dim red light and then placed into opaque tanks of cold, artificial sea water (Marinemix) at 10°C.

Absorbance spectra were successfully obtained from isolated rhabdoms using techniques previously described (4), an example of which is seen in Fig. 1. The visual pigment has a λ_{max} at 498 nm while the resulting long-lived photoproduct has a λ_{max} at 484 nm.

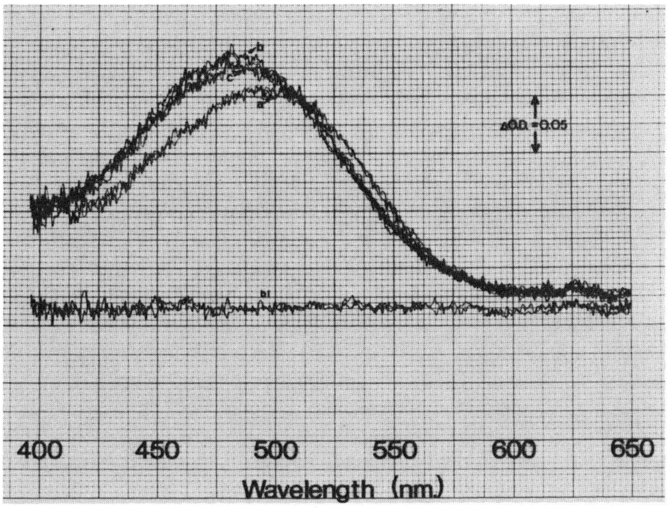

Figure 1. Microspectrophotometer absorbance spectra from an isolated rhabdom. (a) Absorbance spectrum prior to light exposure. The λ_{max} is at 498 nm and the spectrum is fit by a $P_{498}1$ pigment curve constructed from the Dartnall nomogram. (b) The absorbance spectrum of the same rhabdom after 2 min exposure to white light. The λ_{max} is shifted to 484 nm and the extinction is increased. (c) The absorbance spectrum 20 min after exposure. b.l. - baseline.

Attempts to photoregenerate the 498 nm pigment from the 484 nm photoproduct using violet light were unsuccessful. Clearly, the visual pigment system of *Nephrops* was different from that of the closely related lobster, *Homarus americanus*, where photoreversal was easily demonstrable (8). In trying to measure the rate of breakdown of the photoproduct it was noted that there was a steady rise in baseline absorbance above 700 nm which was interpreted as a sign of photoreceptor breakdown. Examination under infrared light confirmed this, but surprisingly it was noted that only those rhabdoms that had been exposed to light showed any signs of physical change! It was further found that if only part of a particular rhabdom was exposed to light, that part alone showed signs of physical change. The character of this change when viewed in white light was one of swelling and vesiculation of the microvillar layers making up the rhabdom. Exposure sufficient to bleach 60% or more of the visual pigment produced the above changes within 30 minutes. However, during this time there was no apparent decrease in the concentration of 484 nm photoproduct in the bleached regions.

In Vivo Experiments

Since the structural changes observed *in vitro* were so dramatic, particularly when compared to the minor structural changes upon bleaching reported for photoreceptors of other animal species (9), a series of *in vivo* experiments was undertaken.

Briefly, lobsters previously exposed only to dim red light were exposed for varying periods of time to white light from an 8 W fluorescent tube placed approximately 30 cm above eye level. The average intensity was 200 cd m^{-2}. Examination of eyes removed and fixed either immediately following light exposure, or after a variable time in darkness indicated that exposure as brief as 20 minutes was sufficient to produce total degeneration of the photoreceptors within 24 hours (Fig. 2). This degeneration was not affected by either a return to darkness following exposure, or exposure to violet light which would have restored visual pigment had photoregeneration been possible.

Microspectrophotometry of rhabdoms isolated from eyes exposed for 2.5 minutes or longer showed only 484 nm photoproduct with no evidence of the 498 nm visual pigment. This was also the finding for all cases in which the animals were kept in darkness following exposure. There was absolutely no evidence for any regeneration of visual pigment in these cases.

For light exposures less than 2.5 minutes there was evidence of some visual pigment in the rhabdoms although this was usually less than 20% of dark-adapted levels and often well below this level. Of the six lobsters exposed within this time range all but two went on to develop photoreceptor degeneration as judged by the quality of the "eyeglow" produced using red light (eyeglow results from the reflection of light out of the eye by the tapetal layer). The two unaffected animals were found to have normal, dark-adapted visual pigment levels when examined 24 hrs after light exposure.

In summary, Loch Torridon *Nephrops* seem to lack any mechanism for the rapid regeneration of visual pigment. In view of the fact that structural damage was always accompanied by very low concentrations of visual pigment and/or high concentrations of non-diffusable photoproduct, and that a period in the dark following even brief light exposure failed to block degeneration or lead to an increase in visual pigment, the conclusion can be drawn that low visual pigment regeneration rate is directly linked to photic damage.

It is of interest to try and relate the intensities of light producing rhabdomeric degeneration in the laboratory with those normally encountered by *Nephrops* in their natural environment. Because the angular distribution and spectral quality of the light used in the laboratory experiments certainly differed from those

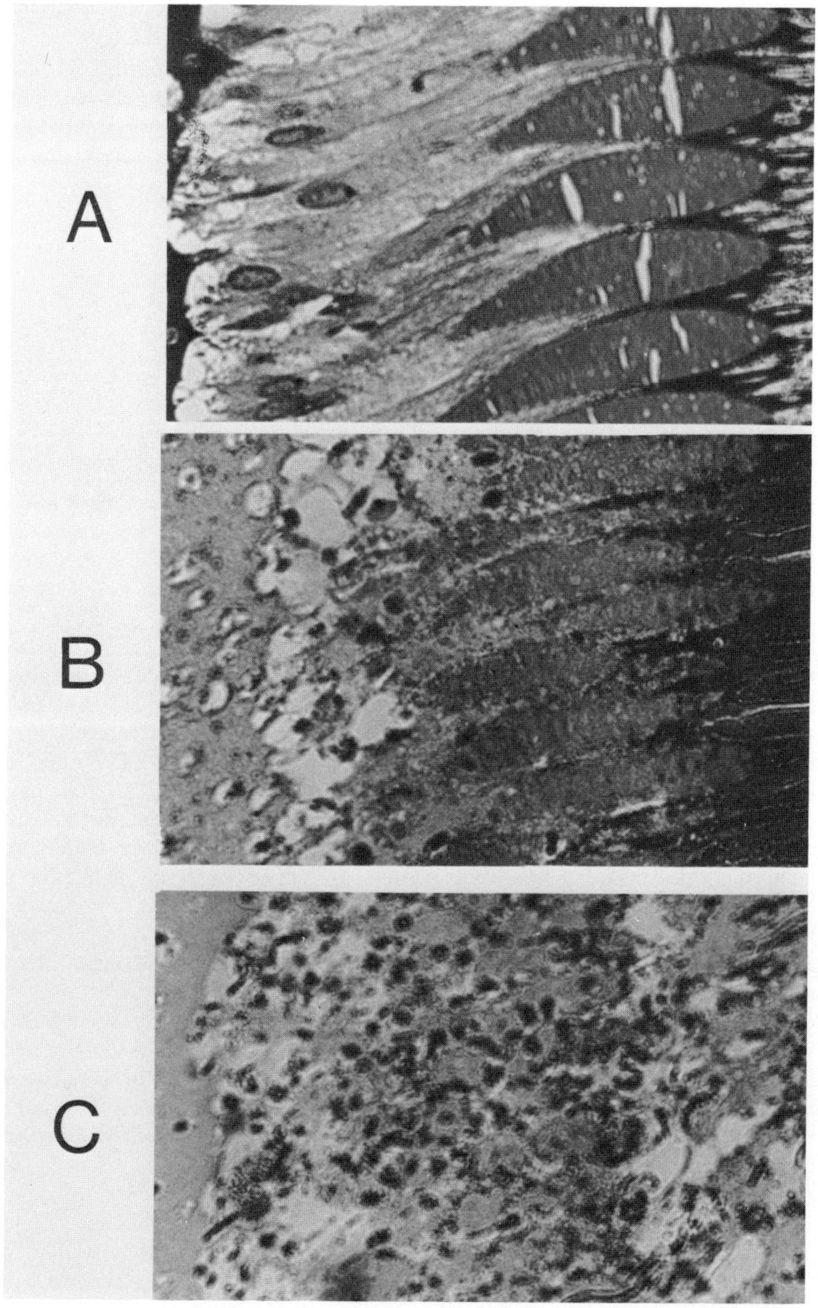

Figure 2. Stages in the light-induced degenerative process. (a) Normal appearance of the rhabdomeric layer. (b) The appearance of the rhabdomeric layer after 2.5 hr of light (200 cd m^{-2}). (c) The rhabdom layer after 14 hr of light.

present in natural waters (10), a direct comparison between field and laboratory light intensities cannot be made. However, in qualitative terms the brightness of the light producing rhabdomeric degeneration is roughly 3 log units above the euphotic range of *Nephrops*, or about the brightness an exposed *Nephrops* would encounter at a depth of 30 m in Loch Torridon at high noon.

Data from other Populations of Nephrops

As mentioned, the above data were gathered from *Nephrops* of the Torridon population. However, other populations show activity at light levels that would certainly lead to photoreceptor degeneration in Torridon lobsters. How do these populations differ?

A schematic diagram of the compound eye of *Nephrops* is seen in Fig. 3. It is of the superposition type and obviously adapted for nocturnal or dim-light vision. The two features relevant to this discussion are the proximal (pp) and distal (dp) screening pigments. The proximal pigment is in a position to shield the tapetal layer (t) thereby greatly reducing the light reflected back through the photoreceptors. The distal pigment can, by changing its distribution, reduce the aperature of the ommatidium, decrease the amount of inter-ommatidial light transfer and absorb more light. In the case of the Torridon lobsters there is no evidence for any movement of either of the screening pigments under the action of light. Both pigments are permanently fixed in the "dark adapted" position. Therefore, even in bright light the eye does not have the capacity to modulate the amount of light actually reaching the rhabdoms. This is not the case for other populations of *Nephrops*. Arechiga and Atkinson (11) have shown that the screening pigments in other populations do move to shield the rhabdoms from over-exposure to light and that this movement occurs over the light intensity range normally encountered by the animal. Exposure to light intensities above the euphotic range "poisons" the photodynamic movement of the pigments.

It is also found that in those species having photodynamic movement, photoreceptor degeneration does not occur at light intensities known to effect Torridon lobsters. This is easily checked by looking for the characteristic changes in the eyeglow accompanying photoreceptor degeneration.

Although it is possible that non-Torridon populations lack a "rapid" visual pigment regeneration mechanism, with screening pigment migration always keeping the light impinging on the photoreceptors within the "euphotic" range, it is more likely that "rapid" visual pigment regeneration does occur. This conclusion is based on the fact that even with the photodynamic mechanism "poisoned" the animals apparently continue to demonstrate a visual

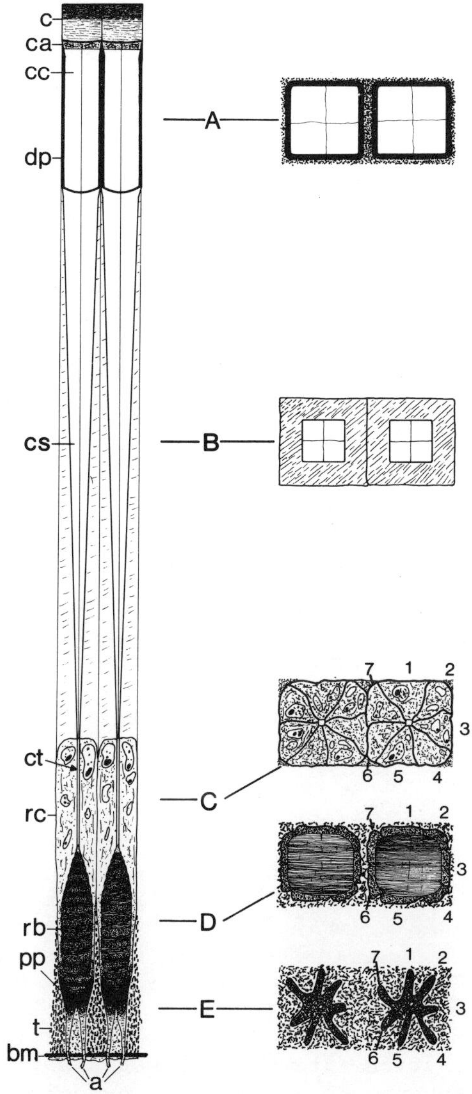

Figure 3. Diagrammatic radial section (1) through several ommatidia of the *Nephrops* compound eye with cross-section (2) at various levels (A-E). c, cornea; ca, corneagen cells; cc, crystalline cone; dp, distal screening pigment; cs, cone stalk; cr, crystalline tract; rc, retinula cells; rb, rhabdom; pp, proximal screening pigment; t, tapetum; bm, basement membrane; a, retinula cell axon.

capacity, and show no signs of photoreceptor damage even when the light exposure is far above the euphotic range. The definitive test, i.e., direct measurement of visual pigment levels in non-Torridon lobsters during and after light exposure, has, unfortunately, yet to be done. However, I feel that these experiments will only support the above conclusion.

Unless one is willing to hypothesize that geographic isolation of the different populations of *Nephrops* has led to sub-speciation with resultant genetic changes in visual pigment regenerative capacity, it must be assumed that conditions subsequent to fertilization, development and/or hatching can affect the inherent regeneration rate. The obvious determining condition is the range of light intensities the lobsters normally encounter. This range would depend on the maximum depth of the environment as well as those behavioral strategies adopted to avoid predation and/or feed. In other words, all lobsters of the species *Nephrops norvegicus* (L) regardless of population would have the same potential regeneration rate, but this rate can be modulated or "set" depending upon the photic stress encountered in early life. Had juvenile lobsters from the Loch Torridon population been used in the above studies instead of only adults, it is very possible that photoreceptor damage would not have been seen. In any case, it would be very interesting to raise Loch Torridon lobsters from spawn in bright light to see if regeneration rate and susceptibility to photic damage could be modulated.

SUMMARY

The following is a summary of the available data for *Nephrops*. For the Loch Torridon population:

1. Exposure to 200 cd m^{-2} white light for more than 2.5 minutes is sufficient to convert all of the visual pigment to photoproduct.
2. No evidence for regeneration of visual pigment following a total bleach is found.
3. As judged from results for less than total pigment bleaches, the inherent regeneration rate must be very low.
4. Total bleaching appears to be a precondition for photoreceptor degeneration.
5. No protection from degeneration can be effected by either a return to darkness following light exposure, or treatment with "isomerizing" violet light.
6. There is no evidence for recovery of the photoreceptors even after a month in the dark following light exposure.
7. The lack of screening pigment movement and the abundant tapetum suggest an eye adapted for operation in very dim light environments.

8. The behavioral adaptations are such that lobsters only encounter light of an intensity not outstripping the capacity of the photoreceptor cell to regenerate visual pigment.

For other populations of Nephrops:

1. Activity in light brighter than that found for the Torridon population is attended by photodynamic movement of screening pigments and probably a faster inherent rate of visual pigment regeneration.

2. Exposure to light producing photoreceptor degeneration in Torridon lobsters produces no damage in lobsters adapted to brighter light conditions.

DISCUSSION

As a working hypothesis for further study I will propose that the different susceptibilities to photic damage among the several populations of Nephrops is a direct result of different inherent visual pigment regeneration rates. It is expected that once kinetic data are available, predictions can be made as to what light intensity will produce photoreceptor degeneration in any Nephrops population. This intensity may also be qualitatively predicted once the euphotic range for the particular population is known.

None of the work on Nephrops to date address the basic physiological/biophysical mechanisms underlying the light-induced photoreceptor degeneration. However, I see no reason why these mechanisms should be any different than those proposed for vertebrates. The mechanisms usually proposed are (see references in (4)):

1. Temperature rise in the photoreceptor producing membrane damage and cell death.

2. Free-radical production with resultant attack on structurally important molecules.

3. Photo-oxidation of structurally important molecules.

4. Production of high concentrations of thermally less-stable photoproduct with loss of the structural integrity of the photoreceptor membrane.

5. "Poisoning" of the visual pigment regeneration pathway by high concentrations of photoproduct leading to reduced concentrations of the structurally more stable visual pigment.

6. "Stressing" of the visual pigment regeneration pathway leading to depletion of precursors, co-factors, etc. This would be both time and intensity dependent.

I believe the most likely mechanism in the case of Nephrops is a combination of 4 and 5 above. I base this on the observation that degeneration only occurs in the presence of high concentrations of photoproduct, and that there is no regeneration of visual pigment in the presence of high photoproduct concentrations even in the dark. Further studies, particularly on the Loch Torridon population, should test the veracity of the above hypothesis.

ACKNOWLEDGEMENTS

I should like to thank Mr. Colin Chapman of the Marine Laboratory, Aberdeen, Scotland for providing me with the opportunity to study such an interesting system. I would also like to thank Dr. J. N. Lythgoe for his support and comments and Professor H. J. A. Dartnall for his help and hospitality. Part of the work mentioned above was supported by a Fight for Sight, Inc. Postdoctoral Fellowship.

REFERENCES

1. Noell, W. K., Walker, V. S., Kang, B. S. and Berman, S. (1966) Invest. Ophthalmol. 5, 450-473.
2. Lawwill, T. (1973) Invest. Ophthalmol. 12, 45-51.
3. Lawwill, T. (1973) In Proc. 10th I.S.C.E.R.G. Symp. (J. T. Pearlman, ed.) Dr. W. Junk B. V., The Netherlands, pp. 65-76.
4. Loew, E. R. (1976) Proc. R. Soc. Lond.-B. 193, 31-44.
5. Chapman, C. J. and Rice, A. L. (1971) Mar. Biol. 10, 321-329.
6. Chapman, C. J., Johnstone, A. D. F. and Rice, A. L. (1975) In Proc. 9th European Mar. Biol. Symp. (H. Barnes, ed.) Aberdeen University Press, Scotland. pp. 33-68.
7. Chapman, C. J. and Howard, F. G. (1979) Mar. Biol. 51, 157-165.
8. Goldsmith, T. H. and Bruno, M. S. (1973) In Biochemistry and Physiology of Visual Pigments (H. Langer, ed.) Springer-Verlag, Berlin. pp. 147-153.
9. Falk. G. and Fatt, P. (1972) In Handbook of Sensory Physiology VII/1 (H. J. A. Dartnall, ed.) Springer-Verlag, Berlin. pp. 200-244.
10. Lythgoe, J. N. (1972) in Handbook of Sensory Physiology VII/1 (H. J. A. Dartnall, ed.) Springer-Verlag, Berlin, pp. 566-603.
11. Arechiga, H. and Atkinson, R. J. A. (1976) Mar. Biol. 32, 66-76.

Selected Topics

IN VITRO EFFECTS OF LIGHT ON THE REGENERATION OF RHODOPSIN

Rosalie K. Crouch

Departments of Ophthalmology and Biochemistry
Medical University of South Carolina
171 Ashley Avenue
Charleston, South Carolina 29403

It is now well established that damage to the retina occurs from visible light at intensity levels below thresholds for thermal burns (1-3). This damage is found to be reversible in the early stages provided the pigment epithelium is intact (4). The mechanism for producing this early reversible light damage is still unknown. This chapter summarizes our preliminary *in vitro* results from studies to ascertain if irradiation damage results in a structural alteration to the rod outer segments (ROS) which would effect the regeneration of the visible pigment and thus lead to a temporary breakdown in the visual response.

The results of several groups have led us to consider that visual pigment itself (or associated lipids) might be damaged and fail to be regenerated in light damaged animals. In the predominantly rod retina of the rat, Noell et al. (1) have demonstrated that the action spectrum for light damage parallels the efficiency of light in bleaching rhodopsin. This result suggests that the selective absorption of energy by photopigments initiates the damaging reactions. Vitamin A deficiency in rats, which reduces the rhodopsin content, has been shown to decrease the light damage (5), again suggesting that the visual pigment is involved in the mechanism.

The electroretinographic sensitivity of rats decreases with increasing exposure to visible light (1, 2). A decrease in rhodopsin content decreases the ERG responses as shown from studies on vitamin A deficiency (6). Recently, Rapp and Williams (7) have shown that the log ERG (b-wave) sensitivity varies directly with the rhodopsin content in light damaged animals even as progressive deterioration is apparent. Organisciak and Noell have shown the rhodopsin extracted from light damaged animals shows a normal absorption spectrum (8). These two results suggest that rhodopsin which is present is structurally unaltered and is functioning normally.

The presence of lipid in the ROS is essential for the regeneration of the visual pigment. Detergents which strip away a high portion of the lipids preserve the spectral properties of the native pigment but, after bleaching, no pigment is regenerated on the addition of the 11-*cis* retinal (9). Farnsworth and Dratz (10) have shown that bovine ROS containing high levels of lipid peroxidation have reduced yields of pigment on regeneration. These workers have correlated the lipid peroxidation in the ROS with α-tocopherol levels. In studies on the isolated frog retina, Shvedova et al., found the induction of lipid peroxidation by a Fe^{+2}-ascorbate system decreases the magnitude of the electroretinograph response (11).

Noell has proposed that rhodopsin photo-products resulting from bleaching initiate reactions leading to light damage (12). The role of vitamin A in lipid peroxidation and hemolysis has been indicated for some time. Increased hemolysis is reported in erythrocytes from rats maintained on an excess of vitamin A (13). Vitamin A has also been reported to be a cause of peroxidation of vegetable oils (14). Delmelle has recently shown that retinal will produce singlet oxygen on irradiation with 365 nm light (15) and that illumination of retinal incorporated in liposomes produces increased membrane fluidity resulting in photodynamic damage (16). We have found that extensive irradiation of ROS *in vitro* leads to the formation of free radicals, the photo-oxidation of the lipid unsaturated fatty acids, and increased fluidity of the membranes. Due to (or in addition to) these alterations in the ROS, the regeneration of the visual pigment is impaired.

All experiments involving the visual pigment or retinals were conducted under dim red light. The retinals were stored in the dark under argon at -70° and were checked periodically by high pressure liquid chromatography for purity; all-*trans* was obtained from Sigma Corp., St. Louis, Mo., and 11-*cis* retinal was the generous gift of Hoffmann-LaRoche, Nutley, N. J. Unless otherwise stated, the buffer used is sodium phosphate, 67 mM, pH 7.4.

Bovine ROS were isolated from frozen, dark adapted retinae (Hormel Co., Austin, Minn.) by the sucrose flotation method of Papermaster and Dreyer (17). Only preparations having a A_{280}/A_{500} ratio of 3.5 or less were used for these studies. Opsin was formed by bleaching the ROS suspended in buffer (1 mg protein per ml buffer) for 90 seconds in the presence of 1 μl hydroxylamine (pH 7.0, 1 M) with white light. The opsin was thrice washed with buffer to remove any residual amine. Pigments were regenerated by combining opsin (1 mg) with 11-cis retinal (5 molar excess in 10 μl ethanol) in buffer and incubated with shaking at 37° for two hours (18). The rod suspensions were spun down at 4° and the pellets dissolved in 2% Ammonyx LO/buffer. Difference absorption spectra were obtained by subtracting the spectra before and after bleaching.

For irradiation experiments, an opsin suspension in buffer (1 mg per ml) was exposed to white light filtered by an ultraviolet filter (Ealing #26-4276) which omitted wavelengths below 420 nm. The suspension was constantly stirred and the temperature maintained at 15°C except for one experiment on temperature effects. Aliquots were withdrawn at various time intervals. For each experiment, control samples were stirred protected from light in red glassware

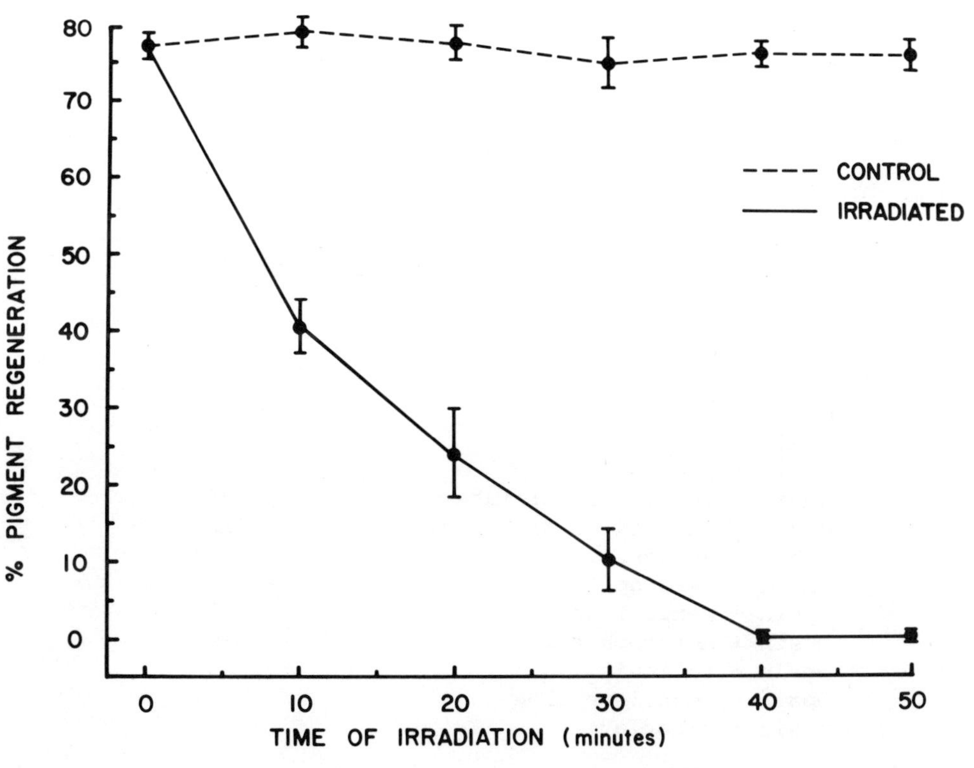

Figure 1: Effect of irradiation of opsin on pigment regeneration. Opsin suspensions (1 mg/ml buffer) were irradiated with visible light at 15°C and aliquots withdrawn at various times. Pigments were regenerated with a 5 molar excess of 11-cis retinal and yields determined by difference absorption spectra.

in the same bath for the maximum periods of irradiation being studied in order to eliminate any possibility of denaturation due to environmental effects other than light. The quantity of damage was determined from the yields of pigment regenerated from the aliquots when combined with 11-*cis* retinal.

Increasing time of exposure of ROS to filtered white light results in decreasing yields of pigment regeneration on the addition of 11-*cis* retinal (Figure 1). The reduction in yields was observed for both irradiated unbleached rod segments and irradiated samples which had been previously bleached in the presence of hydroxylamine and then washed with buffer (19). The pigments regenerated from control and irradiated samples had the same visible absorption spectra, circular dichroism spectra and photosensitivities; therefore the pigments appear to be identical. The addition of excess retinal to the ROS (bleached or unbleached) did not effect the levels of regeneration (Table I). This result suggests that, in our study, free retinal is not acting as the initiator of the photo-damage.

Inhibitors and photosensitizers were added to the opsin suspension before irradiation was initiated. After the irradiations,

TABLE I

Effect of Additives on the Irradiation of Opsin as Measured by Pigment Regeneration

Irradiated Sample	% Pigment Regenerated (±5%)
Opsin	20
Opsin + catalase	20
Opsin + superoxide dismutase	20
Opsin + superoxide dismutase and catalase	22
Opsin + ceruloplasmin	20
Opsin + penicillamine	40
Opsin + α-tocopherol	48
Opsin + retinal	22
Opsin + methylene blue	0
Opsin + riboflavin	3

Samples (1 mg ROS/ml buffer) were irradiated for 30 minutes. Pigments were regenerated with a 10 molar excess of 11-*cis* retinal. Control samples with no irradiation regenerated 82 ± 5% pigment. All additives were at a concentration of 10 mM.

the opsin was washed three times with buffer to remove the additive. Control samples with the additives present were tested to assure the additive itself was not interfering in pigment formation. Likewise, the additives were irradiated, then combined with opsin and pigment regeneration tested to eliminate the possibility that the additive itself yields by-products on irradiation which prevent pigment formation.

The photosensitizers, methylene blue and riboflavin, enhanced the irradiation damage in ROS (Table I). The radical scavengers, penicillamine and α-tocopherol, partially reduced the damaging effect of the irradiation. The enzymes superoxide dismutase, catalase, and ceruloplasmin, which react with the superoxide and hydroxy radicals, were not effective in protecting against the radiation damage _in vitro_ (Table I). Several isozymes of superoxide dismutase have been found in the retina in high concentrations (20, 21). The failure of this enzyme to protect in our system, either alone or in combination with catalase, does not rule out the possibility of the involvement of superoxide ion _in vivo_, as there may be problems with local concentration of the enzyme or the correct isozyme may not have been employed. Likewise, the failure of general radical scavengers, pencillamine and α-tocopherol, to completely prevent damage may be due to problems in obtaining the correct cellular distribution.

Increasing the temperature at which the samples were irradiated decreases the amount of regenerable pigment. On irradiation for 20 minutes, samples at 3° regenerated a 60 ± 7% yield of pigment, samples at 8° regenerated at 45 ± 8% yield, samples at 15° regenerated 23 ± 5% and samples at 32° failed to regenerate any pigment. Light protected samples were stirred at the various temperatures for 20 minutes as controls. No thermal denaturation appeared to be taking place at these temperatures, all controls yielding 75-85% pigment.

Evidence that a structural change has occurred in the ROS on irradiation was obtained in three ways. First, gel electrophoresis was conducted to determine if the major protein was still intact. Disc electrophoresis was performed in 7.5% polyacrylamide gels (22). The opsins were dissolved in 2% Ammonyx LO, mixed with a crystal of bromophenol blue and made 10% with sucrose before application. The gels were stained for protein with Coomassie brilliant blue (23). The gel electrophoresis on the irradiated versus the non-irradiated tissue showed the irradiated tissue to still exist as mainly (∿ 80%) one band but this band had migrated a slightly shorter distance than the control sample. This difference in migration could be explained by a change in charge distribution or protein conformation (24). Extensive lipid peroxidation or sulfhydryl group cross-linking are not reasonable explanations of this phenomenon.

A second line of evidence of structural modification in the ROS is the demonstration of the presence of oxidized lipids in the irradiated ROS. The amount of secondary products of lipid oxidation, i.e., malonyl dialdehyde, was measured in the rod outer segment suspensions by reaction with 2-thiobarbituric acid (TBA) which yields a product absorbing at 535 nm (25). Protein concentrations were determined by the Lowry procedure using bovine serum albumin as the standard (26). Retinal itself reacts with TBA (27) and the nonirradiated bleached rods were also found to yield a slight absorbance at 535 nm on reaction with TBA, due to residual retinal or peroxidized lipids already present. Thus only comparative values could be obtained (Table II). The general free radical inhibitors, α-tocopherol and penicillamine, decreased lipid peroxidation. The photosensitizers methylene blue and riboflavin increased the amount of lipid peroxidation as measured by this method. Ferrous ion, known to promote lipid oxidation (28), increased the levels of malonyl dialdehyde. Likewise, deuterium oxide, which increases the lifetime of singlet oxygen (29) increased the levels of the TBA product. Retinal was found not to have any effect on the TBA reaction in excess of the contribution from the retinal itself, again indicating retinal is not sensitizing the reaction under these conditions.

TABLE II

Effect of Additives on the Irradiation of Opsin as Measured by TBA Adducts

Irradiated Sample	A_{535} Sample/A_{535} Opsin
Opsin	2.0
Opsin + penicillamine	1.3
Opsin + α-tocopherol	1.4
Opsin in D_2O	3.0
Opsin + Fe^{+2}	4.0
Opsin + methylene blue*	2.8
Opsin + riboflavin*	2.5
Opsin + retinal*	2.1

Additives are all 20 μM. Samples irradiated for one hour at 15°. Values normalized to 1 mg protein/ml buffer. All ratios are an average of at least three experiments and are ± 0.2.

*Pigments added to blank to eliminate visible absorbance due to additive itself.

The fatty acid content of the ROS, irradiated and control, has been measured. The phospholipids were extracted by the method of Folch (29). Following hydrolysis, the fatty acids were methylated and analyzed by gas chromatography (in collaboration with Dr. H. Jonsson) (31). Values for the fatty acid content of the ROS were in the same range as these reported by Anderson and Maude (32). In samples that were damaged by irradiation to the extent that $<20\%$ pigment was regenerated, the level of docosahexaenoic acid was reduced by $50 \pm 10\%$ from control values and the total unsaturated acid level was reduced by $60 \pm 8\%$.

The possibility of the reaction of the sulfhydryl groups on the opsin must also be considered. Sulfhydryl enzymes have been reported to be inactivated by lipid peroxidation intermediates (33). Blocking of the sulfhydryl groups on rhodopsin has been shown to prevent regeneration (34). Malonyl aldehyde has been demonstrated to be the agent responsible for the cross-linking of ribonuclease A (35). Therefore, the ROS damage may be more extensive than the decrease of the levels of the unsaturated fatty acids.

Spin labels are sensitive and informative monitors of membrane molecular organization and dynamics (36) and thus have been employed in this study. For these electron spin resonance (ESR) experiments, 100 μl of a 0.1 mM solution of the label in ethanol was placed in a tube and the solvent evaporated with nitrogen. The ROS suspension (1 mg/0.5 ml) was added. The solution was vortexed and allowed to incubate at 30° for 20 minutes. The rod suspension was introduced in a capillary and the spectra recorded with a Varian E-4 spectrometer. The irradiation of ROS in the presence of nitroxide was found to reduce the nitroxide and thus the electron spin resonance (ESR) signal was quenched. The decay of the ESR signal in rods illuminated in the presence of spin labels has been observed previously (37). A similar decay was also observed on the illumination of the light sensitive compounds, retinal, methylene blue, and chlorophyll a, which had been incorporated in lipid bilayers (38). The nitroxide in these reactions is being reduced by the radicals generated on illumination. In a system involving lipid oxidation, Schreier (39) has found the addition of ferrous ions to membranes containing lipid hydroperoxides forms radicals which react with nitroxide spin labels. This quenching of nitroxide signal eliminated the very useful technique for the study of radicals, spin trapping. In spite of an extensive series of experiments varying reaction conditions, no nitroxides were observed by ESR in rod suspensions illuminated in the presence of the nitrones, phenyl-t-butylnitrone and 5,5-dimethylpyrroline-1-oxide.

Our third indication of a photo-dynamic change in the ROS membranes is the changes in fluidity observed on irradiation as monitored by the use of 5-doxylstearic acid (Figure 2). As this nitroxide was reduced on illumination of rods, the label was added

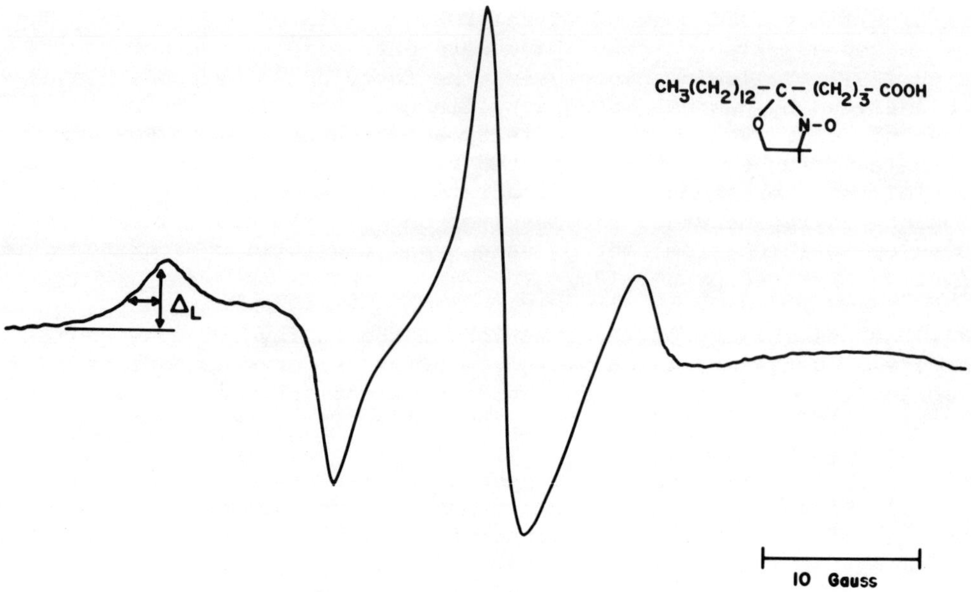

Figure 2: Electron spin resonance spectrum of the 5-doxylstearic acid (100 μM) incorporated in an ROS membrane suspension (1 mg/ml). Δ_1 equals the half width of the downfield peak.

to the rods after irradiation. The amplitude of the downfield peaks were compared. The width parameters for this label decreased with increasing fluidity of the spin-label environment (40). We have observed decreases of (∿ 6%) this peak indicating the membranes have become more fluid. Other workers (41) have obtained similar results on illumination of light sensitive pigments in the presence of other spin-label probes, also attributing the changes to light disorganizations of the lipid films.

Our hypothesis is that the nonthermal reversible photo-damage in the retinae is due, at least partially, to an alteration in the ROS which prevents the visual pigment from regenerating. We have shown in vitro that irradiation of opsin with visible light leads to photo-oxidation of the unsaturated fatty acids and increases in membrane fluidity. Increasing irradiation results in decreasing amounts of regenerable pigments although the pigment formed does not appear to be altered. Our results indicate the photo-oxidation is occurring by a free radical mechanism, possibly involving singlet oxygen. The sensitizer initiating the reaction has not been identified; possibly a retinal Schiff base or minute quantities of a long lived bleaching intermediate surviving the opsin preparation procedure may be involved.

Several mechanisms may be operating in the *in vivo* photo-damaging reaction. It has been shown that peroxidation of the unsaturated fatty acids in the lipids of the ROS effects the electro-retinographic response (11) and it has been demonstrated that an alteration in the ROS unsaturated fatty acids effects regeneration of pigment (10). Our *in vitro* results suggest that one mechanism involved in the photo-damaging process may be the photo-oxidation of the unsaturated fatty acids in the ROS which prevents the normal regeneration of the visual pigment.

Acknowledgement is made to the Research Coporation and the Donors of the Petroleum Research Fund, administered by the American Chemical Society, for support of this research.

REFERENCES

1. Noell, W. K., Walker, V. S., Kang, B. S. and Berman, S. (1966) *Invest. Ophthal.* 5, 450-473.
2. Gorn, R. S., Kuwabara, T. (1967) *Arch. Ophthal.* 77, 115-118.
3. Ham, W. T. H., Muller, A. and Sliney, D. H. (1976) *Nature* 260, 153-154.
4. Kuwabara, T. (1970) *Amer. J. Ophthal.* 70, 187-198.
5. Noell, W. K., Delmelle, M. C. and Albrecht, R. (1971) *Science* 172, 72-75.
6. Dowling, G. E. (1960) *Amer. J. Ophthal.* 50, 875-889.
7. Rapp, L. M. and Williams, T. P. (1977) *Nature* 267, 835-836.
8. Organisciak, D. T. and Noell, W. K. (1977) *Invest. Ophthal.* 16, 188-190.
9. Henselman, R. A. and Cusanovich, M. A. (1974) *Biochem.* 13, 5199-5203.
10. Fransworth, C. and Dratz, E. (1976) *Biochem. Biophys. Acta* 443, 556-570.
11. Shvedora, A. A., Sidorov, A. S., Novikov, K. N., Galushchenko, I. V. and Kagan, V. E. (1979) *Vision Res.* 19, 49-55.
12. Noell, W. F. and Albrecht, R. (1971) *Science* 172, 76-79.
13. Lucy, J. S. and Dingle, J. T. (1964) *Nature* 204, 156-160.
14. Bhattacharya, S., Chowdhury, N. K. and Basu, U. P. (1954) *J. Indian Chem. Soc.* 31, 231-240.
15. Delmelle, M. (1978) *Photochem. Photobio.* 27, 731-734.
16. Delmelle, M. (1978) *Photochem. Photobio.* 28, 357-360.
17. Papermaster, D. S. and Dreyer, W. J. (1974) *Biochem.* 13, 2438-2444.
18. Zorn, M. and Futterman, S. (1971) *J. Biol. Chem.* 246, 881-886.
19. Crouch, R. and Coffman, M. (1976) *Biochem. Biophys. Res. Commun.* 73, 428-433.
20. Hall, M. O. and Hall, D. O. (1975) *Biochem. Biophys. Res. Commun.* 67, 1199-1209.
21. Crouch, R., Priest, D. and Duke, E. (1978) *Exp. Eye Res.* 27, 503-509.

22. Maurer, H. R. (1971) Disc Electrophoresis, Walter deGruyer, New York, pp. 44-45.
23. Weber, K. and Osborn, W. (1969) J. Biol. Chem. 244, 4406-4412.
24. Orstein, L. (1964) Ann. N. Y. Acad. Sci. 121, 321-349; Davis, B. J., Ibid., 404-427.
25. Ohkawa, H., Ohishi, N. and Yagi, K. (1978) J. Lipid Res. 19, 1053-1057.
26. Lowry, O. H., Rosebrough, N. J., Farr, A. L. and Randall, R. J. (1951) J. Biol. Chem. 193, 265-275.
27. Futterman, S. and Saslaw, L. D. (1961) J. Biol. Chem. 236, 652-657.
28. Wills, E. D. (1965) Biochim. Biophys. Acta 98, 238-251.
29. Nilsson, R., Merkel, P. B. and Kearns, D. R. (1972) Photochem. Photobiol. 16, 117-124.
30. Folch, J., Lees, M. and Sloane Stanley, G. H. (1957) J. Biol. Chem. 226, 497-509.
31. Ackman, R. G. and Burgher, R. D. (1965) J. Amer. Oil Chem. Soc. 42, 38-42.
32. McKnight, R. C. and Hunter, F. E. (1966) J. Biol. Chem. 241, 2757-2765.
33. Wald, G. and Brown, P. K. (1952) J. Gen. Physiol. 35, 797-821.
34. Chio, K. S. and Tappel, A. L. (1969) Biochem. 8, 2827-2832.
35. Anderson, R. E. and Maude, M. B. (1970) Biochem. 9, 3624-3628.
36. Schreier, S., Polnaszek, C. F. and Smith, I. C. P. (1978) Biochem. Biophys. Acta 515, 365-546.
37. Verma, S. P., Berliner, L. J. and Smith, I. C. P. (1973) Biochem. Biophys. Res. Comm. 55, 704-709.
38. Verma, S. P., Schneider, H. and Smith, I. C. P. (1972) FEBS Letters 25, 197-200.
39. Schreier, S. (1978) Biophys. J. 21, 205a.
40. Wey, C-L, Cone, R. A. and Gaffney, B. J. (1979) Photochem. Photobiol. 29, 707-712.
41. Verma, S. P., Schneider, H. and Smith, I. C. P. (1974) Arch. Biochem. Biophys. 162, 48-55.

SOLAR RETINOPATHY AS A FUNCTION OF WAVELENGTH: ITS SIGNIFICANCE FOR PROTECTIVE EYEWEAR

William T. Ham, Jr., H. A. Mueller, John J. Ruffolo, Jr., and DuPont Guerry, III
Department of Biophysics
Virginia Commonwealth University
Richmond, Virginia 23298

INTRODUCTION

The paper of Verhoeff and Bell (1) in 1916 became a landmark in the history of radiation injury to the retina. These authors demonstrated that high intensities of light on the retina produced burn lesions. In their own words, "our experiments show with the utmost distinctness that the effects known as eclipse blindness are wholly thermic, due to the intense concentrations of the solar energy upon the retina by the refracting system of the eye itself forming an image of destructive energy intensity---". This emphasis on thermal injury, particularly for solar retinitis and eclipse blindness, tended during the ensuing years to obfuscate other possible effects of light on retinal tissue.

During World War II the syndrome of foveomacular retinitis was investigated by Cordes (2) who adopted this nomenclature to describe an eye disease first noted in naval personnel. It consisted of a retinal pathology which began as a macular edema with loss of foveal reflex, developing later into a foveal hole or cyst. The original syndrome was confined primarily to personnel serving in the Hawaiian Islands or the Pacific combat zone. Only two etiological factors received series consideration: solar retinitis and angiospastic retinopathy. Over the past several decades the etiology of foveomacular retinitis has been controversial, opinions ranging all the way from psychic disturbance to deliberate sun gazing with intent to avoid military service. The status and epidemiology of foveomacular retinitis among military personnel, including the 1969 outbreak at San Diego, was reviewed by Marlor et al. (3) in 1973. They concluded that solar retinitis could not be excluded as a major etiology for foveomacular retinitis.

Smith (4), while conducting an ophthalmological clinic on a tropical island in the Pacific theatre during World War II, had occasion to examine 150 servicemen whose complaints were similar and characteristic: "I just haven't been able to see sharp outlines---"; "the targets on the range have recently become blurred"; "my vision is definitely worse than when I arrived here". Fundoscopic examination revealed constant and typical pigmentary disturbance in the macula. There was faint but definite macular edema, fading gradually into normal appearing tissue in the paramacular region. The edematous area was about two-thirds that of the optic disk. The foveal reflex was blurred. Smith noted gradual decrease in visual acuity and pigment changes in the macula as characteristic for patients who had served four months or longer in the tropics. There was no evidence of this syndrome among indoor military personnel (yeomen and pharmacist mates). He postulated that the observed retinal pathology might be caused by the abiotic rays in the solar spectrum and he noted that the effects appeared similar to the earlier stages of senile macular degeneration. There was no mention of future prognosis probably because he was unable to perform follow-up studies on these patients.

Although Czerny (5) as early as 1867 had reported on the production of lesions in the rabbit eye by rays of the sun, Eccles and Flynn (6) were unable to produce a lesion in the fully dilated rabbit eye, even after long exposures to direct sunlight. They succeeded in inflicting retinal lesions by using a telescope which produced a magnification of ten for the image of the sun on the rabbit retina. These experiments demonstrated the hazard of viewing the sun through optical instruments which magnify the image on the retina. It is well known that no optical device can increase the power density on the retina as long as the retinal image is not diffraction-limited. However, Ham et al. (7) had shown that the production of lesions in the rabbit retina was a strong function of image size. A retinal irradiance on the mammalian retina which is harmless for short exposures and small image sizes may produce a thermal lesion for large image diameters because the temperature at the center of the image increases more than linearly with image diameters, even though the irradiance remains constant, Ham et al. (8).

Reports in the medical literature on solar retinitis and eclipse blindness after a solar eclipse have been frequent. For example, the eclipse in this country in 1970 produced 145 cases of eye injury, Hatfield (9). Permanent impairment of visual function resulted in 55 per cent of these cases, while the other 45 per cent eventually recovered to their previous level of visual function. Penner and McNair (10) reported remarkable recovery in patients suffering from eclipse blindness. Ocular injury from solar radiation has also been reported among anti-aircraft lookouts, plane spotters and gunners in World War II who received severe retinal lesions while tracking planes attacking from the direction of the sun. Religious devotees, psychotic patients, lifeguards and sunbathers have also been victims of solar retinitis.

Up until about 1970 it was commonly believed that solar retinopathies were the result of thermal injury to the retina. For example, Duke-Elder (11) writes as follows: "Photic energy is known to damage the neural retina in two distinct ways--[1] Photochemical Damage. Prolonged exposure to a bright light is thought to cause damage by a photochemical process due to bleaching an excessive amount of the visual pigment so that the regeneration cycle is inhibited. There is no thermal element in this type of damage since the maximum rise in temperature would be in the order of $3^{o}C$ (12), and the damage occurs in the absence of melanin (13). [2] Thermal Damage. Such lesions are therefore caused indiscriminately by infra-red or visible light, the reaction being unrelated either quantitatively or qualitatively to any particular wavelength, but depending merely on the concentration of energy incident in this region--".

The reference to Friedman and Kuwabara (12) concerns their finding that exposure of the rhesus retina to the light of an indirect ophthalmoscope for 15 minutes produced a lesion, even though the maximum temperature rise in the retina as measured with a thermocouple was $3^{o}C$. They recognized that prolonged exposure of a limited area of the retina to a white light source (tungsten bulb, $2900^{o}K$) at low levels of retinal irradiance, ($0.27W/cm^2$), such that the retinal temperature rise did not exceed $3^{o}C$, produced a lesion which could not be attributed solely to thermal effects. In their own words, "the precise mechanism responsible for this retinal damage was not ascertained but was felt to be a light effect which was potentiated by heat".

The reference to Noell et al. (13), relates to the environmental exposure within cages of albino rats to constant fluorescent illumination. Such environmental exposure for extended periods of time results in an almost total loss of rod receptors in the albino rat. Pigmented rats show the same effects but to a lesser degree. The phenomenon is extremely temperature sensitive and the action spectrum seems to follow the absorption spectrum for rhodopsin. It is not thought that the effects of environmental exposure of nocturnal rodents to fluorescent lights bear any direct relation to the type of discrete or limited retinal lesions produced in primate and man by solar radiation. However, possible similarities should not be overlooked since environmental exposure effects have been noted in birds and several other species of mammals, (14).

Research since 1970 on photic damage from a variety of optical sources, focused as discrete or limited images on the retina, has shown that photochemical effects from short wavelength light play a major role in many photoretinopathies, especially those involving extended exposures. Vassiliadis (15) found that 1000 s exposures of the rhesus retina to the He-Ne (633 nm) laser produced thermal damage, whereas the argon laser (514.5 nm) produced a lesion at much lower levels of retinal irradiance, indicating a photochemical nature of injury. Harwerth and Sperling (16) produced permanent blue blindness in the rhesus monkey exposed to a retinal irradiance of 3 x 10^{-4}W/cm^2 of blue light (463 nm) for 3 hours per day for 21 days. These levels of irradiance are not far above the maximum permissible exposure (MPE) level for continuous sources, (17). Tso (18) exposed rhesus monkeys to the indirect ophthalmoscope for periods of one hour, in a set of experiments similar to those of Friedman and Kuwabara (12). Damage to the retina was moderately severe and probably represented a combination of photochemical and thermal insult. Histological and ultrastructure studies were performed at 5 hours, 24 hours, 40 hours, one week, 3 weeks, 4 weeks, 3 months and 5 months postexposure. It was concluded that gradual recovery of the photoreceptor cells over a period of months provided an explanation of recovery of vision in patients suffering from solar retinitis as reported by Penner and McNair (10). Lawwill (19) using ERG techniques found that the short wavelengths in the visible spectrum were primarily responsible for photic damage to the retina of rabbits and rhesus monkeys exposed over a four hour period of time.

Ham et al. (20) exposed rhesus monkeys to a simulated solar spectrum focussing an image diameter of 159 μm on the rhesus retina corresponding to the image diameter of the sun on the human retina. They found that a 3 minute exposure produced a retinal lesion, even though the maximum temperature rise in the retina was less than 3°C. Since a lesion of this type could not be attributed solely to thermal injury, they postulated that thermally enhanced photochemical processes must be responsible. Then, Tso and La Piana (21) exposed three volunteer patients, scheduled for monocular enucleation (because of malignant melanoma) to solar radiation. These patients agreed to view directly the sun for an hour. One had a pupillary diameter of 4 mm induced by a mydriatic, while the other two gazed with a pupillary diameter estimated at 1.5 mm. Maximum retinal temperature rise during exposure, as estimated from the mathematical model of White et al. (22), was 5° for the 4 mm pupil and less then 1°C for the two volunteers with 1.5 mm pupils. Only the volunteer with a 4 mm pupil sustained a loss of visual acuity at 24 hours postexposure and this was minor. All three eyes were enucleated between 38-48 hours postexposure. Histological examination revealed damage to the retinal pigment epithelium (RPE) with some clumping of melanosomes and depigmentation within the exposed area. The outer segments (OS) of the photoreceptor cells appeared normal, thus explaining the retention of visual acuity.

To further investigate the effect of wavelength on photic injury to a limited area of the retina, Ham et al. (23) exposed the rhesus retina to 8 monochromatic laser lines extending from 1064 nm to 441 nm. Retinal image diameter was 500 μm to the e^{-2} points of the Gaussian distribution. The criterion for injury was the funduscopic appearance of a minimal lesion at 48 hours postexposure. The results demonstrated a definitive dependence of photic damage upon wavelength. For extended exposures (1000 s), blue light (441 nm) was three orders of magnitude more effective in producing a minimal lesion than near infra-red (1064 nm) radiation. The infra-red lesion was a characteristic retinal burn produced by temperatures 23°C above ambient, while the blue lesion resulted from a level of retinal irradiance which was so low that temperature rise effects were completely negligible ($< 0.1^{o}$C). Only photochemical processes could account for such a lesion. Moreover, the funduscopic appearance and the characteristic properties of the blue light lesion were different from a burn lesion. To define these differences more definitively, Ham et al. (24) examined histologically over 3000 sections from 20 eyes at postexposure times ranging from one hour to 90 days. They also investigated the effects of minimal, moderate and moderately severe photochemical lesions produced by blue light (441 nm) on the visual acuity of rhesus monkeys trained to perform a visual task (25).

Both histological and visual acuity data confirmed the funduscopic findings by demonstrating that the blue light lesion differed fundamentally from a burn lesion. The initial site of injury appeared primarily to involve the RPE and more specifically the melanosomes in the apical portion of the RPE. Phagocytosis of the melanosomes by macrophages appearing in the sub-retinal space at 48 hours postexposure resulted in hypopigmentation of the RPE. This explained why minimal blue light lesions required 48 hours to become funduscopically visible. There was no histological evidence of major damage to the outer segments (OS) of the photoreceptors at 48 hours postexposure. This finding agreed with the visual function until the 5th day postexposure. It also corroborated the histological findings of Tso (18) on the rhesus monkeys exposed to the indirect ophthalmoscope and with the histological data of Tso and LaPiana (21) on the three volunteers who gazed at the sun for an hour, thereby suggesting that the etiology of solar retinitis might be closely related to the blue light lesion. Specimens examined histologically at 5 days postexposure showed extensive damage and disorientation in the OS. This observation confirmed the visual acuity tests on rhesus monkeys which had shown a sharp decline in visual function on the 5th day after macular exposure to blue light.

Recent research provides convincing evidence that the etiology of solar retinitis is closely related to the photochemical effects of the short wavelengths in the solar spectrum rather than to thermal effects from an indiscriminate mixture of visible and infrared radiation. The development of mathematical models of heat conduction in the retina by White et al. (22) and the experimental measurements of retinal temperatures by Cain and Welch (26) have made it untenable to explain eclipse blindness and solar retinitis in terms of thermal injury since it can be shown that the maximum temperature induced in the retina by gazing at the sun does not exceed a few degrees Celsius. This also suggests that solar retinitis and eclipse blindness are predominantly photochemical, rather than thermal phenomena. On this basis it becomes possible to understand the clinical literature on solar retinopathies.

Solar retinitis is a general term for inflammation of the retina caused by overexposure to the sun. It does not imply necessarily, severe and irreparable damage comparable to that from a burn lesion which generally constitutes permanent damage to the non-regenerative photoreceptor cells of the neural retina, resulting in an absolute scotoma, i.e. an area in the visual field where no photoreceptor response can be elicited. A relative scotoma, on the other hand, involves a partially impaired area in the visual field where the response is below normal, but still functioning. This is in keeping with the findings of Penner and McNair (10) and Tso (18, 21) and relates quite well with the histological and visual acuity data on the blue light lesion by Ham (24) and Moon et al. (25) indicating that minimal to moderate photochemical lesions are reparable with time. The outer segments, while damaged, are able to recover if injury to the RPE is not too severe. Thus, many cases of solar retinitis and eclipse blindness exhibit complete recovery of visual function after a time interval which depends upon the severity of the exposure. Severe photochemical damage also may result in photoreceptor death and, thus, in an absolute scotoma, if injury to the RPE is great enough to completely impair its supportive role in the metabolic maintenance of the photoreceptor cells. In the case of the rhesus monkey, a radiant exposure to 90 J/cm^2 of blue light (441 nm) administered in 1000 s produced a permanent loss of 20/20 vision (25).

The principal objectives of the research reported in this paper were to demonstrate that the short wavelengths in the solar spectrum are primarily responsible for eclipse blindness and solar retinitis and to analyse the protective features of several ophthalmic glasses having different transmission characteristics. To accomplish these objectives the retinae of rhesus monkeys are exposed to a simulated solar spectrum (SSS) at sea level through 5 different glasses for periods of time ranging from 1 to 1000 s. The criterion for comparison of these glasses is the production of a funduscopically visible minimal lesion at 48 hours postexposure. While sunglasses are neither designed nor intended for direct viewing of the sun, this worst case criterion is adopted as a valid method of comparing the protective features of the different glasses. The image size of the SSS at sea level (300-1400 nm) on the monkey retina is made to correspond to the image size of the sun on the human retina, namely 159 μm in diameter. To test the relative effects of visible vs.

near-infra-red solar radiation on the mammalian retina the SSS is further divided into two separate bandwidths, 400-800 nm and 700-1400 nm, for comparison as to effectiveness in producing a minimal lesion.

EXPERIMENTAL DETAILS

The optical system used to simulate the solar spectrum at sea level is a 2500 W xenon lamp with associated optics similar to that described by Ham et al. (20), except that quartz optics are used throughout the system; this produces an output 5 times more powerful than the original system. Current through the xenon arc is maintained constant at 95 A by a servo-mechanical control system. This

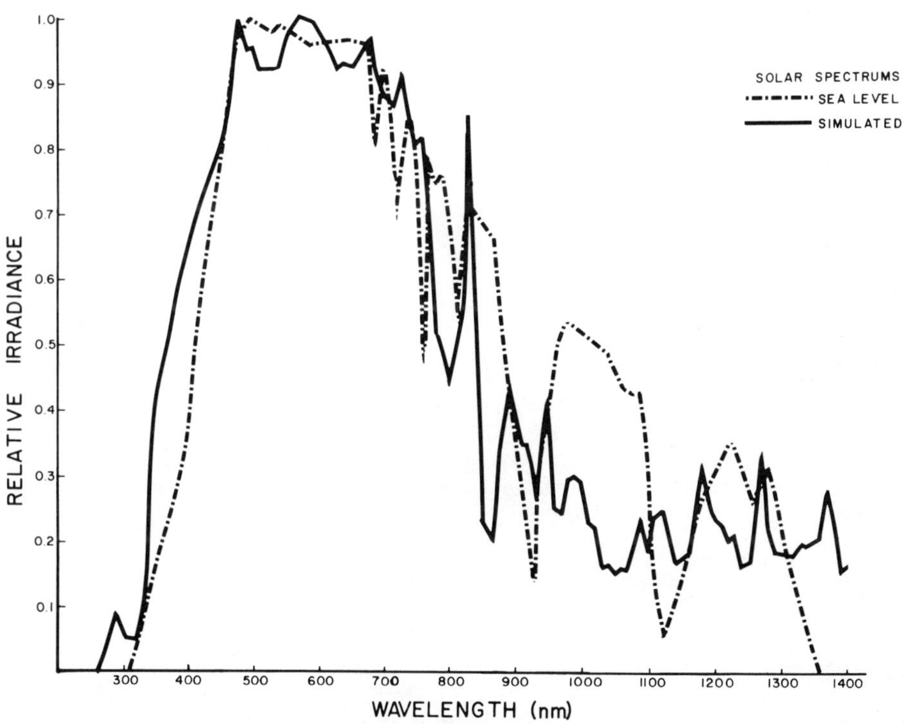

Figure 1: A comparison of the solar spectrum at sea level with the simulated solar spectrum produced by the xenon 2500W lamp system with quartz optics and the Aerospace Control Corp. filter ER 7187. Relative irradiance is plotted along the ordinate vs. wavelength in nm along the abscissa.

insures that the spectral quality of the xenon lamp output will remain constant at all times. The focussed beam provided 5 W over the spectral range 300-1400 nm. Although the ocular media of the human and rhesus eye do not transmit wavelengths below 400 nm, inclusion of the near ultraviolet is necessary to maintain the transmission characteristics of the photochromic glasses used in this study. The simulated solar spectrum, SSS, is obtained by passing the focussed beam from the xenon arc through a specially designed reflection filter (Aerospace Control Corp. filter, ER 7187). A comparison of the SSS spectrum with the solar spectrum at sea level is shown in Fig. 1. The solar spectrum was taken from Gast (27) and from Thekaekara (28); it represents optimum irradiance at sea level with the sun at zenith on a clear day. The SSS was measured at 10 nm bandwidths with a high intensity Bausch and Lomb monochromator provided with 3 separate gratings, ruled for the near ultraviolet, visible and near infra-red respectively. The monochromator was calibrated with a spectral standard of irradiance as a source and an Eppley black body receiver as a detector. The latter was calibrated against a standard of irradiance from the Bureau of Standards. Aside from some excess in the near ultraviolet and a small deficiency in the near infra-red, the SSS spectrum agrees reasonably well with the solar spectrum.

The sun subtends an angle of 0.00931 radians at the surface of the earth. When the human eye is focussed on the sun, the retinal image has a diameter, D_r, of 159 μm. To provide the same diameter, D_r, on the rhesus retina, it is necessary to increase the beam divergence entering the pupil from 9.31 to 11.8 milliradians, mr, corresponding to the difference in focal length between human and primate eye, namely, 17.1 vs. 13.5 mm, respectively.

The maximum solar irradiance at sea level on a clear day with the sun at zenith is 0.0788 W/cm^2 for the spectral range 300-1400 nm. Irradiance, E, on the retina as a function of pupillary diameter, D_p, can be calculated if the integrated transmittance, T, for the ocular media (cornia, aqueous, lens and vitreous) is known for a given spectral distribution. Thus,

$$E = \frac{0.0788 \times T \times D_p^2}{D_r^2} \quad \text{in } W/cm^2 \qquad [1]$$

T is 0.726 for the human ocular media and for the solar spectral distribution at sea level. For the rhesus ocular media and the SSS spectral distribution, T is 0.698. Table 1 gives the retinal irradiance as a function of pupillary diameter for both the human retina exposed to the sun at sea level and for the monkey retina exposed to the SSS spectrum. Estimates of maximum temperature at the center of the lesion are also given for both species, utilizing the calculations of White et al. (22) which predict a temperature rise of $1^{o}C$ above ambient for a retinal irradiance of 6.77 W/cm^2.

TABLE 1

Power entering both human and monkey eye as a function of pupillary diameter for a solar irradiance at sea level of 78.8 $mW \cdot cm^{-2}$. Human retinal irradiance is based on an integrated transmittance through the ocular media for the solar spectrum of 0.726, while the monkey irradiance is based on an integrated transmittance through the ocular media of 0.698 for the SSS spectrum. Maximum retinal temperature is estimated from the mathematical model of White et al. (22). Spot size on the retina 0.159 mm.

Pupil Diameter (D_p in mm)	Power entering Cornea (P_c in mW)	Retinal Irradiance Human (E in $W \cdot cm^{-2}$)	Temp. (^oC)	Retinal Irradiance Monkey (E in $W \cdot cm^{-2}$)	Temp. (^oC)
8	39.6	146.7	21.7	141.2	20.9
7	30.3	112.3	16.6	108.1	16.0
6	22.5	83.3	12.3	79.4	11.7
5	15.5	57.3	8.5	55.2	8.2
4	9.9	36.7	5.4	35.3	5.2
3	5.6	20.6	3.0	19.9	2.9
2	2.5	9.1	1.3	8.8	1.3
1.5	1.4	5.1	0.75	5.0	0.74

Animals scheduled for retinal exposures are first tranquilized with 1 mg/kg (i.m.) phencyclidine hydrochloride; pupillary dilation to 8 mm or more is accomplished with cyclopentolate hydrochloride and phenylephrine; deep anesthesia follows intravenous injection of sodium pentobarbital (15 mg/kg). Hypothermia under anesthesia is prevented by enveloping the animal in a thermal blanket controlled electronically by a rectal thermometer. The lids are kept open by means of a speculum, while frequent applications of physiological saline prevent drying of the cornea.

Exposure of the retina at any desired location is accomplished by maneuvering the animal, secured firmly to a stereotaxic platform with 6 degrees of freedom for manipulation. A fundus camera is fixed rigidly at 90° and aligned with the optic axis of the radiation beam, which passes through a hole in the center of a mirror aligned at 45° to the fundus camera and the exposed eye. This allows observation of the fundus during exposure. The entire radiation beam enters the eye with a divergence of 11.8 mr or approximately 0.68 degrees. The power entering the eye is measured before and after exposure with a Scientech calorimeter calibrated at the Bureau of Radiological Health and power output is monitored and recorded during exposure. Preselected exposure times from 1 to 1000 s are made with an electronic timer which controls an electronic shutter in the beam.

The procedure employed to determine a minimal lesion has been described elsewhere (20) in some detail. Briefly, it consists of reducing the power input by known amounts until no lesion is funduscopically visible. At least two more exposures are made at known levels of power below the previous levels. When the fundus is examined again at 24 and 48 hours postexposure, lesions will range from mild to minimal to complete absence. Interpolation makes it possible to estimate accurately the power needed to produce a minimal lesion. Typically, it requires from 5 to 7 exposures to determine a minimal lesion; because the lesion size is small it is possible to place from 50 to 70 exposures in a single eye. This method avoids the large expenditure of animals needed to do a Probit analysis. Furthermore, it provides an accurate estimate of variation in pigmentation from animal to animal. The corneal power needed to produce a minimal lesion at each exposure time is the average of determinations in 16 eyes from 8 animals.

Once the corneal power P_c to produce a lesion has been determined, the retinal irradiance, E, and radiant exposure, H, can be calculated, provided the transmittance, T, through the ocular media and the spot diameter, D_r, of the beam on the retina are known.

$$E = \frac{4P_c \times T}{\pi D_r^{2}} \quad \text{in W/cm}^2 \qquad [2]$$

$$\text{and} \qquad H = E \times t \qquad \text{in J/cm}^2 \qquad [3]$$

where it is the exposure time in seconds. Transmittance through the ocular media, as measured in this laboratory on 9 enucleated eyes from 5 rhesus monkeys, is shown in Fig. 2 where the standard deviation for each 10 nm bandwidth from 350-1400 nm is given. In all cases these transmittances are higher than those measured by Boettner and Wolter (29) and by Maher (30). A comparison of the mean transmittances through the ocular media of man (32 eyes), monkey (9 eyes) and rabbit (48 eyes) over the spectral range 350-1400 nm as measured in this laboratory are shown in Fig. 3. The spot diameter on the retina is determined by the product of the angular divergence in radians (0.0118) and the focal length of the rhesus eye (13.5 mm), namely, 0.159 mm or 159 μm. The xenon arc source provides a remarkably flat profile of power density normal

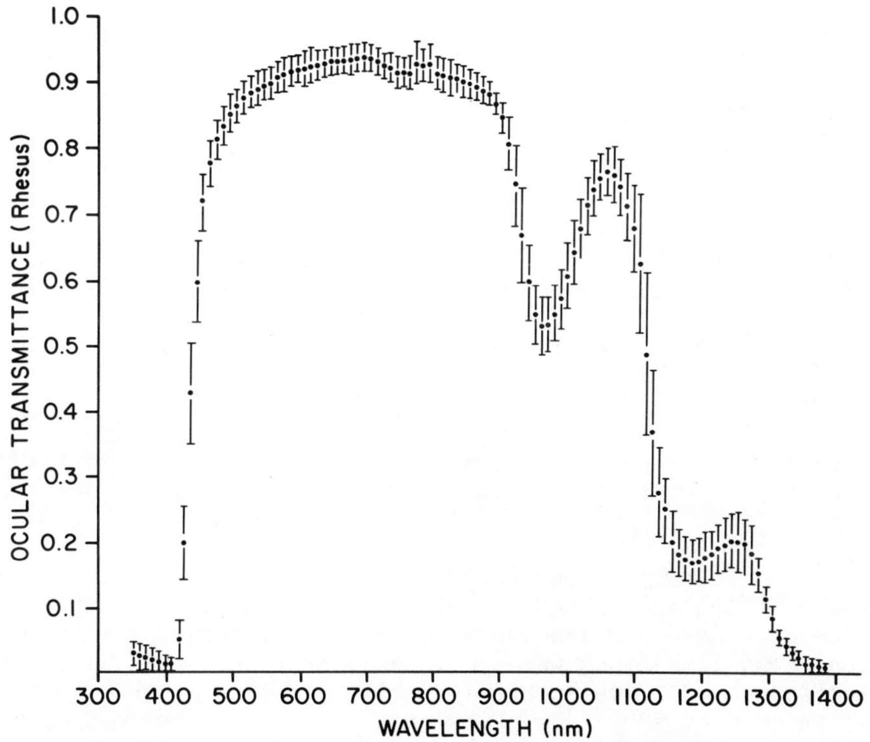

Figure 2. Transmission through the ocular media (cornea, aqueous, lens and vitreous) of the rhesus monkey as measured on 9 eyes from 5 animals. Transmittance is plotted along the ordinate vs. wavelength in nm on the abscissa. Standard deviation for each 10 nm bandwidth is represented by the vertical bar.

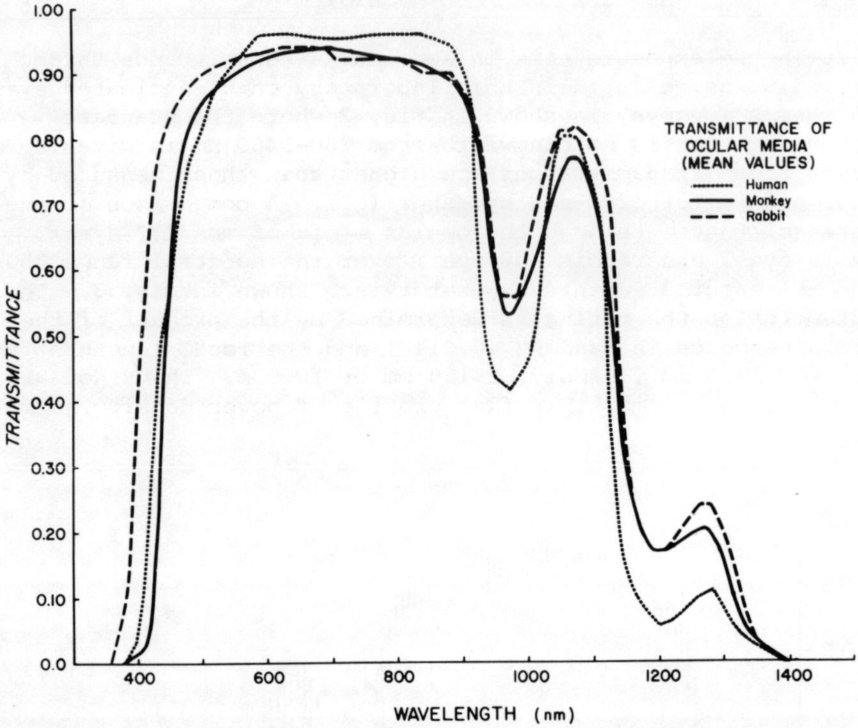

Figure 3. A comparison of the transmittance through the ocular media (cornea, aqueous, lens and vitreous) vs. wavelength as measured for 32 human eyes, 9 rhesus monkey eyes and 48 rabbit eyes. Transmittance is plotted along the ordinate vs. wavelength in nm along the abscissa.

to the beam, resulting in uniform retinal irradiance across the spot diameter, rather than a peaked or Gaussian distribution of energy.

It was necessary to investigate the spectral characteristics of 5 types of Corning ophthalmic glass during exposure to the SSS source before these glasses could be tested for their protective features. The five types were designated as W (crown #8361), N (neutral #8364), P-1 (photochromic, Aton Grey #8101), P-2 (photochromic Photo Grey #8097) and P-3 (photochromic, Photosun Grey #8098). The W and N glasses which transmit 90.4 and 13.5 percent of sunlight respectively have fixed spectral characteristics which do not vary with radiation exposure or temperature. However, the photochromic glasses, P-1, P-2 and P-3, transmitting under optimal conditions 45.1, 58.7 and 47.0 percent sunlight, have transmittances which depend markedly upon previous exposure history, the intensity and spectral distribution of the SSS source and the temperature of the glass during exposure.

To control these variables it was necessary to impose a strict schedule of pre-exposure history and temperature regulation. Eight samples of each glass were exposed for 2 hours per day for 10 days to black light bulb (BLB) radiation at room temperature before exposure to the SSS source. A temperature of 77°F was chosen as representative of average summer conditions and maintained during exposure to the SSS source by means of a water-cooled holder regulated by an attached thermocouple. Each glass sample was allowed to come to minimal transmittance during exposure to the SSS source before the spectral transmittance was measured from 300-1400 nm in 10 nm bandwidths by means of the monochromator using an Epply cell as a black body detector. After exposure each glass sample was placed in boiling water for one hour to fade out all previous exposure history. The sample was then subjected again to the 10 day BLB regimen before another exposure to the SSS source. In this manner the average spectral transmittance of 8 samples of each photochromic glass and the W and N glasses was determined. In Figs. 4, 5 and 6 spectral distribution of the solar spectrum on the human to that of the SSS source on the monkey retina before and after filtration by the 5 ophthalmic glasses furnished by the Corning Glass Works. The close correspondence of the two spectral distributions warrants reasonable confidence in the scientific validity of the SSS source as a laboratory substitute for the solar spectrum.

RESULTS

The Simulated Solar Spectrum (SSS), 300-1400 nm.

Knowing the spectral distribution of the SSS source and the transmission characteristics of each glass for the SSS source, it now was possible to measure and compare the protective features of each glass. This was done by determining the corneal power needed to produce a minimal lesion on the rhesus retina after attenuation by each of the 5 glasses. Exposure times ranged from 1 to 1000 s. Although the SSS source was capable of producing a greater power density than the solar radiation at sea level, the power entering the eye was limited to that for direct sun-gazing with an 8 mm pupillary diameter. For the spectral range 300-1400 nm this was 78.8 mW/cm^2 multiplied by the area of an 8 mm pupil, or 39.6 mW.

It requires complete dark adaptation for at least half an hour or the use of a mydriatic drug to produce an 8 mm pupil in normal human subjects. Eccles and Flynn (6) measured the pupillary diameter on ten individuals exposed to bright sunlight. The average

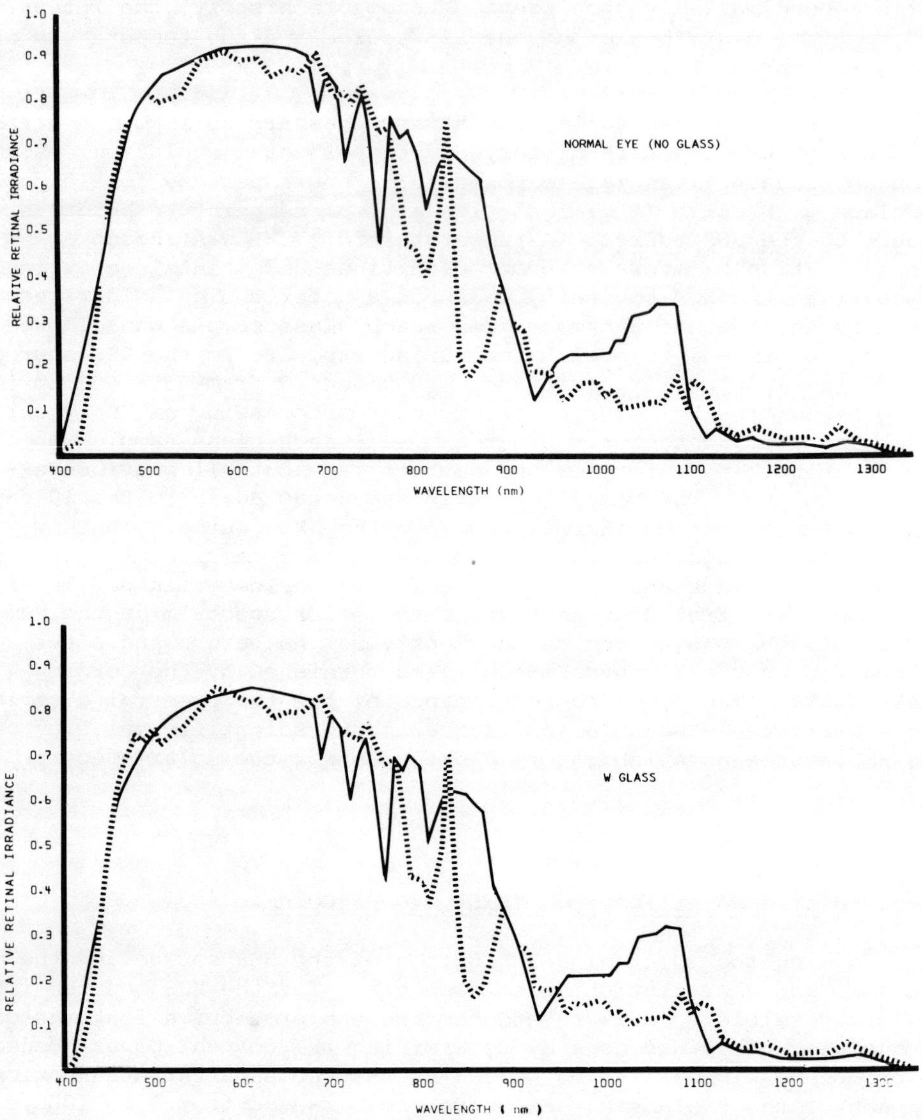

Figure 4. A comparison of the solar spectral distribution on the human retina with the SSS distribution on the rhesus monkey retina. The solid lines represent the solar radiation at sea level, the broken lines the SSS source. The upper figure is a comparison with no filtration. The bottom figure compares the two distributions after filtration by the W glass. Relative irradiance is plotted along the ordinate vs. wavelength along the abscissa.

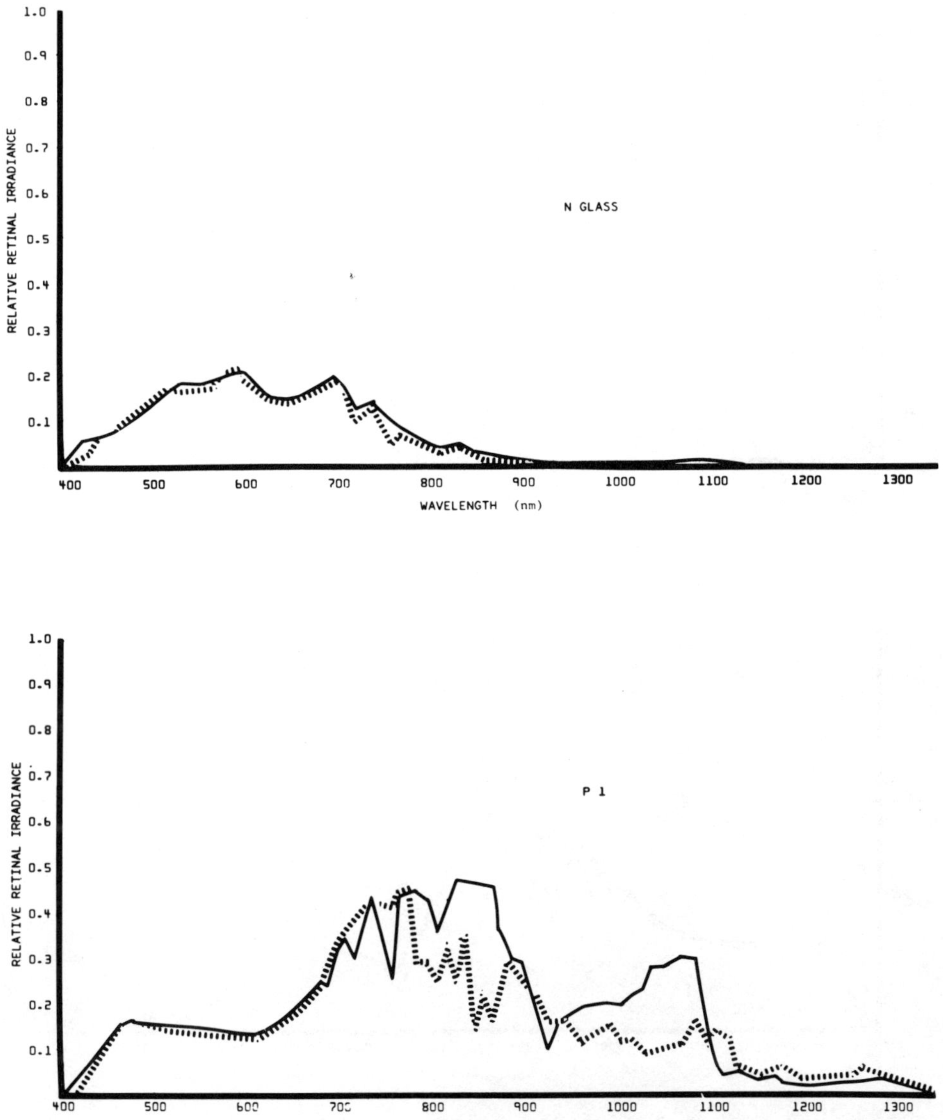

Figure 5. A comparison of the solar spectral distribution on the human retina with the SSS distribution on the rhesus monkey retina. The solid lines represent the solar radiation at sea level, the broken lines the SSS source. The upper figure is a comparison after filtration by the N glass, the lower figure after filtration by the P-1 glass. Relative retinal irradiance is plotted along the ordinate vs. wavelength in nm along the abscissa.

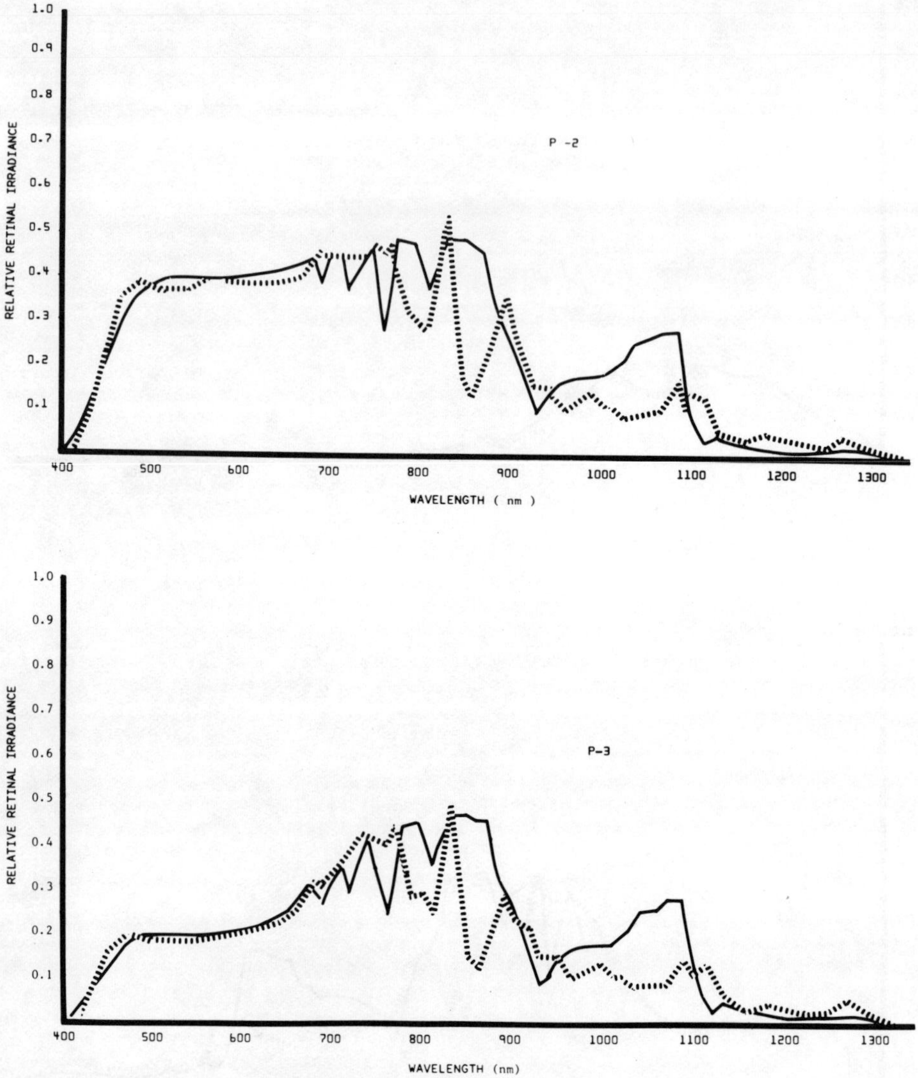

Figure 6. A comparison of the solar spectral distribution on the human retina with the SSS on the rhesus monkey retina. The solid line represents the solar radiation at sea level, the broken lines the SSS source. The upper figure is a comparison after filtration by the P-2 glass, the lower figure after filtration by the P-3 glass. Relative retinal irradiance is plotted along the ordinate vs. wavelength in nm along the abscissa.

diameter was 1.6 mm and the range was 1.5 to 1.8 mm. They also measured the pupillary diameter through welders' goggles while viewing the sun. The average pupillary diameter was 2.6 mm and the range 2.0-3.0 mm. To further test the effect of filters on pupillary diameter, 10 individuals varying in age from 3 to 49 years were measured while wearing each of the 5 glasses in bright sunlight. In no case did the pupillary diameter exceed 3 mm.

The results of exposing 8 monkeys (16 eyes) to the SSS source through the 5 glasses at exposure times ranging from 1 to 1,000 seconds are summarized in Table 2. The corneal power, P_C, and retinal irradiance, E, required to produce a minimal lesion are listed with standard deviation in mW and $W \cdot cm^{-2}$ respectively. Each power and irradiance listed is the mean of 8 determinations, each on a different animal. The maximum temperature at the center of the irradiated area during exposure is given also as estimated from the mathematical model of White et al. (22).

The relationship between the data given in Table 2 and irradiance on the human retinal when viewing the sun is shown in Fig. 7 where retinal irradiance in $W \cdot cm^{-2}$ is plotted against pupillary diameter in mm using the data in Tables 1 and 2. For example, the maximum energy from the sun at sea level that can enter the eye through a 8 mm pupillary diameter is 39.6 mW. This is reduced by the transmittance of a specific glass for the SSS source and by the integrated transmittance of the monkey ocular media for the SSS spectral distribution after passage through the glass. Thus, knowing the power which reaches the retina and the area of the radiation beam on the retina, the retinal irradiance in $W \cdot cm^{-2}$ can be calculated. This has been calculated for each individual glass and plotted on Fig. 7 for comparison with the human retinal irradiance vs. pupillary diameter for the solar spectrum at sea level. The crosses superimposed on the curve of each glass represent the irradiance required to produce a minimal lesion in a given exposure time. Minimal lesions could not be produced in one second through any of the glasses except the W (91.5% transmittance) without exceeding the retinal irradiance from the sun through a 8 mm pupil. Exposures longer than 100 do not appear on the curves because the retinal irradiances are below those passing through a 1.5 mm pupillary diameter.

The Simulated Solar Spectrum (SSS), 400-800 nm.

The SSS source between 400 and 800 nm is obtained by inserting a "hot" mirror at normal incidence to the beam and reflecting the transmitted beam into the monkey eye with a "cold" mirror. The "hot" mirror reflects wavelengths greater than 800 nm and the "cold" mirror reflects wavelengths between 400 and 800 nm but not the near ultraviolet. The photochromic glasses intercepted the radiation beam

TABLE 2

Corneal power P_c in mW and retinal irradiance E in $W{\cdot}cm^{-2}$ (mean ± Std. Dev.) on 8 eyes from 8 different monkeys required to produce a minimal lesion as observed funduscopically 48 hours postexposure are listed in columns 1 and 2 respectively. Column 3 gives the exposure times in seconds to SSS source (300-1400 nm) filtered by the 5 different glasses which are designated in column 4. Column 5 gives the estimated maximum temperature rise in ^{o}C during exposure as calculated from White et al. (22). Retinal image diameter, 159 μm in all cases.

Power Entering Cornea (P_c in mW)	Retinal Irradiance (E in $W{\cdot}cm^{-2}$)	Exposure Time (seconds)	Glass Type	Temperature (^{o}C)
30.5 ± 2.6	109.1 ± 9.3	1	W	16.4
17.7 ± 1.8	63.1 ± 6.4	10	W	9.5
21.8 ± 2.7	76.0 ± 9.4	10	P-2	11.4
8.0 ± 2.3	27.8 ± 8.0	30	P-2	4.2
11.3 ± 1.2	37.0 ± 4.5	30	P-1	5.7
13.1 ± 1.9	45.7 ± 6.6	30	P-3	6.9
4.4 ± 0.56	15.5 ± 2.0	30	N	2.3
4.0 ± 1.1	13.8 ± 3.8	60	P-2	2.1
6.9 ± 1.2	23.1 ± 4.8	60	P-1	3.5
7.0 ± 1.6	24.1 ± 5.6	60	P-3	3.7
2.1 ± 0.11	7.4 ± 0.4	60	N	1.1
2.0 ± 0.17	7.1 ± 0.6	100	W	1.1
2.0 ± 0.94	7.1 ± 3.3	100	P-2	1.1
3.0 ± 0.7	10.0 ± 2.7	100	P-1	1.5
2.7 ± 0.6	9.4 ± 2.1	100	P-3	1.4
1.2 ± 0.38	4.2 ± 1.3	100	N	0.6
0.23 ± 0.04	0.8 ± 0.1	1000	W	0.11
0.23 ± 0.09	0.8 ± 0.3	1000	P-2	0.11
0.24 ± 0.06	0.8 ± 0.2	1000	P-1	0.11
0.28 ± 0.1	0.91 ± 0.3	1000	P-3	0.13
0.09 ± 0.02	0.29 ± 0.07	1000	N	0.04

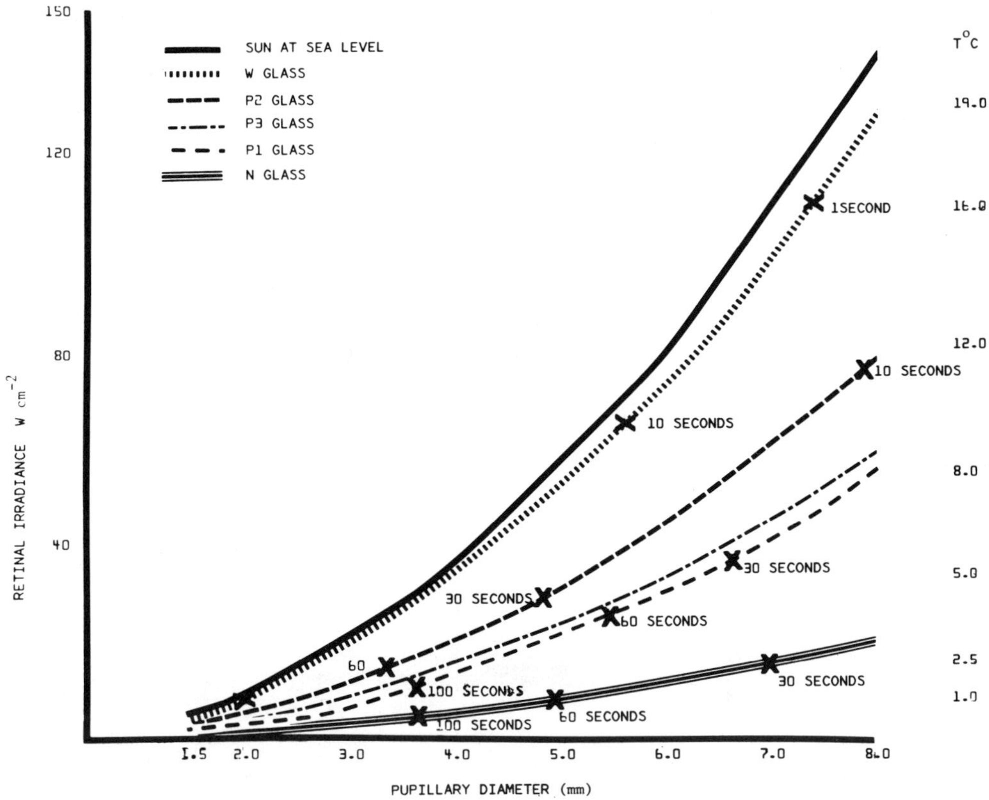

Figure 7. Retinal irradiance in W•cm^{-2} is plotted on the left ordinate vs. pupillary diameter in mm on the abscissa for the sun at sea level (300-1400 nm) on the human retina and for the SSS source (300-1400 nm) on the monkey retina after transmission through each of the 5 glasses. Estimated temperature rise in °C above ambient for a given retinal irradiance is plotted on the right ordinate. Crosses represent retinal irradiance required to produce a minimal lesion by the 300-1400 nm SSS source after transmission through each of the 5 glasses. Exposure times in seconds are indicated beside each cross.

before it encountered the "hot" mirror. This was necessary to maintain minimum transmittance through the photochromic glasses. The ideal situation would have been to divide the SSS source into the spectral bands 300-700 nm and 700-1400 nm. Unfortunately, this was not possible because both "hot" and "cold" mirrors reflect the near ultraviolet.

The results from the exposure of 8 monkeys (16 eyes) to the 400-800 nm SSS source are given in Table 3 which is similar in

TABLE 3

Corneal power P_c in mW and retinal irradiance E in $W \cdot cm^{-2}$ (mean ± Std. Dev.) on 8 eyes from 8 different monkeys required to produce a minimal lesion as observed funduscopically 48 hours postexposure are listed in columns 1 and 2 respectively. Column 3 gives exposure times in seconds to visible spectrum (400-800 nm) filtered by the 5 different glasses designated in column 4. Column 5 gives estimated maximum temperature in °C during exposure as calculated from White et al. (22). Retinal image diameter, 159 μm in all cases.

Power Entering Cornea (P_c in mW)	Retinal Irradiance (E in $W \cdot cm^{-2}$)	Exposure Time (Seconds)	Glass Type	Temperature (°C)
17.7 ± .96	71.7 ± 3.9	1	W	10.6
11.2 ± .89	45.3 ± 3.6	10	W	6.7
3.84 ± .14	14.6 ± 0.57	30	P-2	2.2
3.13 ± .18	12.5 ± 0.72	30	P-1	1.8
3.40 ± .14	14.0 ± 0.58	30	P-3	2.1
3.52 ± .08	13.5 ± 0.31	30	N	2.0
1.99 ± .64	7.6 ± 2.6	60	P-2	1.1
1.70 ± .33	6.8 ± 1.3	60	P-1	1.0
1.81 ± .14	7.9 ± 0.58	60	P-3	1.2
2.14 ± .20	9.4 ± 0.77	60	N	1.4
0.92 ± .04	3.7 ± 0.16	100	W	0.55
1.18 ± .10	4.5 ± 0.41	100	P-2	0.7
0.99 ± .06	3.9 ± 0.24	100	P-1	0.6
0.95 ± .04	3.9 ± 0.16	100	P-3	0.6
1.44 ± .11	5.5 ± 0.42	100	N	0.8
0.18 ± .03	0.73 ± 0.12	500	W	0.11
0.22 ± .02	0.84 ± 0.08	500	P-2	0.11
0.18 ± .03	0.73 ± 0.01	500	P-1	0.11
0.13 ± .01	0.53 ± 0.04	500	P-3	0.11
0.15 ± .03	0.59 ± 0.11	500	N	0.11

format to Table 2. Corneal power, retinal irradiance, exposure time, type of glass and temperature are given in the same units as in Table 2. Exposures beyond 500 seconds in duration were not made for any of the glasses because the retinal irradiances required to produce a minimal lesion were less than those permitted by a 1 mm pupil when gazing directly at the solar waveband between 400-800 nm. Again, 1 and 10 second exposures produced minimal lesions through the W glass but not after filtration by the photochromic glasses or the N glass.

The data in Table 3 are shown graphically in Fig. 8 which is a plot of retinal irradiance vs. pupillary diameter for each glass

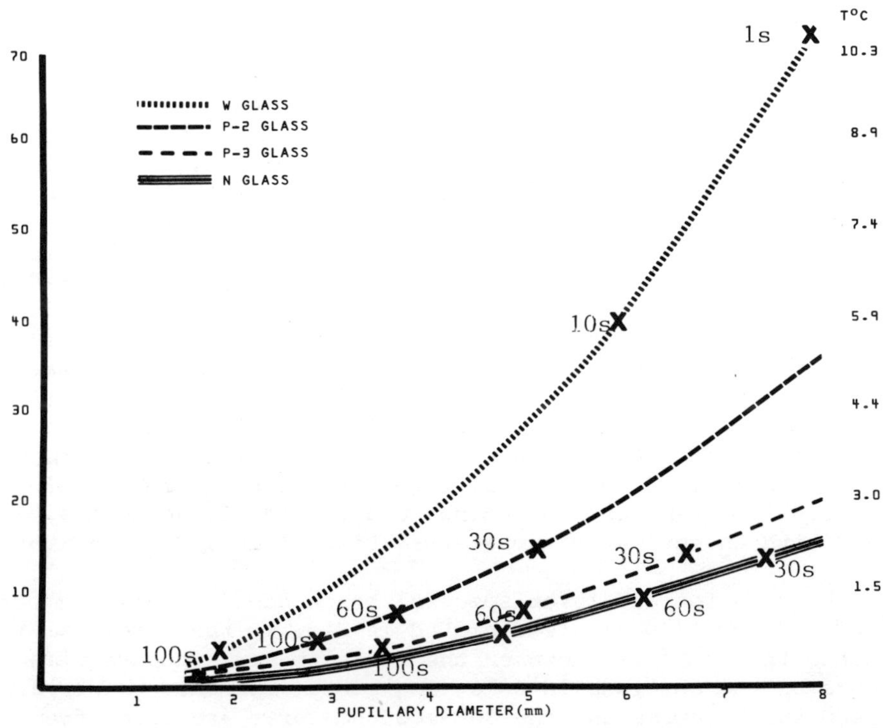

Figure 8. Retinal irradiance in W•cm^{-2} is plotted on the left ordinate vs. pupillary diameter in mm on the abscissa for the SSS source between 400-800 nm on the monkey retina after transmission through the 4 glasses W, P-2, P-3 and N. Estimated temperature rises are plotted on the right ordinate. Crosses represent retinal irradiance required to produce a minimal lesion by the 400-800 nm SSS source after transmission through each of the 4 glasses. Exposure times in seconds are indicated beside each cross.

except the P-1 glass. The data for the P-1 glass are given in Table 3 but not plotted on Fig. 8 because the P-1 curve overlaps the N curve. The irradiance scale is reduced to half the scale in Fig. 7 because the irradiances needed to produce a minimal lesion are approximately one-half those required by the 300-1400 nm spectrum. Crosses on each curve represent irradiances needed to obtain a minimal lesion in a given exposure time. Minimal lesions were obtainable only through the W glass for 1 and 10 second exposures without exceeding the retinal irradiances from the sun through a 8 mm pupil for the waveband 400-800 nm. The 500 second exposure data do not appear on the graph because the irradiances are below those passed by a 1 mm pupil exposed directly to the solar waveband between 400 and 800 mm.

The Simulated Solar Spectrum (SSS), 700-1400 nm.

The SSS source between 700 and 1400 nm was produced by inserting a RG-715 Jena glass filter in the radiation beam. The spectral distribution of this near-infra-red source is shown in Fig. 9 where relative irradiance is plotted on the ordinate vs. wavelength in nm on the abscissa. Also shown is the spectral distribution on the rhesus retina after filtration by the ocular media.

Attempts to produce a minimal lesion with this infra-red source were completely negative for exposure times up to 500 seconds in duration. The power incident on the cornea was limited to 19.4 mW which corresponds to the solar power at sea level for the waveband 700-1400 nm passed through a 8 mm pupil. An exposure time of 1000 seconds produced a minimal lesion in 3 out of 8 animals. In order to obtain lesions in the other 5 animals it was necessary to go slightly above 19.4 mW. The average corneal power for 8 animals was 19.78 ± 0.36 mW corresponding to a retinal irradiance of 69.1 $W \cdot cm^{-2}$ with an estimated temperature rise of 10.3°C above ambient.

The data for all three spectral bandwidths, 300-1400 nm, 400-800 nm and 700-1400 nm are depicted in Fig. 10 where retinal irradiance in $W \cdot cm^{-2}$ is shown on the ordinate vs. exposure time in seconds on the abscissa in a log-log plot. The data points along the 400-800 nm curve and the 300-1400 nm curve are taken from exposures through the W glass which is Crown glass with a flat transmittance of 0.915 throughout the visible and near infra-red spectrum. The 4 crosses representing exposure times of 10, 30, 60 and 180 seconds plotted on the 700-1400 nm curve are data taken from Ham et al. (20). The data point at 1000 seconds represents the mean irradiance for 8 animals exposed to the spectral bandwidth 700-1400 nm.

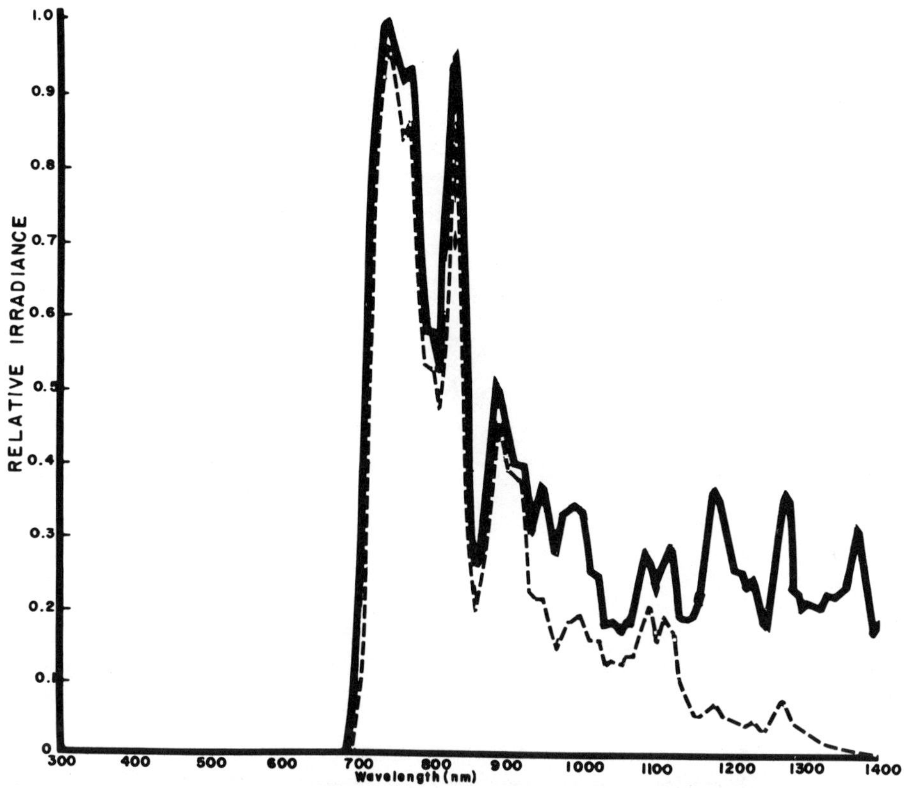

Figure 9. Relative irradiance is plotted on the ordinate vs. wavelength in nm on the abscissa for the SSS infra-red source (700-1400 nm) produced after transmission through an RG-715 Jena glass filter. The solid curve represents the spectral distribution at the cornea, the dotted curve the spectral distribution on the rhesus retina.

DISCUSSION

Table 4 provides a convenient comparison of the data reported in the results under A and B for the spectra 300-1400 nm vs. 400-800 nm. Here are listed the radiant exposures in J/cm^2 required to produce a minimal lesion in exposure times of 10 seconds or longer for each of the 5 glasses. The upper set of data give radiant exposures for the SSS distribution 300-1400 nm, while the lower set of data provides the same for the SSS distribution 400-800 nm. The average or mean radiant exposure for each glass is given at the base of each column. It is immediately apparent that with the exception of the N glass the radiant exposures required by the 300-1400 nm spectral distribution are about twice those required by the 400-800 nm distribution. Approximately 50% of the energy between

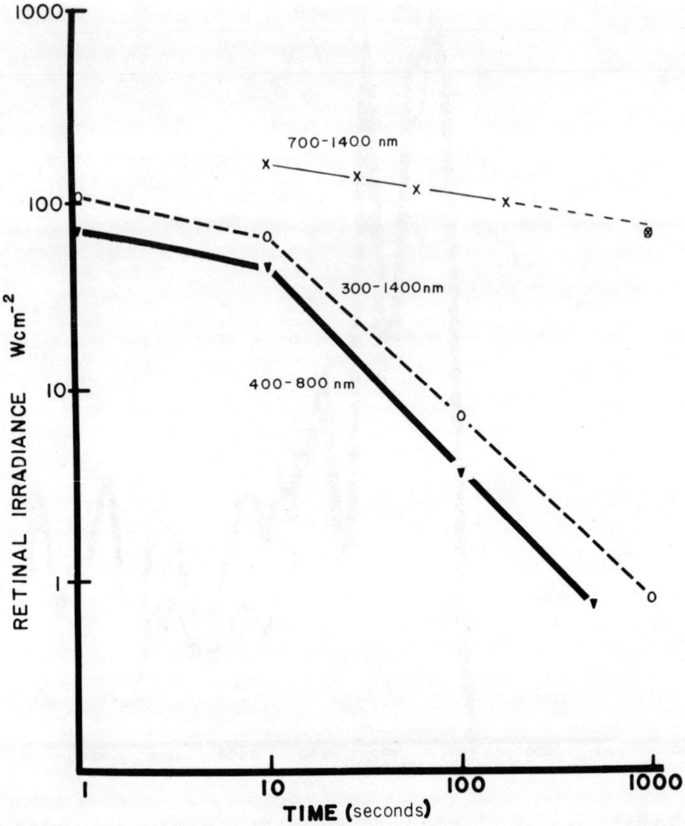

Figure 10. A log-log plot of retinal irradiance in $W \cdot cm^{-2}$ on the ordinate vs. exposure time in seconds on the abscissa for minimal lesions produced by the three spectral distributions of the SSS source, 300-1400 nm, 400-800 nm and 700-1400 nm.

400-1400 nm resides in the wavelength interval 400-800 nm, while the other 50% is in the 800-1400 nm waveband. The conclusion is that the long wavelength component in the 300-1400 nm spectrum produces negligible damage to the rhesus retina.

A glance at Fig. 5 explains why the N glass has approximately equal radiant exposures for both spectral distributions. Only a small amount of near infra-red radiation reaches the rhesus retina through the N glass. In Fig. 6 it can be seen that, with the exception of the W crown glass and the photochromic P-2 glass, exposures of 100 seconds or less would require pupillary diameters greater than 3 mm to produce a lesion when wearing the N glass or the P-1 and P-3 photochromic glasses. Pupil diameters greater than

TABLE 4

Radiant exposure in $J \cdot cm^{-2}$ required to produce a minimal lesion for exposure times of 10 seconds or longer are listed for each glass for the SSS source, 300-1400 nm and the SSS source, 400-800 nm. The mean radiant exposure is given at the bottom of each column.

Exposure Time	SSS Source, 300-1400 nm Radiant Exposure For Each Glass in $J \cdot cm^{-2}$				
(Seconds)	W	P-1	P-2	P-3	N
10	633		760		
30	---	1170	834	1371	465
60	---	1386	824	1464	444
100	710	1000	710	940	420
1000	800	800	810	910	290
Avg.	714	1074	788	1171	405

	SSS Source, 400-800 nm				
	W	P-1	P-2	P-3	N
10	453				
30	---	375	438	420	405
60	---	408	456	474	564
100	370	390	450	390	550
500	---	365	420	265	295
1000	365	---	---	---	---
Avg.	396	385	441	387	454

3 mm are unlikely for subjects in sunlight even when wearing sunglasses. To receive damage through the N glass in 30 seconds would require a 7 mm pupil. It should be emphasized, however, that no sunglass provides protection from direct viewing of the sun for extended periods of time. Sunglasses were never designed nor intended for sun gazing. It would require an optical density of at least 3 to gaze safely at the sun.

A comparison of Fig. 7 and Fig. 8 or Tables 2 and 3 shows that temperatures in the retina during exposure are much higher for the 300-1400 nm spectrum than for the 400-800 nm spectrum. For example, in Table 3, temperature rises never exceed 2.2°C for exposures of 30 seconds or longer. This can only mean that extended exposures to the 400-800 nm bandwidth of solar radiation produce photic damage which is photochemical in nature. The solid line in Fig. 7 shows, that for pupillary diameters of 3 mm or less, the temperature rise in the retina will not exceed 4°C even if no protective filter is used. Again it can be concluded that solar retinitis is a photochemical phenomenon having little or nothing to do with thermal injury. It should be noted, however, that a temperature rise of 3° to 4°C above ambient in the retina may produce thermally enhanced photochemical damage.

Exposure of the rhesus retina to the near infra-red spectrum 700-1400 nm shown in Fig. 9 provides conclusive evidence that infrared radiation in the solar spectrum cannot produce a retinal lesion unless one gazes directly at the sun for 1000 seconds with a 8 mm pupil. If the wavelengths below 700 nm in solar radiation are removed with a filter like the RG-715 Jena glass filter, direct sun gazing can be tolerated for appreciable periods of time. In Fig. 10 the sharp bend in the two lower curves can be interpreted as denoting different mechanisms of retinal injury. For exposures through the W crown glass of 10 or less, temperature rises in the retina exceed 10°C for both the 300-1400 nm spectrum and the 400-800 nm spectrum. The type of retinal damage at these short exposure times is primarily thermal in nature though there is a significant contribution by the shorter wavelengths to photochemical damage as well. This has been noted in a previous publication by Ham et al. (31) where one second exposures to 1064 nm radiation required a retinal temperature of 55°C to produce a minimal lesion, whereas 514.5 nm light required only 25°C to produce a minimal lesion. Both lesions are thermal in character but there must be a photochemical contribution to the damage by 514.5 nm light to account for the differences in temperature. In Fig. 10, the sharper slope for exposure times beyond 10 seconds indicates a different mechanism which is presumably photochemical in nature. Note, however, that the graph for the 700-1400 nm spectrum shown for exposure times of 10, 30, 60, 180 and 1000 seconds is a straight line entirely in the thermal region where the temperature rises on the retina during exposure are well above 10°C.

In summary, solar retinitis and eclipse blindness are photochemical phenomena and near infra-red solar radiation makes only a negligible contribution to retinal damage. Sunglasses can provide protection to both the retina and the lens for the long-term chronic effects of exposure to sunlight but they do not provide protection from direct sun gazing.

ACKNOWLEDGEMENTS

This research was supported jointly by the Corning Glass Works and the U.S. Army Medical Research and Development Command. In conducting the research described in this report, the investigators adhered to the "Guide for Laboratory Animal Facilites and Care" of the Committee on the Guide for Laboratory Animal Resources, National Academy of Sciences - National Research Council.

REFERENCES

1. Verhoeff, F. H. and Bell, L. (1916) Proc. Acad. Arts Sci. 51, 629-759.
2. Cordes, F. C. (1948) Am. J. Ophthalmol. 31, 101-107.
3. Marlor, R. L., Blair, B. R., Preston, F. R. and Boyden, D. G. (1973) Invest. Ophthalmol. 12, 5-16.
4. Smith, H. E. (1944) U. S. Navy Med. Bull. 42, 675-680.
5. Czerny (1867) K. Akad. d. Wissensch. 6, Oct.
6. Eccles, J. C. and Glynn, A. J. (1944) Med. J. Australia 1, 339-342.
7. Ham, W. T., Jr., Wiesinger, H., Guerry, D., III, Schmidt, F. H. Williams, R. C., Ruffin, R. S. and Shaffer, M. C. (1957) Am. J. Ophthalmol. 43, 711-718.
8. Ham, W. T., Jr., Williams, R. C., Mueller, H. A., Guerry, D., III, Clark A. M. and Geeraets, W. J. (1966) Trans. N. Y. Acad. Sci. 28, 517-526.
9. Hatfield, E. M. (1970) Sight-Saving Rev. 40, 79-86.
10. Penner, R. and McNair, J. N. (1966) Am. J. Ophthalmol. 61, 1452-1457.
11. Duke-Elder, Sir Stewart and MacFaul, Peter A. (1972) in System of Ophthalmology, Vol. XIV, part 2, p. 885, C. V. Mosby Co.
12. Friedman, E. and Kuwabara, T. (1968) Arch. Ophthalmol. 80, 265-279.
13. Noell, W. K., Walker, V. S., Kang, B. S. and Berman, S. (1966) Invest. Ophthalmol. 5, 450-473.
14. Lanum, J. (1978) Survey of Ophthalmol. 22, 221-248.
15. Vassiliadis, A. (1972) Contract RCS, Project 1649, Stanford Research Institute, Menlo Park, Ca.
16. Harwerth, R. S. and Sperling, H. G. (1971) Science 174, 520-522.
17. Sliney, D. H. and Freasier, B. C. (1973) Appl. Optics 12, 1-24.
18. Tso, M. O. M. (1973) Invest. Ophthalmol. 12, 17-34.
19. Lawwill, T. (1973) Invest. Ophthalmol. 12, 45-51; Lawwill, T., Crockett, S. and Currier, G. (1977) Documents Ophthalmologica, 44, 379-385.
20. Ham, W. T., Jr., Mueller, H. A., Williams, R. C. and Geeraets, W. J. (1973) Appl. Optics 12, 2122-2129.
21. Tso, M. O. M. and LaPiana, F. G. (1975) Tr. Am. Acad. Ophthalmol. 79, 788-795.
22. White, T. J., Mainster, M. A., Wilson, P. W. and Tips, J. H. (1971) Bull. Math. Biophysics 33, 1-17.
23. Ham, W. T., Jr., Mueller, H. A. and Sliney, D. H. (1976) Nature 260, 153-155.
24. Ham, W. T., Jr., Ruffolo, J. J., Jr., Mueller, H. A., Clarke, A. M. and Moon, M. E. (1978) Invest. Ophthalmol. and Vis. Sci. 17, 1029-1035.
25. Moon, M. E., Clarke, A. M., Ruffolo, J. J., Jr., Mueller, H. A. Ham, W. T., Jr. (1978) Vis. Res. 18, 1573-1577.
26. Cain, C. P. and Welch, A. J. (1974) Invest. Ophthalmol. 13, 60-70.

27. Gast, P. R. (1965) in Handbook of Geophysics and Space Environments, McGraw-Hill Book Co., chpt. 16.1-16.11.

28. Thekaekara, M. P. (1972) Optical Spectra, March, pp. 32-35.

29. Boettner, E. A. and Wolter, J. R. (1962) Invest. Ophthalmol. 1, 776-783.

30. Maher, E. F. (1978) Report SAM-TR-78-32, USAF School of Aerospace Medicine, Brooks AFB, TX, 78235.

31. Ham, W. T., Jr., Mueller, H. A., Ruffolo, J. J. and Clarke, A. M. (1979) J. Photochem. and Photobiol. 29, 4, 735-743.

IN VIVO DENSITY OF CONE PIGMENTS AFTER REPEATED COMPLETE BLEACHES

Howard D. Baker and Thomas K. Kuyk

Department of Psychology
Florida State University
Tallahassee, FL 32306

Even short durations of intense light may cause retinal damage, of the sort discussed by Dr. Ham and Dr. Lawwill in their papers for this symposium. (For a general review of light damage see Lanum, 1). One of the authors of the present paper, HDB, has a long history of repeated exposures to intense lights, that occurred during a program of measurements of his cone pigments by the technique of retinal reflection densitometry. In 1963, and again in 1968 he estimates his left eye was totally bleached over 100 times (2, 3). Then, beginning in the early 1970's the regimen became more regular with approximately 100 total bleaches of his left eye each year; between September and April this year's count already exceeds 100.

During the course of data collection for the research program we observed two features of this subject's data that led us to suspect that successive bleaching may have damaged his left eye, which was the eye that had always been bleached. First, we found that cone pigment density as measured by reflection densitometry appeared to decline during successive bleaches in a single session. Second, the regeneration time-constant was considerably longer during the past year compared to measurements made on the same eye in 1963 and in 1968. It is also longer compared to the time constant of other subjects measured recently on our device.

Apparatus

All of the pigment measurements for this paper were made on an apparatus that was a modification of Rushton's Florida State densitometer (4). The basic design can be seen in Fig. 1.

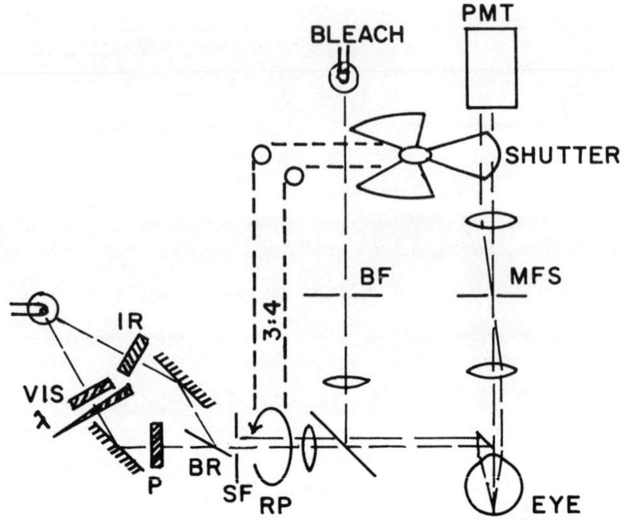

Figure 1. Plan of the modified Florida State retinal densitometer used in the present experiment. S, source of visible and infra-red comparison beams; P, sheet polarizer; BR, Brewster-angle plate; SF, field stop; RP, rotating polarizer to alternate the visible and infra-red beams; BF, bleaching field stop; PMT, photomultiplier tube; MFS, measuring field limiting stop.

Sinusoidally alternating visible test and infra-red reference beams originate from a tungsten-halide source, S, and eventually enter the pupil of the subject's eye. Both beams pass through the retina, are reflected off the back of the eye, pass back through the retina, and exit out of the pupil where they are picked up by a photomultiplier, PMT. Baseline determinations are made prior to bleaching by adjusting the intensity of the reflected visible beam to match the photomultiplier's response to the reflected infra-red reference.

Then the eye is bleached with 20,000,000 trolands of light from a tungsten-halide source for one minute. This bleaching light is intense enough to bleach away all measureable cone pigment, and it is of the same order of magnitude as that of indirect ophthalmoscope lamps, which have been suspected of causing retinal light damage (1). The bleaching light subtends a visual angle larger than the test beams, and all beams are presented in Maxwellian view.

Following the bleach, the visible beam is always found to be too bright and density must be added in that beam to rebalance the photomultiplier's response to the infra-red reference. The infra-red response is unchanged of course, since it is not affected by the visual pigment. We believe the density that must be added to the visible beam is that of the cone pigments, multiplied by two since the light passes through the retina twice.

The Problem

As mentioned previously, two features of the subject's data led us to consider that his history of repeated left-eye bleaching might have damaged that retina. In the first place, the 2.0 minute time-constant of regeneration in 1963 had extended to 3.24 minutes in 1979.

Second, the rule of thumb of allowing at least the equivalent of three time constants for the pigment to return to baseline (in this case 9.72 minutes), just didn't seem to be appropriate here. Instead we observed successive shifts in baseline level of pigment even though 12-15 minutes always were allowed for recovery between each bleach. These shifts can be seen in Fig. 2.

Baseline Shifts

The data in Fig. 2 were gleaned from a large number of experiments, and represent only those occasions when successive bleaches were made in one session under the same experimental conditions, with the usual 12-15 minutes between bleaches. By "conditions", we mean size, wavelength and location of the measuring stimulus; these variables affect the measured total density of the visual pigment. All of the data reported here are for centrally fixated

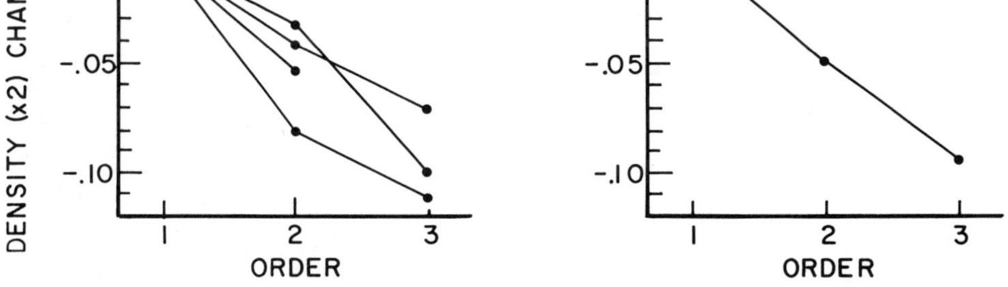

Figure 2. Shift of baseline density (x2) of foveal cone pigment during 3 successive bleach-and-recovery cycles under 4 conditions, with 12-15 minutes of recovery between each bleach. Significant: $p = .0078$.

test lights, and represent cone pigments. For ease of comparison the averages of the first baselines taken in a session are set to zero in Fig. 2, to illustrate better the change that was occurring. The direction of the shift is the important consideration, not the size, since each curve was taken under different conditions.

In the four conditions represented, the second and third baselines were always successively lower in terms of pigment density. That all comparisons are in one direction, down, is unlikely to be by chance. The probability of such a chance occurrence is 0.0078.

In deference to the growing literature on light damage, and after observing the results in Fig. 2, we began last summer requiring 30 minutes of recovery time between each bleach. During the last 15 minutes the subject always remained in a very dim light below cone threshold. This procedure resulted in the data of Figs. 3 and 4.

Figure 3 shows baseline measurements for 5 conditions in which successive bleaches were made on the same day, but with 30 minutes (not 12-15) for recovery between bleaches. The downward trend may still be present, but is much reduced. It is also not consistently down for all conditions. The downward trend is not statistically significant, because such a combination of upward and downward changes would be expected with a probability of 0.15 chance alone.

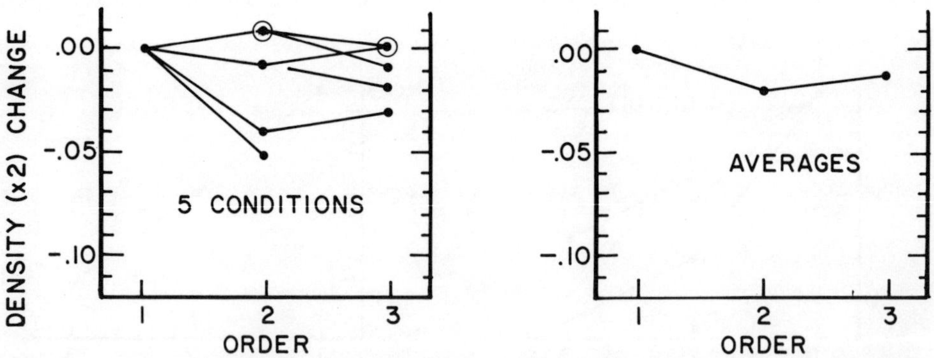

Figure 3. Shift of baseline density (x2) of foveal cone pigment during 3 successive bleach-and-recovery cycles under 5 conditions, with more than 30 minutes of recovery between each bleach, and with each series measured on same day. Not significant: P = .15.

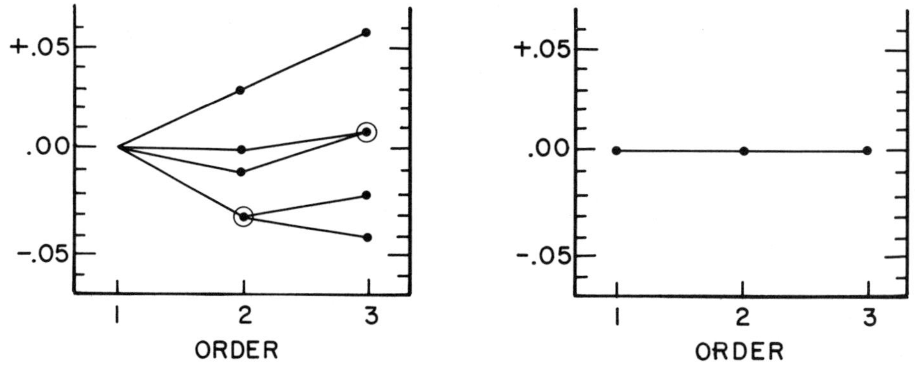

Figure 4. Shift of baseline density (x2) of foveal cone pigment before first, second or third bleach of the day, but not necessarily in succession on the same day. More than 30 minutes of recovery before second and third bleaches. No consistent trend.

Figure 4 shows successive baseline measurements for the five conditions without the requirements that successive bleaches for any one condition be taken on the same day. This figure therefore represents considerably more data than the previous figures, and we find on the whole that the baseline after 30 minutes is as likely to increase as to decrease on successive bleaches. The unchanging averages on the right reflect this. A 30 minutes recovery period lets the pigment all come back.

We conclude that cone pigments in this subject are not completely restored within usual recovery periods (10 minutes). However, if sufficient time is allowed for recovery between bleaches, there is no difference in the amount regenerated that can be assigned to frequent, successive bleaching.

Pigment Regeneration

We next investigated the matter of the longer time constant found for the left eye. Does the extension of the time constant over the years, from 2.0 to 3.24 minutes reflect a gradually damaged retina? Since the subject's right eye had never been bleached we made one set of measurements of baseline, total pigment and recovery in this eye for comparison with the left eye data. If repeated bleaching had damaged the left eye but not the unused right eye, this should show up as a difference in the time constants of regeneration for the two eyes.

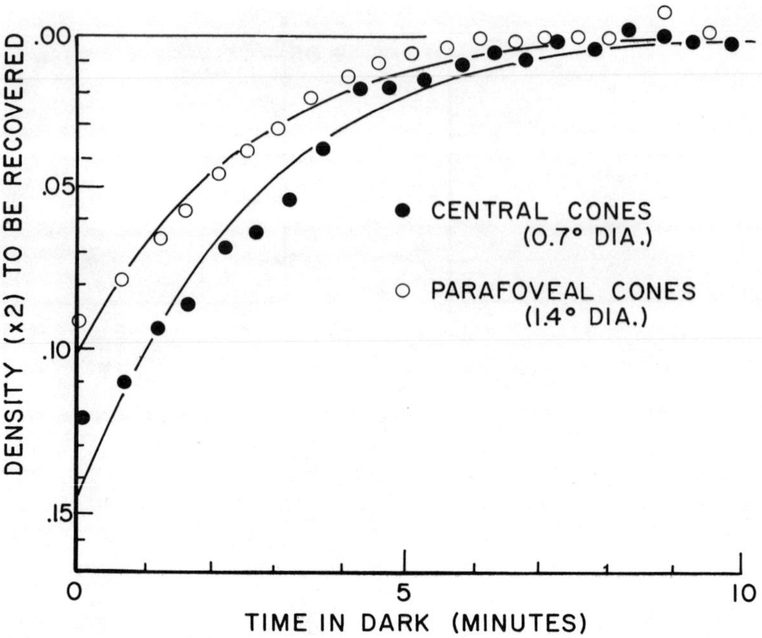

Figure 5. Pigment recovery in left eye, after many bleaches.

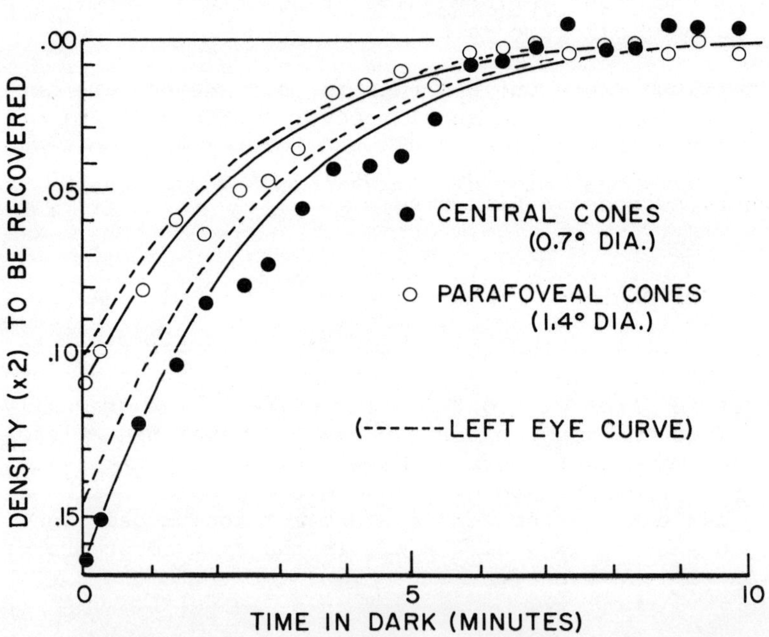

Figure 6. Pigment recovery in right eye, previously unbleached.

Figure 5 shows the return of pigment bleached away in the subject's left eye using the tungsten halide light that subtended 10° of visual angle. The measurements were made with two sizes of test beams, both of 589 nm wavelength. The curve in each case describes recovery of the pigment and represents the same exponential with 3.24 minute time constant. Each curve is an average of six individual curves. This eye has been bleached hundreds of times.

Figure 6 shows the single set of measurements in the previously unused right eye, for conditions similar to those of Fig. 5. The time constant of these curves is the same, 3.24 minutes. For comparison, the dashed lines are the curves from Fig. 5. There appears to be about one hundredth of a double density more pigment in the right eye for both conditions. But this amount is small; such a difference is within the day-to-day variability of the measurements in one eye.

Summary

We find that cone pigments may not be completely restored within usual cone recovery periods following complete bleaches. But when the time between bleaches is extended to 30 minutes the deficiency disappears. Although the time constant for pigment recovery in our subject's left eye has increased from 2.0 minutes to 3.24 minutes during sixteen years of repeated complete bleaching, the increase appears related to age rather than damage, since the recovery time constant was found to be similar in the subject's right eye, which was never bleached until this experiment. We conclude that repeated bleaching away of all cone pigments over a long period of time can be done with little or no damage to the mechanism of pigment recovery.

REFERENCES

1. Lanum, J. (1978) Survey Ophthal. 22, 221-249.
2. Baker, H. D. and Rushton, W. A. H. (1965) J. Physiol. 176, 56-72.
3. Rushton, W. A. H., Fulton, A. B. and Baker, H. D. (1969) Vision Res. 9, 1473-1479.
4. Hood, C. and Rushton, W. A. H. (1971) J. Physiol. 217, 213-229.

Molecular Aspects of Photoreceptor Physiology

EYE PIGMENTATION AND CONSTANT LIGHT DAMAGE IN THE RAT RETINA

Matthew M. LaVail

Department of Anatomy
School of Medicine
University of California, San Francisco
San Francisco, CA 94143

Photoreceptor destruction is known to occur as a result of exposure for various lengths of time to low (less than about 1000 lux) (1, 2), moderate (about 1000-3000 lux) (3, 4) and higher (5, 6) levels of visible light, as well as by the radiant energy from lasers of different wavelengths (7, 8). In the case of high levels of visible light (6) and laser-induced damage to the retina, the mechanism appears to be the absorption of energy by melanosomes in the pigment epithelium and, indeed, the extent of laser-induced retinal destruction is directly related to the degree of retinal pigmentation (7, 8).

In contrast, retinal damage by low and moderate levels of visible light is mediated by the absorption of light energy by the visual pigment of the photoreceptor outer segments (3, 4, 9). Furthermore, the presence of melanosomes in the eye seems to slow the rate of photoreceptor destruction (3). Most of the studies dealing with low and moderate levels of visible light have used albino rats as experimental animals. In an early study, Noell and co-workers (3) compared the light damage (1200-2500 lux) in albino and pigmented rats. In this case they dilated the eyes of the pigmented rats and found that the time to produce equivalent lesions required somewhat more than twice the time in pigmented rats than in albino rats.

In our animal quarters several years ago, the lights were inadvertantly left on continuously for many months. The retinas of the albino rats showed almost complete degeneration of the photoreceptor cells as expected from the work of O'Steen (1), but the pigmented rats had histologically normal retinas. It seemed that the difference in degeneration rates between the albino and pigmented rats was far greater than the factor of two found by Noell et al. (3). However, our levels of in-cage illuminance were much lower and more variable (5-350 lux) than those of Noell et al., and the rats did not have dilated pupils. The fact that ocular pigment apparently protected the retina from constant-light damage for many months stimulated a more systematic comparison of constant-light damage in pigmented and albino rats, and the results of that study are presented here. In addition, the protective role of melanosomes in the retinal pigment epithelium, specifically, has been examined by analyzing the effect of constant light on the eyes of two genetic rat mutants with different patterns of ocular pigmentation. The first, red-eyed yellow (gene symbol, _r_), has a mostly pigmented eye, except that the retinal pigment epithelium lacks melanosomes. The second, pink-eyed dilution (gene symbol, _p_; also called pink-eyed yellow), has an albino-like eye, except that it has a few lightly melanized melanosomes, particularly in the retinal pigment epithelial cells of the far peripheral retina.

MATERIALS AND METHODS

Rats of four phenotypes, albino, fully-pigmented (black-eyed), red-eyed and pink-eyed (Fig. 3), were arranged into three groups. Each group had black-eyed rats and one or two of the pigmentation dilution mutants, and each group was exposed to different illuminance levels of constant light for different lengths of time. One to three rats were examined at each exposure time. The various parameters are indicated in Tables I, II and III, and only general comments will be made here.

Animals

Red-eyed/CDF Group (Group I; Table I). The red-eyed yellow (_r/r_) rats were obtained from Dr. W. J. Dodds-Laffin of the New York State Department of Health. They have been maintained by inbreeding in our laboratory, and the rats used were at our F6 generation. The rat strain was called "fawn-hooded" because of its "fawn-colored" coat. However, the strain proved to be homozygous for _r_ in linkage testing with _c_ and _p_ on linkage group I, wild-type (+) for pink-eyed dilution and homozygous for non-agouti (_a_) (LaVail, unpublished observations). CDF is an abbreviation for inbred caesarian-derived Fischer albino rats that were originally obtained from Charles River Laboratory (Wilmington, MA) and maintained by inbreeding in our laboratory. The breeding scheme for this group is shown in Fig. 1. It provides littermates that are either albino (_c/c_), red-eyed (_r/r_) or black-eyed (+/+, +/_c_ or +/_r_).

Table I

Red-eyed/CDF Group (Group I)

Rats:
Albino, red-eyed and black-eyed
Produced by intercrossing red-eyed X CDF F1 rats

Age into constant light:
14 months

Illuminance levels:
First 90 days: 50-110 lux reflected from bedding
200 lux maximal in-cage incident illuminance
After 90 days: 385-500 lux reflected from bedding
1650 lux maximal in-cage incident illuminance

Duration in constant light:
35, 126, 140, 247 and 338 days

Table II

ACI/RCS Group (Group II)

Rats:
Pink-eyed and black-eyed
All +/*rdy*
Produced by backcrossing ACI x RCS F1 to RCS

Ages into constant light:
One of each phenotype at 9, 11.5 and 15 months

Illuminance levels:
90-175 lux maximal in-cage incident illuminance

Duration in constant light:
77 days

Table III

Albino, Pink-eyed and Black-eyed Group (Group III)

Rats:
Albino: Non-inbred CD strain, Charles River
Pink-eyed: RCS-*rdy*+ strain
Black-eyed: RCS-*rdy*+ *p*/+ strain

Ages into constant light:
Albino and pink-eyed: 5 months
Black-eyed: 6.5 months

Illuminance levels:
200-500 lux reflected from bedding
1650 lux maximal in-cage incident illuminance

Duration in constant light:
43, 58 and 112 days

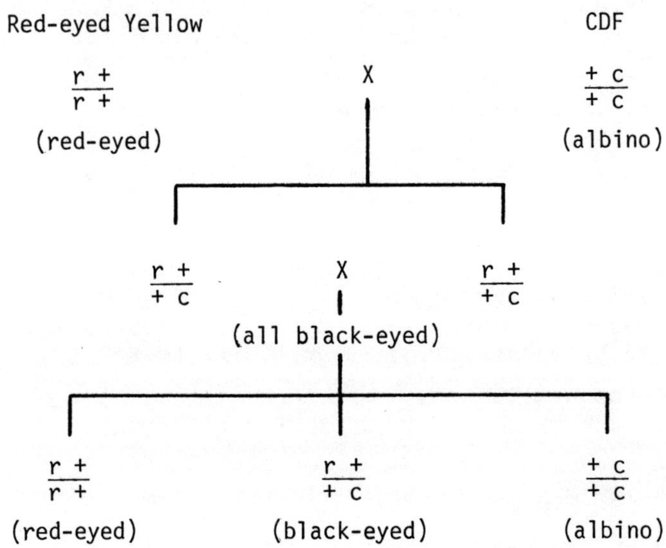

Figure 1. Breeding scheme for the red-eyed/CDF (Group I) rats.

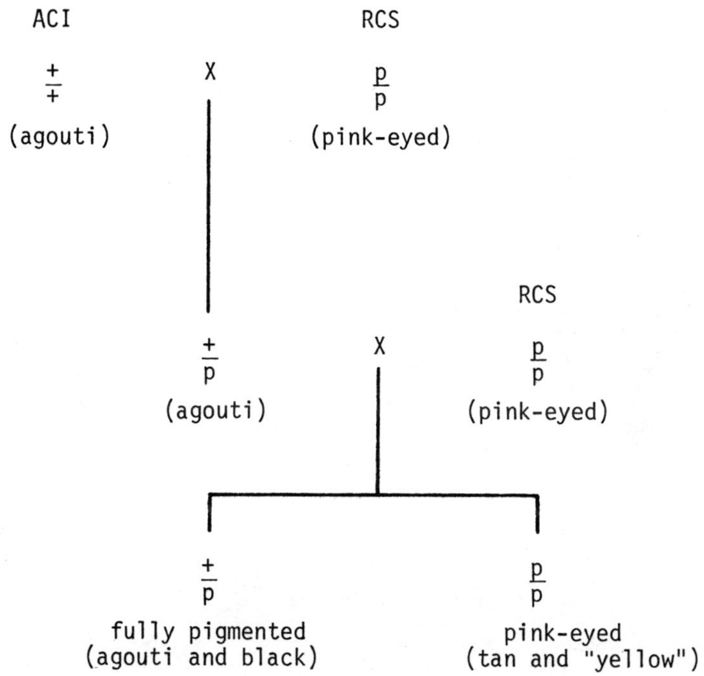

Figure 2. Breeding scheme for the ACI/RCS (Group II) rats.

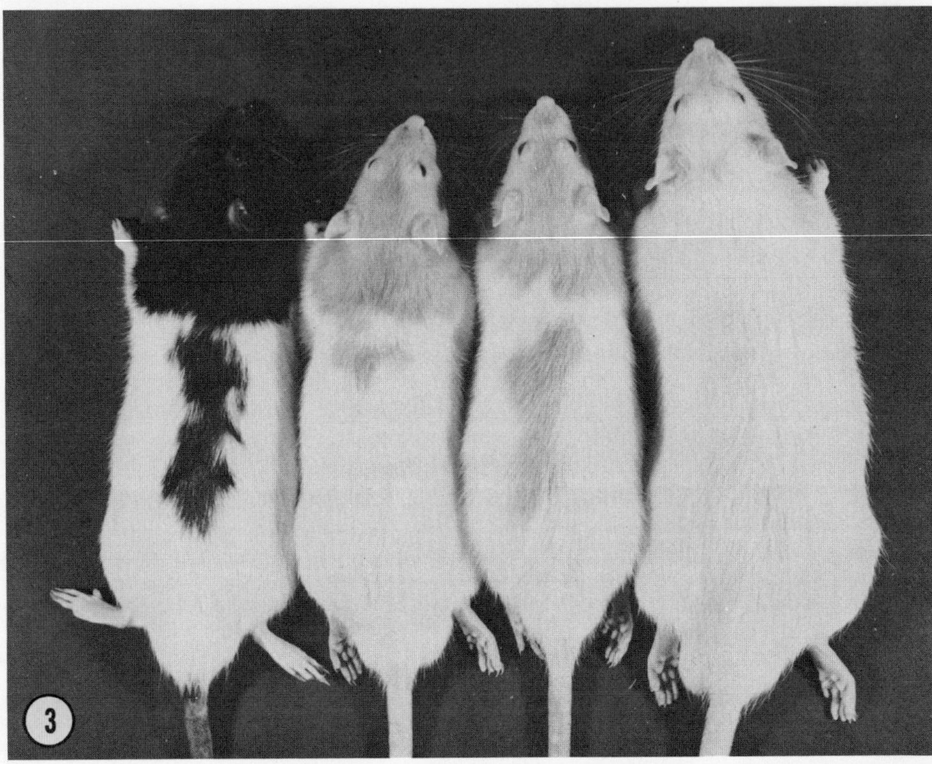

Figure 3. Four pigmentation types of rats examined. From left to right, fully pigmented (+/+, wild-type), red-eyed yellow (r/r), pink-eyed dilution (p/p) and albino (c/c). The age-matched albino is larger than the others because it is a non-inbred animal.

ACI/RCS Group (Group II; Table II). Inbred ACI rats were obtained from Microbiological Associates, Inc. (Bethesda, MD) and were bred to RCS rats according to the scheme shown in Fig. 2. The backcross progeny were of four coat colors (Fig. 2) due to the interaction of the agouti (A), non-agouti (a) and pink-eyed dilution (p) genes, but both the agouti and black rats have fully pigmented eyes and both the tan and "yellow" rats have pink eyes (Fig. 2). Since RCS rats are homozygous for retinal dystrophy (rdy), some of the backcross progeny are +/rdy and others are rdy/rdy. The first eye of each rat was taken using ether anesthesia at 1-2 months of age. Only those with histologically normal retinas (+/rdy) were used for the constant light study.

Albino, Pink-eyed and Black-eyed Group (Group III; Table III). Non-inbred albino rats (CD strain) were purchased from the Charles River Laboratories. Pink-eyed and black-eyed rats were of the RCS-rdy+ and RCS-rdy+ p/+ strains, respectively. These strains are congenic with the retinal dystrophic RCS strain, but they are both +/+ at the rdy genetic locus, and they differ from one another in eye and coat pigmentation.

Cages and Light Environment

The rats were maintained in transparent polycarbonate cages with stainless steel wire-bar covers. A water bottle and food were kept in the appropriate depressions in the cage covers, so that a light shadow fell across a small portion of the cage. Most of the animals frequently huddled together in the region of the shadow with some animals burying their heads under the other animals to differing degrees. This resulted in a quite variable amount of light exposure as described by Birch and Jacobs (10), but most of the animals probably were in the lower levels of illuminance given in Tables I-III for most of the time. The bedding was a light birch cube ("Betta-chip", Northeastern Products Corp., Warrenburg, N. Y.).

Constant light was provided from banks of GE "Cool-white"® fluorescent bulbs. The cages were placed on the top of animal cage racks directly under the lights in the case of Groups II and III. Cages of rats from Group I were placed on partially shaded rows two and three of cage racks, which were mostly shaded from direct light, for the first 90 days, and then they were moved to the tops of the racks. Before being placed into constant light, the rats were maintained in a 12:12 light:dark cycle with an in-cage illuminance of 5-200 lux for either one month (CD rats in group III) or their entire life (remainder of rats).

Histological Procedures

Most of the rats were killed by vascular perfusion, their eyes were removed and bisected in the vertical meridian, and they were embedded in a mixture of Epon-Araldite. Some eyes were enucleated and then immersed in fixative rather than fixing by perfusion. The eyes were sectioned at 1-2 μm thickness in the vertical meridian so that single sections included a full length of retina passing through the optic nerve head and the ora serrata in both the superior and inferior hemispheres of the eye. The sections were then stained with toluidine blue. For the analysis of eye pigmentation, the eyes were embedded in polyester wax, sections were cut at 8 μm and examined without staining. Details of each of these procedures are presented elsewhere (11).

RESULTS

Coat and Eye Pigmentation

The coat pigmentation of the four rat phenotypes is illustrated in Fig. 3, and the anterior and posterior aspects of their gross eye pigmentation is shown in Figs. 4 and 5, respectively. The red-eyed yellow mutation results in a reddish-black eye, whereas the pink-eyed dilution mutation produces an albino-like eye. Both mutations dilute the black-hooded coat to a tan or slate color.

The iris pigmentation of the four phenotypes in shown in Figs. 6-9. The iris of the red-eyed rats is pigmented, but not quite as heavily as that of black-eyed, wild-type rats (cf. Figs. 6 and 7). The stroma of the red-eyed iris is moderately pigmented, whereas the posterior iris epithelium is only lightly pigmented (Fig. 7). The iris of the pink-eyed rats (Fig. 8), is almost as free of pigment as that of the albino rats (Fig. 9), but a few melanocytes are present in the iris stroma (Fig. 8).

The choroidal and pigment epithelial cell pigmentation of the four phenotypes is shown in Figs. 10-15. In the fully pigmented, wild-type rat, most regions of the eye show heavy pigmentation of the choroid and pigment epithelium (Fig. 10). In the dorsal or superior retina, in the region extending from the equator posteriorly to about halfway to the optic nerve head, the pigment epithelium shows a lighter pigmentation pattern (Fig. 11). This is similar to the lighter pigmentation that Deol (12) has described in the superior or dorsal aspect of the gross mouse eye, although the pattern appears to be more consistent in the rat than the mouse.

Figures 4 and 5. Anterior (Fig. 4) and posterior (Fig. 5) views of freshly enucleated whole eyes of rats in Fig. 3. From left to right, +/+, *r*/*r*, *p*/*p* and *c*/*c*.

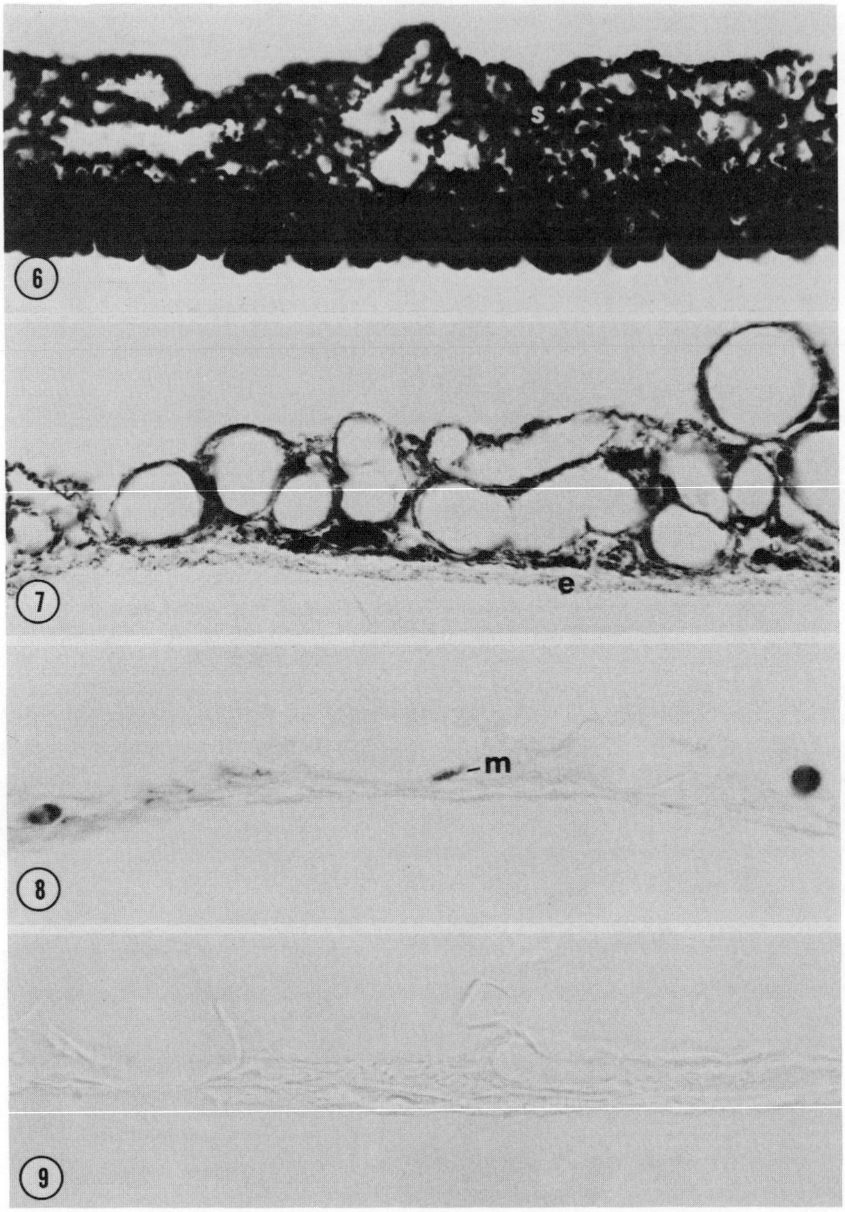

Figures 6-9. Unstained sections of the iris of +/+ (Fig. 6), r/r (Fig. 7), p/p (Fig. 8) and c/c (Fig. 9) rats. The blood vessels in the stroma (s) of the +/+ iris (Fig. 6) are not as round and distended as in the others, because the +/+ eye was fixed by immersion, whereas the other rats were fixed by vascular perfusion. e, iris epithelium; m, melanocyte. Polyester wax, 8 μm. All x810. (Reduced 20% for purposes of reproduction.)

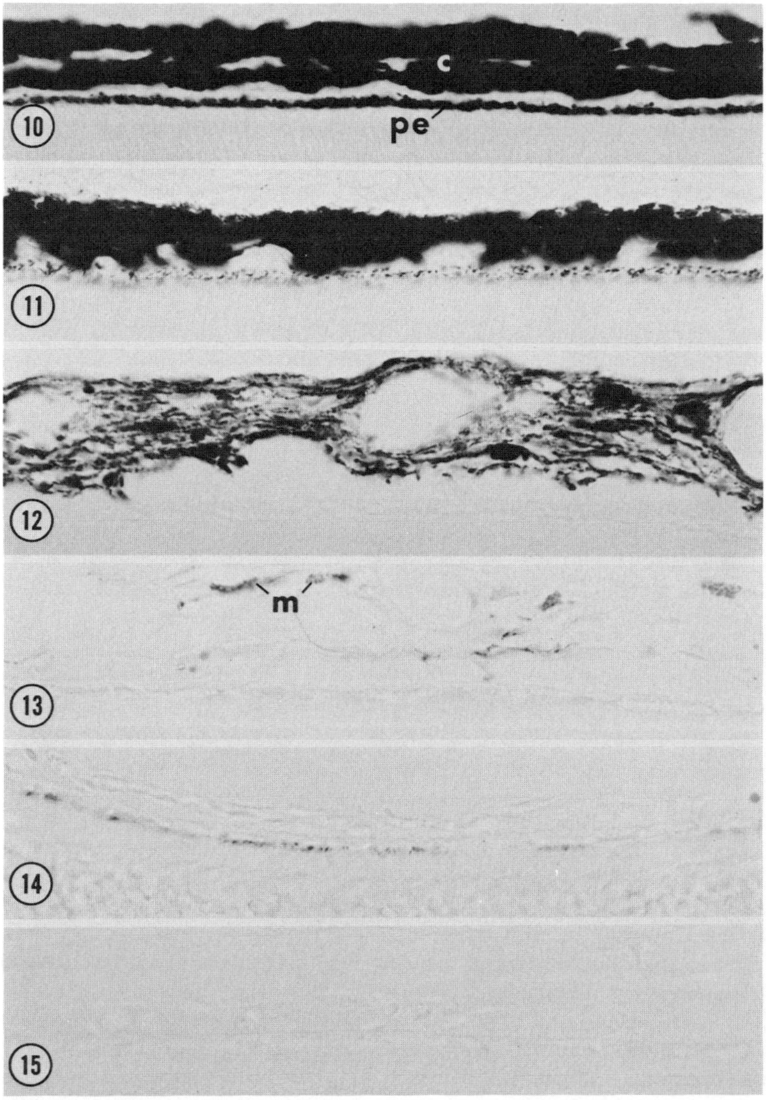

Figures 10-15. Unstained sections of the choroid (c) and pigment epithelium (pe) of the different rat phenotypes. Figs. 10 and 11 are from a fully pigmented +/+ rat, and they illustrate the differences in pigmentation of the pigment epithelium in the superior (Fig. 11) and inferior (Fig. 10) hemispheres of the retina. Fig. 12, r/r. Fig. 13, p/p, posterior retina illustrating the maximal number of choroidal melanocytes (m) seen in a given field. There are often none seen, as in Fig. 14. Fig. 14, p/p, peripheral retina showing some melanosomes in the pigment epithelium. Fig. 15, c/c. Polyester wax, 8 µm. All x810.
(Reduced 20% for purposes of reproduction.)

The red-eyed rat has a moderately pigmented choroid that is somewhat more lightly pigmented than that of wild-type rats (cf Figs. 12 and 10). The pigment epithelium, however, shows a remarkable absence of melanosomes in red-eyed rats (Fig. 12). The _r/r_ rat has been described as having a slightly pigmented retinal pigment epithelium (13) but Wise and Lund (14) found both in unosmicated and unstained 1 μm plastic sections by light microscopy and in thin sections by electron microscopy that the _r/r_ pigment epithelium lacks melanosomes. In the present study, the _r/r_ pigment epithelium does show a light brown stippling in toluidine blue-stained sections of both plastic and polyester wax-embedded eyes. Unstained wax-embedded sections of _r/r_ eyes, however, do not show melanosomes, in confirmation of Wise and Lund. The lightly staining structures apparently are the relatively abundant lipofuscin granules and/or residual bodies in the pigment epithelium of _r/r_ eyes (LaVail, unpublished observations).

The pink-eyed rats have almost no pigment in the choroid and pigment epithelium in the posterior regions of the eye (Fig. 13). In the peripheral retina, from the ora serrata extending posteriorly several hundred μm, the pigment epithelium contains a number of moderately pigmented melanosomes (Fig. 14).

Constant-Light Damage in Fully Pigmented Eyes

Different regions of a retina from a 9-month-old wild-type control rat reared in cyclic light are shown in Figs. 16-18. The outer nuclear layer consists of 7-8 rows of nuclei at this age, and it is thinner than at younger ages due to continued growth of the eye (15) without the addition of new cells. The superior (Fig. 16) and inferior (Fig. 17) regions of the eye show the pigmentation differences indicated above, and rod outer segments are also found to be longer in the superior than in the inferior hemisphere of the eye, as described previously (16).

No apparent loss of photoreceptor cells was found in the fully pigmented rats of Group III that were kept in constant light for up to 112 days. Examples of retinas from rats kept 58 and 112 days in constant light are shown in Figs. 20 and 21, respectively. The only difference between these rats and the cyclic light controls is that the rod outer segments are somewhat shorter in the constant light. The illuminance levels for this group were sufficient to destroy most of the photoreceptor cells in albino rat eyes by 43 days in constant light (Fig. 19). In the albinos, some rod and cone photoreceptor nuclei with their surrounding cytoplasm remain in the absence of outer segments, and although the cells are progressively reduced in number with continued exposure to constant light, some may remain for the life of the animal (17).

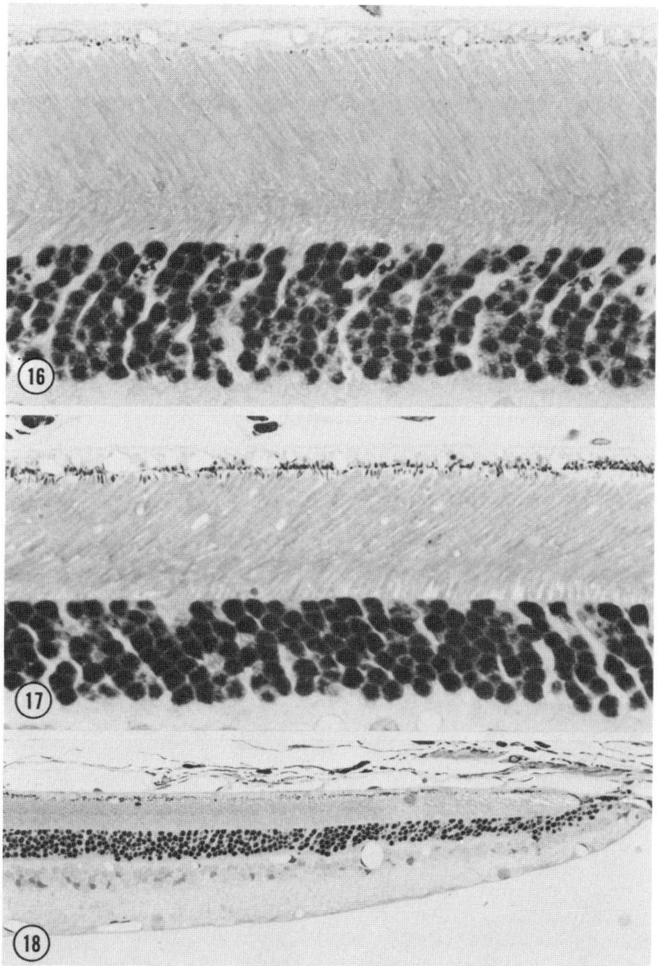

Figures 16-18. Light micrographs from one retina of a 9-month-old RCS-rdy+p/+ control rat maintained in cyclic light. Approximately equatorial regions are shown in the superior (Fig. 16) and inferior (Fig. 17) hemispheres, and the ora serrata is illustrated in Fig. 18. Note the longer rod outer segments and lighter pigmentation of the pigment epithelium in the superior than in the inferior hemisphere. The large apparent vacuoles within the pigment epithelium are actually spaces between the pigment cells and at the bases of the cells. They are due primarily to excessive perfusion pressure, and they can be seen in many of the following micrographs. These and all of the following figures are toluidine blue-stained 1-2 μm thick sections of Epon-Araldite. Figs. 16 and 16, X1000. Fig. 18, X250. (Reduced 30% for purposes of reproduction.)

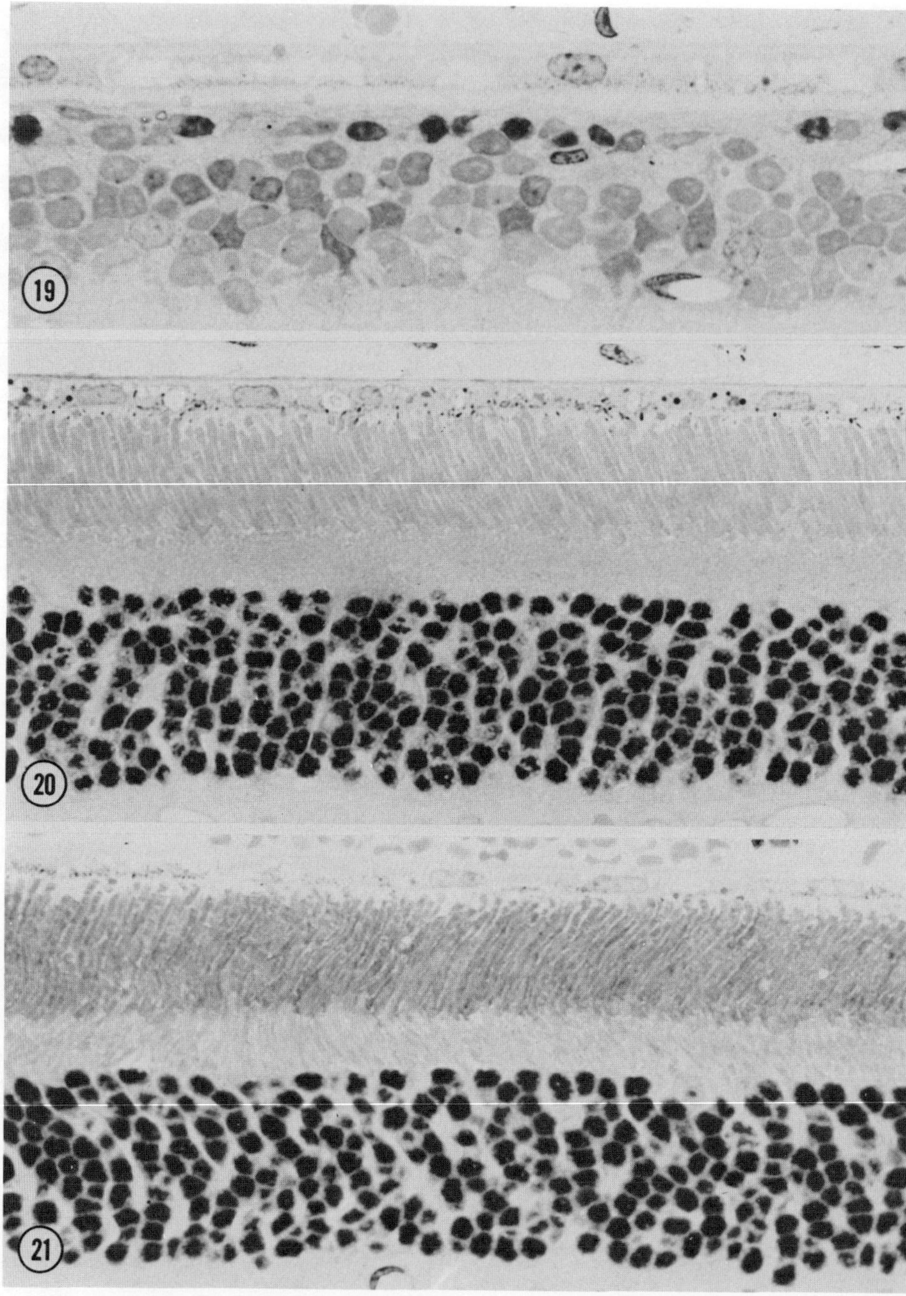

Figures 19-21. Retinas from rats in Group III. Fig. 19, c/c, 43 days of light exposure. Figs. 20 and 21, fully pigmented, 58 and 112 days, respectively, of light exposure. All X1000. (Reduced 10% for purposes of reproduction.)

The results were virtually the same with the Group II rats. After 77 days of constant light exposure, the pink-eyed rats had almost complete loss of photoreceptor cells, whereas the fully pigmented rats showed no apparent loss.

The fully pigmented rats in Group I also showed no obvious loss of photoreceptor nuclei through 140 (Fig. 22) and 247 (Figs. 23 and 24) days of constant light exposure, since the outer nuclear layer thicknesses were within normal limits for the respective ages. However, some cell death was beginning by 247 days in constant light, because occasional pyknotic photoreceptor nuclei were present, as were darkly staining inner and outer segment portions of some photoreceptor cells (Figs. 23 and 24). In addition, a reddish (metachromatic) staining was present at or near the apical surface of the pigment epithelium at this time, whereas none was seen at shorter exposure intervals.

By 338 days of constant light, the pigmented rat eyes showed some regions of almost complete photoreceptor degeneration, but no eye had this degree of degeneration throughout the retina. For example, Figs. 25-28 illustrate four regions within one section of an eye where there is complete degeneration from the ora serrata to the equatorial region in the superior hemisphere (Figs. 27 and 28), yet the retina in the inferior hemisphere (Figs. 25 and 26) appears almost completely normal, with the loss of perhaps 1-2 rows of photoreceptor nuclei. In these regions the rod outer segments appear normal, albeit somewhat shorter than controls in cyclic light as noted earlier. Even in the regions where the outer nuclear layer grades from almost normal to fully degenerated and fewer than normal nuclei are present (Figs. 27), the outer segments still mostly retain their integrity, although they and the inner segments may be conspicuously shorter and less well-organized than controls (Fig. 46).

Constant-Light Damage in Red-eyed Rat Eyes

For presentation of the constant light experiments on this (*r*/*r*) pigmentation mutant, the fully pigmented littermates in Group I will not be shown. It should be noted that it is only necessary to discuss up to 140 days in constant light, and at that time the retinas of the fully pigmented rats showed no apparent loss of photoreceptor cells (Fig. 22).

Examples of retinas from albino and red-eyed yellow rats that had been kept in constant light for 126 days are shown in Figs. 29-33. The albino retinas (Figs. 29 and 30) show the pattern of degeneration expected of 30-40 days in direct, constant light (Figs. 19 and 38-40). (It should be noted that after the initial 35 days in low levels of light, the albinos showed only a slight thinning

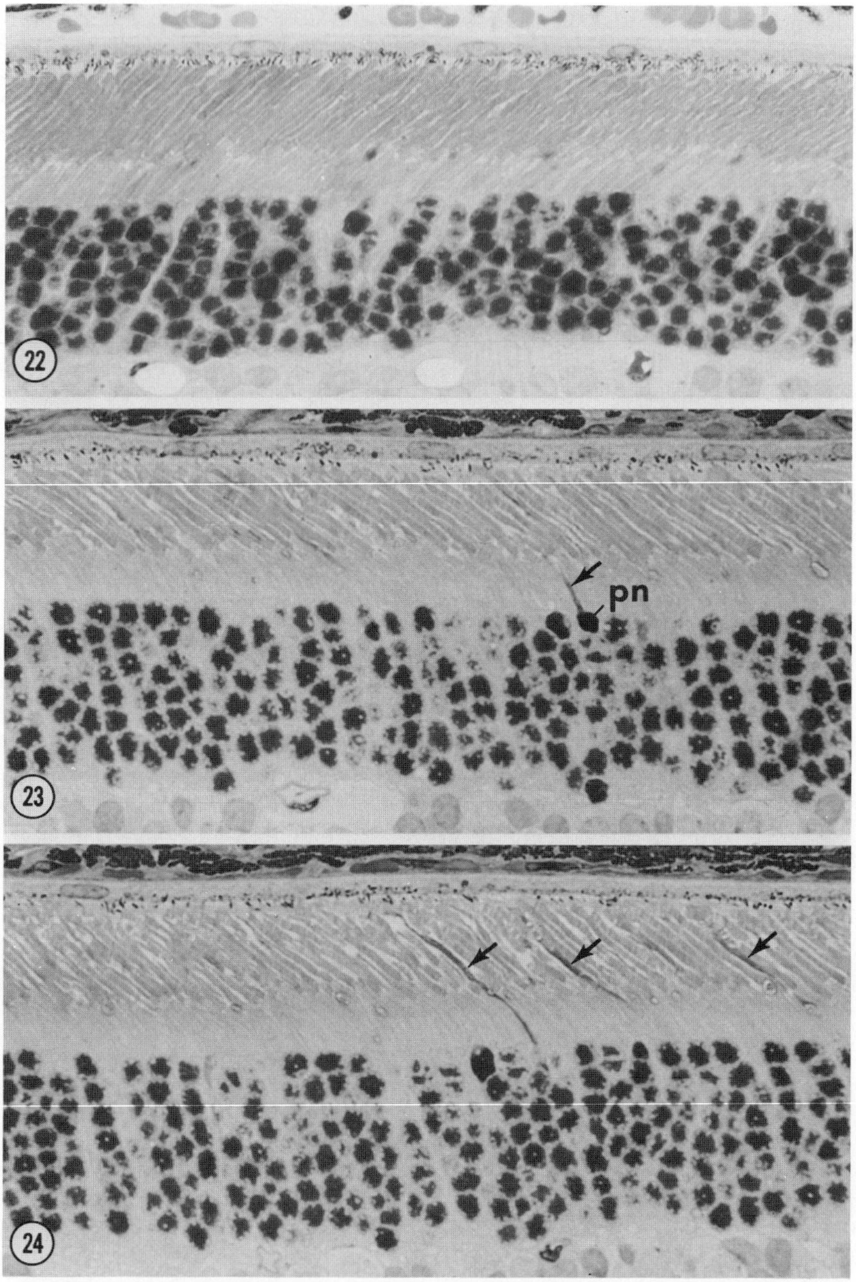

Figures 22-24. Fully pigmented retinas from rats in Group I after 140 (Fig. 22) or 247 (Figs. 23 and 24) days of light exposure. pn, pyknotic nucleus; arrows, dark inner and outer segments. All X1000. (Reduced 20% for purposes of reproduction.)

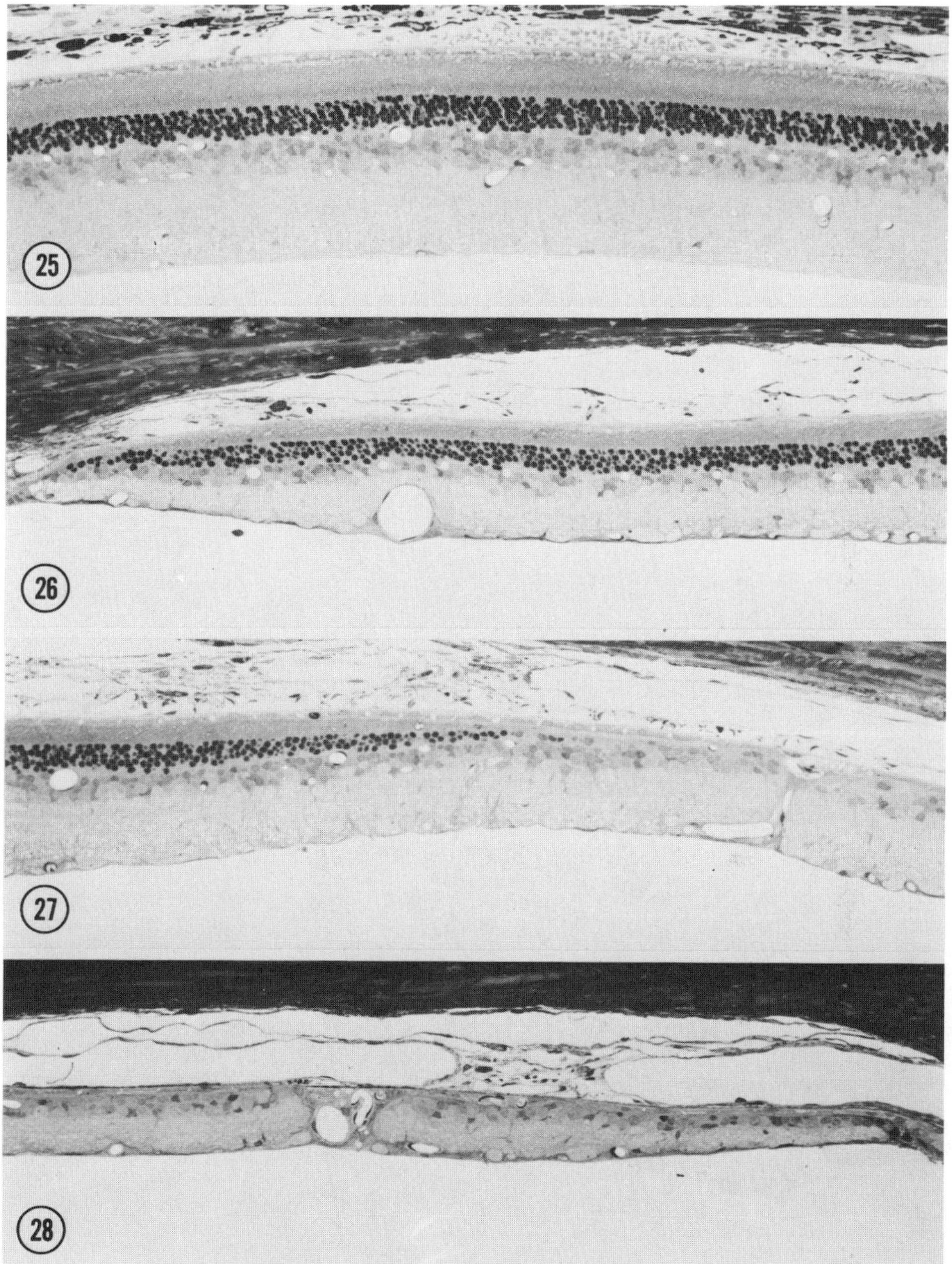

Figures 25-28. Micrographs from one retinal section from the eye of a fully pigmented rat in Group I after 338 days of constant light exposure. Figs. 25 and 26, inferior hemisphere; Figs. 27 and 28, superior hemisphere. All X250. (Reduced 10% for purposes of reproduction.)

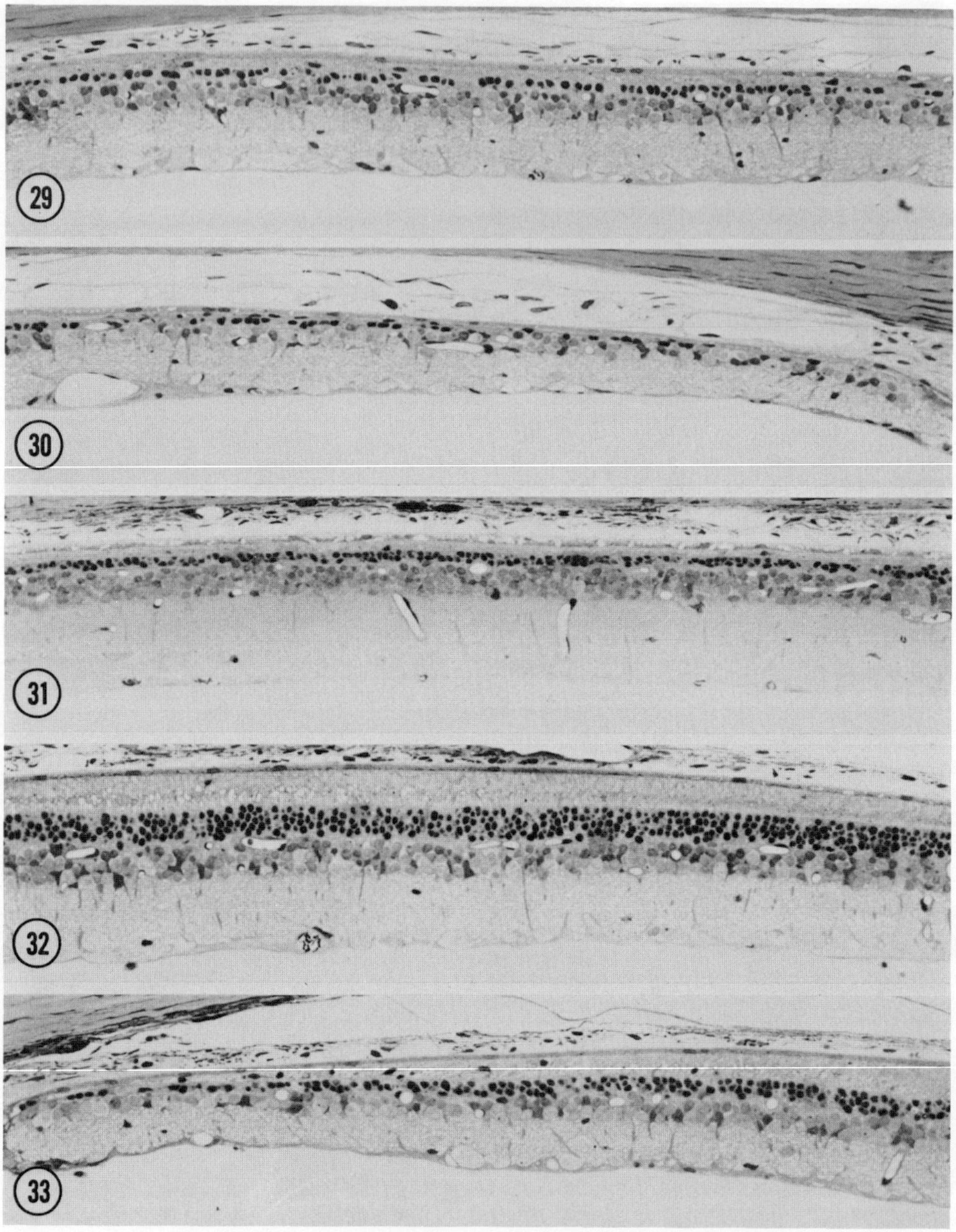

Figures 29-33. Retinas from *c/c* (Figs. 29 and 30) and *r/r* (Figs. 31-33) Group I rats after 126 days of light exposure. Fig. 31, posterior retina; Figs. 29 and 32, equatorial retina; Figs. 30 and 33, ora serrata. All X250.(Reduced 10% for purposes of reproduction.)

of the outer nuclear layer and that only the last 36 days of the 126 days were in direct light at moderate levels of illuminance; see Materials and Methods.) In some regions of the eyes of red-eyed yellow rats, the amount of photoreceptor degeneration was almost equal to that of albino rats (Figs. 31 and 33), despite the eye being mostly pigmented. In other regions of the same eyes, many more photoreceptor nuclei were still surviving, some regions of the outer nuclear layers showing 3-4 rows (Fig. 32). However, no part of the eye had the full complement of photoreceptor nuclei, and in all regions the outer segments were grossly disrupted (Figs. 32 and 33).

The variability in degeneration in the *r/r* rat eyes at 126 days of constant light was also seen at 140 days. Some *r/r* rats showed outer nuclear layers almost as thick as those of normal, fully pigmented rats, but only in some regions of the eye (Fig. 48). In all cases where outer segments were present, they were quite disrupted (Fig. 48). Most remarkably, however, some *r/r* rats showed photoreceptor degeneration as extensive as that in the albino littermates (Figs. 34-37).

Constant-Light Damage in Pink-eyed Rat Eyes

The albino rats of Group III kept in constant light for 43 days showed an almost complete loss of photoreceptor nuclei in the posterior retina (Fig. 38) and in some far peripheral regions near the ora serrata (Fig. 40). In far peripheral regions in other retinal hemispheres, 1-2 rows of nuclei still remained, but no surviving outer segments were evident (Fig. 39). In age-matched pink-eyed rats kept in adjacent cages, the pattern of degeneration in the posterior retina was virtually identical to that of the albino retinas (Fig. 41). In the far peripheral retina, however, in both the superior and inferior hemispheres of the eyes, the degeneration was substantially less than that in the albino eyes. Either about two (Fig. 43) or three (Fig. 42) rows of photoreceptor nuclei were still present. In addition, some outer segment membranes were present, although they were mostly disorganized (Figs. 42 and 47).

The difference between the degree of degeneration in the peripheral retina of albino and pink-eyed rats was still evident after 58 days in constant light (Figs. 44 and 45), but not after 112 days.

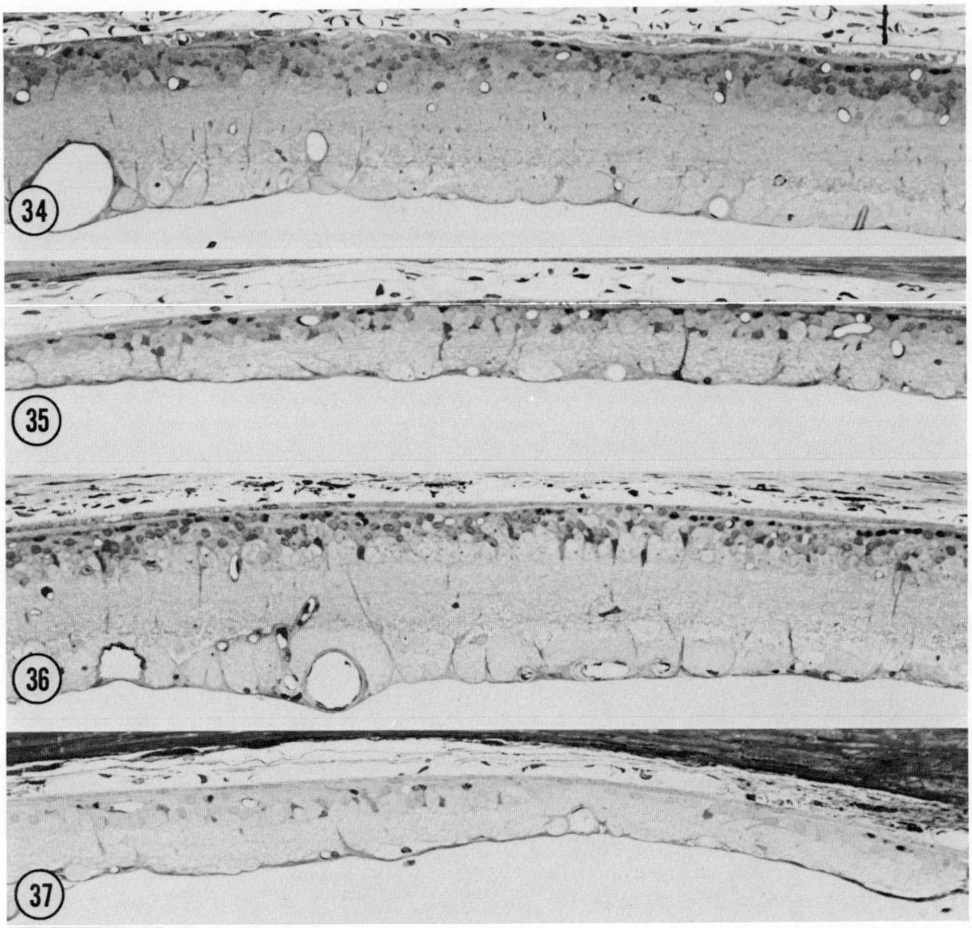

Figures 34-37. Retinas from c/c (Figs. 34 and 35) and r/r (Figs. 36 and 37) Group I rats after 140 days of light exposure. The retinas from these rats are fully degenerated and virtually indistinguishable, whereas their fully pigmented littermate shows a completely normal retina (Fig. 22). All X250.
(Reduced 10% for purposes of reproduction.)

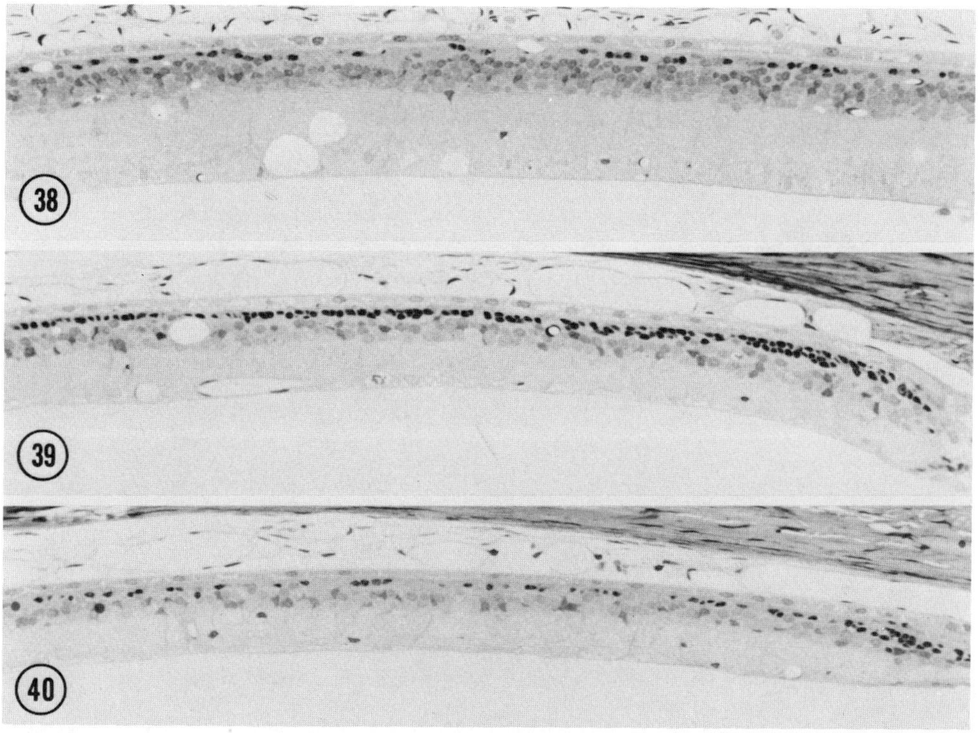

Figures 38-40. Micrographs from one retinal section from the eye of an albino rat in Group III after 43 days of light exposure. Fig. 38, posterior retina; Figs. 39 and 40, ora serrata. All X250. (Reduced 10% for purposes of reproduction.)

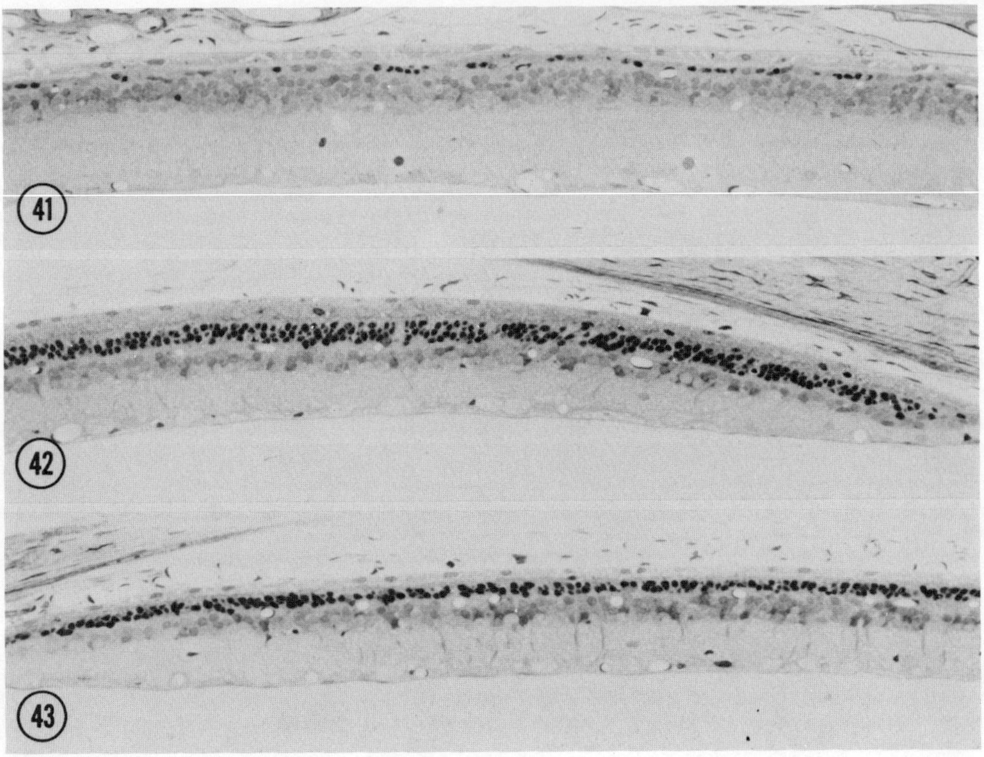

Figures 41-43. Micrographs from one retinal section from the eye of a *p/p* rat in Group III after 43 days of light exposure. Fig. 41, posterior retina; Figs. 42 and 43, ora serrata. All X250. (Reduced 10% for purposes of reproduction.)

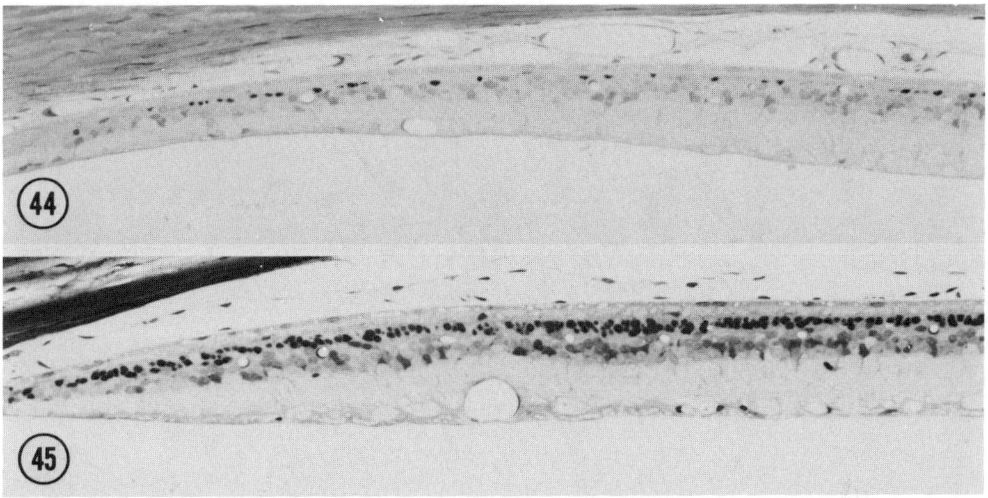

Figures 44 and 45. Ora serrata regions of the retinas from c/c (Fig. 44) and p/p (Fig. 45) rats in Groups III after 58 days of constant light exposure. Both X250. (Reduced 10% for purposes of reproduction.)

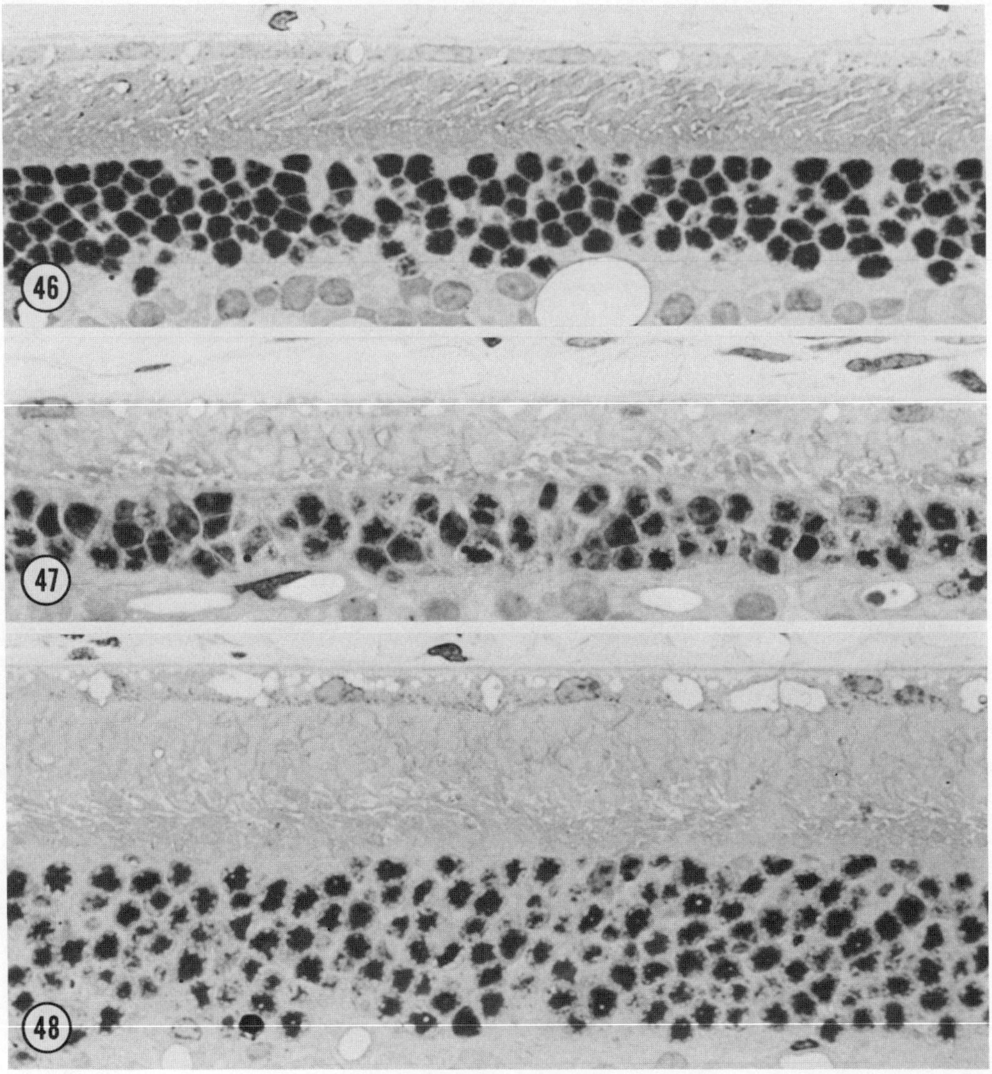

Figures 46-48. Retinas from different pigmentation phenotypes after different exposure times in constant light illustrating varying degrees of outer segment disruption. Fig. 46, +/+, 338 days exposure; Fig. 47, *p/p*, 43 days exposure; Fig. 48, *r/r*, 140 days exposure. All X1000. (Reduced 10% for purposes of reproduction.)

DISCUSSION

Constant Light Damage in Fully Pigmented Eyes

The present work demonstrates that full eye pigmentation in untreated rats at normal body temperature provides substantial protection from constant light damage to photoreceptors. The amount of time to produce a constant-light lesion in pigmented rats equivalent to that in albino rats is a factor greater than 4 and probably as great as 8-10 (or higher). For example, in the Group III rats the albinos showed almost full degeneration of the outer nuclear layer after 43 days of constant light, and the work of others (1, 2) indicates that this degree of degeneration was probably reached by 30 days. In fully pigmented rats, the retinas were normal (except for somewhat shorter than normal rod outer segments) after 112 days in constant light. In the Group I rats which received the longest exposure to constant light, the retinas showed some regional loss of the outer nuclear layer after 338 days of light exposure, but large (hemispheric) expanses of retina showed the loss of only 1-2 rows of photoreceptor nuclei, and photoreceptor outer segments were still present in these relatively intact regions. This alone would suggest a factor of greater than 11 for the time to produce equivalent lesions in albino and pigmented rats. In this group, however, a lower illuminance level was used for the first 90 days, which did not produce full degeneration in the albinos after 35 days of exposure. On the other hand, 36 days after the illuminance level was increased (126 days total exposure), the albinos in this group did show the almost complete degeneration expected after 30-40 days of exposure. Therefore, if day 90 is taken as day zero, the maximal exposure period would be reduced from 338 days to 248 days. Since the typical 1-month degeneration pattern seen in albinos was not seen in pigmented rat retinas after 8 months of constant light, a factor of greater than 8 between albinos and pigmented rats is suggested. The reason one cannot be more precise in the estimation of this factor is the possible complicating factors of 1) not knowing the additional exposure period that would be required to produce truly equivalent lesions in the two phenotypes (i.e., to destroy the remainder of the relatively intact outer nuclear layer in the pigmented rats already exposed for 338 days) and 2) the possible protective effect that might have been provided by the initial 90-day exposure to low levels of illumination (4).

As noted in the Introduction, Noell and co-workers (3) found that with pupil dilation the protective effect of full eye pigmentation is somewhat greater than a factor of 2 in the time to produce equivalent lesions in pigmented and albino rats. This estimate was made by electroretinography on the percent of control a-waves in animals exposed for different lengths of time. Even after the longest exposure time, the complete loss of a a-wave was not achieved.

In this Symposium, Rapp et al. (18) have also shown that a factor of about 2 is obtained between albino and pigmented rats with dilated pupils. The criterion level of degeneration used by Rapp et al. was also less than full degeneration (35% of outer nuclear layer cell loss). In both of these studies, it is not clear that further degeneration in the pigmented eyes would proceed at exactly the same rate as at earlier stages. Therefore, the factor of 2 might be somewhat different, perhaps greater, if complete degeneration were used as the criterion as in the present study (although the methods used by Noell et al., and by Rapp et al. are probably more precise for a given stage of degeneration). Nevertheless, it is difficult to imagine the factor increasing to the level found in the present work with undilated pupils. This difference in the two other studies and the present work reinforces the well-known fact that iris pigmentation is an important factor in reducing retinal illumination.

One must be cautious in comparing sensitivity to constant-light damage in different species, particularly when considering the human retina and safety factors related to light exposure. Nevertheless, the findings in the present study do warrent a comment in this regard. When compared to the Rhesus monkey, rabbit and other mammalian retinas, the rat retina has been said to be by far the most sensitive to light damage (4, 9). Nevertheless, in illuminance levels as great as virtually any office fluorescent lighting, pigmented rats subjected to many months of constant light still have almost entirely normal retinas. Since the human eye is also pigmented and since the human retina may be less sensitive to light damage than the rat retina, it seems that the human eye should be in little danger of photoreceptor damage in low to moderate illuminance levels for continuous periods of time up to at least several weeks, if not longer.

Retinal Pigment Epithelium in Protection from Constant-Light Damage

Constant-light experiments with the red-eyed and pink-eyed dilution mutants were carried out to determine whether the melanosomes in the retinal pigment epithelium, specifically, are important in the protection that eye pigmentation affords the retina from damage by constant light. In the red-eyed mutant, the eye is mostly pigmented, yet degeneration can occur in these animals almost as rapidly as in albinos, and the most significant difference in pigmentation from fully pigmented eyes is the absence of melanosomes in the retinal pigment epithelium. Although the eyes of the red-eyed rats are not quite as fully pigmented as those in wild-type, pigmented rats, the reddish-black eye must have a substantially lower retinal illumination than albinos due to the iris pigment and the probable reduction in scattered and reflected light by the pigmented choroid.

In the pink-eyed mutant, the eye is almost free of pigment and appears grossly to be almost an albino eye. As might be expected, most of the retina shows virtually the same rate of degeneration as that of albino eyes. Most remarkably, however, there is a substantially retarded rate of degeneration in the far peripheral retina, and this is precisely where there is an abundance of lightly melanized melanosomes in the retinal pigment epithelium.

These data from the pigmentation mutants leads to the conclusion that melanosomes in the retinal pigment epithelium play an important role in protecting photoreceptors from damage by constant light at low to moderate illuminance levels.

An alternative explanation of the data on the red-eyed rats is that the iris, which is not quite as fully pigmented as that of wild-type-rats, reduces the retinal illumination just slightly from that of the retinal illumination of albinos. This seems unlikely, but it cannot be excluded as a possibility. In consideration of this alternative explanation however, the data on the pink-eyed rats must also be taken into account. In these almost pigment-free eyes, there is no barrier to light reaching all parts of the retina, and, therefore, no reason for the peripheral retina to be preserved except for the relatively small amount of melanin in the peripheral pigment epithelial cells.

Another alternative explanation for the different patterns of degeneration in the mutants is strain differences. This is probably not a factor because Noell et al. (3) found no substantial susceptibility to constant light damage in 4 albino strains. Furthermore, the breeding schemes and congenic animals used in the present study hopefully minimized such genetic factors, if they do exist.

Another series of experiments may also support the hypothesis that melanosomes in the retinal pigment epithelium play an important role in the protection of retinal photoreceptors from constant light damage. Rapp (18) and Rapp et al. (19) report that in constant-light experiments on pigmented rats, the retina, in a region of the superior hemisphere extending from about 500 μm from the optic nerve head to about 2000 μm from it, is more severely affected than is the retina of the inferior hemisphere. Their more affected region corresponds precisely to the region of reduced pigmentation of the retinal pigment epithelium in fully pigmented rat eyes (Figs. 10-11 and 16-17). There are alternative explanations, however, for the greater susceptibility of the superior retina to constant light damage. One is that the rod outer segments are longer in the superior retina (Figs. 16 and 17 and ref). More rhodopsin being present there might lead to greater degeneration, since visual pigment absorption of light appears to be the first event in constant light damage (3, 4, 9). Another possible explanation is that intrinsic differences exist in the susceptibility to constant light damage in the superior and inferior hemispheres of the eye. It is clear that superior/inferior hemispheric differences do exist, for example, in rod outer segment length in normal rats (16) and in the rate of inherited retinal dystrophy in RCS rats (11).

It should be noted that in the rabbit retina, Lawwill (9) found the greatest area of damage after a 4-hour light exposure to be the area of greatest concentration of photoreceptor cells. He also found no significant differences in the threshold for damage in lightly and darkly pigmented eyes or in lightly pigmented areas of dark fundi. On the one hand, this might argue against the interpretation of the rat data presented here (i.e., a protective role of pigment epithelial cell melanosomes). On the other hand, it is difficult to make direct comparisons between the two species and experimental protocols. Furthermore, even the very light pigmentation of the pink-eyed rat pigment epithelium is enough to retard degeneration somewhat. Therefore, one might not see a difference in fully and lightly pigmented regions of rabbit retina, depending upon the intensity and duration of light exposure.

Possible Mechanism of Protection from Constant-Light Damage by Pigment Epithelial Cell Melanosomes

It is generally accepted that melanin acts as a sunscreen in that it absorbs visible and ultraviolet light and dissipates this energy as heat. Melanin in the iris, therefore, acts to screen out light and reduce retinal illumination. Once light passes through the photoreceptor cells, it may be absorbed by the melanin in the pigment epithelial cells and choroid. In this way light is prevented from being diffusedly scattered and reflected within the eye. The analysis of the pigmentation dilution mutants in the present study suggests that the melanin in the pigment epithelial cells may have a role in addition to that of direct absorption of light energy. In the case of the red-eyed mutant, substantial absorption of light energy must occur by the iris and choroid, yet the retina can degenerate in constant light almost as readily as in an albino retina. This must be due in large part to the lack of melanin in the pigment epithelium. In the case of the pink-eyed mutant, the small amount of melanin in the peripheral pigment epithelial cells clearly cannot reduce the retinal illumination much, if at all, in this albino-like eye, yet the region of lightly melanized pigment granules is precisely where the photoreceptors show a slowed degeneration. Presumably the melanin is playing a role in this partial protection from light damage.

The protective role melanin plays (other than its sunscreening role) may result from its being a powerful electron acceptor. McGinness and Proctor (20) have suggested that melanin can act as a sink for free radicals and excited species such as singlet oxygen.

Crouch has demonstrated lipid peroxidation in light damage to bovine opsin *in vitro* (21) and in preliminary experiments to rat opsin *in vivo* (Crouch, personal communication), implicating free radical generation in the destructive process. Organisciak et al. (22) have also shown lipid peroxidation of rat outer segment membrane following an intense *in vivo* light exposure. It therefore seems plausible that melanin in the pigment epithelium acts as a sink for excited species that are generated on exposure of photoreceptor outer segment membranes to light. In this regard, melanin would act in concert with superoxide dismutase in the pigment epithelium. Such a role for melanin would explain 1) at least in part, why albino animals have a greater susceptibility to constant-light damage than do pigmented animals and 2) the findings in each of the pigmentation dilution mutants in the present study.

A role of free radical sink for melanin in the pigment epithelium might also explain why melanin is present in the retinal pigment epithelium in the first place. For this role, it would be important for melanin to be located near the photoreceptor outer segments. If diffuse-light absorption were the only role of melanin behind the retina, the melanin in the more external choroid might be adequate, since there clearly is more melanin in the multilayered choroid than in the single-layered pigment epithelium.

Possible Differences in Cytopathological Mechanisms in Different Pigmentation Phenotypes

In an early description of light damage in rats, Noell et al. (3) pointed out that different patterns of degeneration may occur in short exposures to moderate levels of illumination and longer exposures to low levels of illumination. More specifically, pigment epithelial cell involvement is not apparent in the latter case.

In the present study, there was no obvious abnormality in the pigment epithelium. There was, however, a conspicuous difference in the appearance of rod outer segments in degenerating retinas in fully pigmented eyes as compared to that in the pigmentation dilution mutants. In the fully pigmented eyes, most outer segments retained their integrity virtually until the outer segments disappeared (Fig. 46). The outer segments of pink-eyed and red-eyed rats, on the other hand, largely disintegrated into masses and whorls of outer segment debris membranes (Figs. 47 and 48). This also occurs in early stages of photoreceptor degeneration in albino rat eyes (see figures in references 18 and 23). In these cases, the question arises whether the pigment epithelium is involved in the cytopathological process and/or whether the phagocytosis mechanism of the cells is compromised. It is possible that the pigment epithelium is not affected and that the presence of whorls of membranes simply reflects a change in the membrane structure so that it is not readily recognized and ingested. Alternatively, the pigment epithelial cells may be affected by a factor from the degenerating membranes (perhaps free radicals) in a subtle way such that the metabolism of the cells is affected, but not in a way (or to an extent) that kills the cells. Resolution of this issue awaits further work.

ACKNOWLEDGMENTS

I wish to thank Bonnie A. P. Lord, Douglas Yasumura, Nancy Lawson and Patricia Ann Ward for technical and photographic assistance and Dave Akers for additional photographic assistance. This investigation was supported in part by USPHS Research Grant EY 01919 and Research Career Development Award EY 70871 from the National Eye Institute, Biomedical Research Grant RR 05535 and funds from the Academic Senate Committee on Research, University of California.

REFERENCES

1. O'Steen, W. K. (1970) Exp. Neur. 27, 194-205.
2. O'Steen, W. K. and Anderson, K. V. (1971) Exp. Neur. 30, 525-534.
3. Noell, W. K., Walker, V. S., Kang, B. S. and Berman, S. (1966) Invest. Ophthal. 5, 450-473.
4. Noell, W. K. and Albrecht, R. (1971) Science 172, 76-80.
5. Gorn, R. A. and Kuwabara, T. (1967) Arch. Ophthal. 77, 119-123.
6. Ham, W. T., Jr., Mueller, H. A., Ruffolo, J. J., Jr., and Clarke, A. M. (1979) Photochem. Photobiol. 29, 735-743.
7. Marshall, J. (1970) Invest. Ophthal. 9, 97-115.
8. Borland, R. G., Brennan, D. H., Marshall, J. and Viveash, J. P. (1978) Exp. Eye Res. 27, 471-493.
9. Lawwill, T. (1973) Invest. Ophthal. 12, 45-51.
10. Birch, D. and Jacobs, G. H. (1977) Physiol. Behav. 19, 255-259.
11. LaVail, M. M. and Battelle, B. A. (1975) Exp. Eye Res. 21, 167-192.
12. Deol, M. S. (1971) J. Embryol. Exp. Morph. 26, 123-133.
13. Harkins, R. K., Allison, J. E. and Macy, R. M. (1974) J. Hered. 65, 273-276.
14. Wise, R. P. and Lund, R. D. (1976) Exp. Neur. 51, 68-77.
15. Braekevelt, C. R. and Hollenberg, M. J. (1970) Amer. J. Anat. 127, 281-302.
16. Battelle, B. A. and LaVail, M. M. (1978) Exp. Eye Res. 26, 487-497.
17. LaVail, M. M. (1976) Invest. Ophthal. 15, 64-70.
18. Rapp, L. M. (1979) in The Effects of Constant Light on Visual Processes, this volume (Williams, T. P. and B. N. Baker, eds.)

19. Rapp, L. M., Howell, W. L., and Williams, T. P. (1979) Invest. Ophthal. Vis. Sci. 18 (Suppl.), 44.
20. McGinness, J. and Proctor, P. (1973) J. Theor. Biol. 39, 677-678.
21. Crouch, R. K. (1979) in The Effects of Constant Light on Visual Processes, this volume (Williams, T. P. and B. N. Baker, eds.).
22. Organisciak, D. T., Mitrosky, S. L. and Swafford, T. A. (1979) Invest. Ophthal. Vis. Sci. 18 (Suppl.), 112.
23. Anderson, K. V. (1979) in The Effects of Constant Light on Visual Processes, this volume (Williams, T. P. and B. N. Baker, eds.).

MEMBRANE TURNOVER IN ROD OUTER SEGMENTS: CHANGING PATTERNS OF SHEDDING AFTER EXTENDED PERIODS OF CONSTANT LIGHT AND DARKNESS

Joe G. Hollyfield, Mary E. Rayborn and
Scott F. Basinger
Cullen Eye Institute
Baylor College of Medicine
Houston, Texas 77030

The suggestion of Bairati and Orzalesi (1) that the dense bodies in the pigment epithelium resembled fragments of outer segment material was experimentally established in the autoradiographic studies of Young and Bok (2). The latter investigators followed packets of radioactive outer segment membranes as they were shed from the outer segment tips and phagocytized by the pigment epithelium. Their studies clearly demonstrated that the dense bodies were of rod origin and pointed to an important role for the pigment epithelium in the rod outer segment renewal process.

Renewed interest in rod shedding has followed the discovery by LaVail (3, 4) that the loss of outer segment membranes occurs in a near synchronous manner shortly after the onset of the light cycle in rats maintained under diurnal conditions of cyclic lighting. The timing of rod shedding has now been evaluated in amphibia (5, 6, 7), goldfish (8), rabbit (9) and squirrel (10). In all these animals, rod shedding is initiated shortly after the beginning of the light cycle. In contrast to rods, cone shedding in goldfish (8), lizard (11) and the chick (12) takes place after the beginning of the dark cycle.

The close correlation in the time of rod and cone shedding with the time of lighting transitions has prompted questions as to the consequences of photoreceptor shedding when diurnal lighting cycles are altered. When rats entrained to cyclic lighting failed to undergo a normal lighting change in the morning, rod shedding occurred at about the usual time each day even though the onset of the light cycle did not take place (4). This finding indicates that rod shedding is entrained by the light cycle but follows a circadian pattern in darkness. Similarly, when

Xenopus laevis conditioned to cyclic lighting were retained in darkness for three days, rod shedding, though somewhat reduced in amount, occurred at approximately the correct time each day (7). In contrast to the rat and *Xenopus*, when *Rana pipiens* entrained to cyclic lighting were kept in darkness, rod shedding did not take place for the first few days (5). The rod shedding response is abolished by constant light treatment in rats, *Xenopus* (7) and *Rana* (13). In the latter animal shedding has been inhibited for up to 10 days by constant light treatment.

Autoradiographic studies of photoreceptor membrane renewal in amphibia maintained in constant light or darkness indicate that although changes in the rates of membrane addition occur under these altered lighting conditions, new membranes continue to be added even though the process of membrane loss through shedding either ceases or is greatly reduced (7, 14, 15). The net result of this imbalance is an increase in rod outer segment length to approximately 1½ times that observed in cyclic light.

To date most of the observations of rod shedding patterns in altered lighting conditions have been made after relatively short periods of treatment and have utilized relatively few animals. In this study we evaluate rod shedding in adult *Rana pipiens* at three hour intervals on the first, seventh and twentieth day in constant light and darkness by monitoring the phagosome population of the pigment epithelium. Our analyses of dark maintained frogs indicate that rod shedding does not take place during day 1, but on day 7 and 20 shedding occurs at levels higher than observed in cyclic lighting. In constant light rod shedding is virtually abolished on days 1 and 7, but occurs spontaneously in some animals on day 20. When shedding occurs in constant light maintained frogs at this time, it involves virtually all rods in the retina but can proceed independently in either eye.

METHODS

Adult *Rana pipiens* used in this study were of the Northern variety purchased from West Jersey Biological Supply, Wenonah, New Jersey. Animals were conditioned for several months to a rigidly controlled lighting cycle of 14 hours light followed by 10 hours darkness per day (fluorescent ceiling lights providing 40-60 lux at the level of the animals containers). Frogs were housed in leucite holding bins with a constant flow of tap water to remove debris from the floor of the animal chambers. Frogs were fed each week with crickets during this period.

Phagosome counts from animals taken during a normal day were analyzed in frogs sampled on 3 different days with a total of nine animals per sample time. Each frog had received an injection of

^{3}H-mannose, ^{3}H-leucine or ^{3}H-uridine three hours prior to fixation (50-100 μCi per frog). An autoradiographic study of the diurnal metabolism of these compounds will be reported elsewhere (16).

Conditions of constant light or darkness were provided as follows: The 1 and 7 day constant light frogs were housed in shallow plastic pans containing 2-3 cm of water in constant temperature incubators (Freas, 25°C) fitted with fluorescent lights identical to those providing room illumination for the cyclic light maintained frogs. The 20 day constant light frogs were kept in similar pans in a room in which the fluorescent lighting remained on during the 20 days of treatment. The room temperature was 23-25°C.

Constant dark maintained frogs sampled on day 1 and day 7 of treatment were kept in incubators at temperatures identical to those used for the constant light experiments but in total darkness. The 20 day constant dark frogs were held in shallow pans of water inside a light proof box which in turn was kept inside a darkened room at 23-25°C. Water was changed daily throughout the period of treatment.

Three frogs were killed every 3 hrs throughout a 27 hr period. Each animal had received 3 hours earlier an intraperitoneal injection of ^{3}H-uridine (100 μCi/animal, Amersham Searle) for an analysis of RNA synthesis to be reported elsewhere (16). Eyes were removed and slit along the anterior segment with a razor blade and were placed in 4% glutaraldehyde in 0.087 M phosphate buffer at pH 7.2. After a few hours in a coldroom the anterior segment was removed and the posterior hemisphere bisected with a cut along the dorsoventral axis, passing through the optic nerve head. Small rectangular pieces containing approximately 1 mm of retina on either side of the optic nerve were cut from the two halves. After washing in buffer, the tissue was post-fixed in 1% osmium in the same buffer, then dehydrated and embedded in Epon. The tissue pieces were oriented in flat embedding molds prior to polymerization so that sections could be taken from the block face on either side of the optic nerve.

1 μm thick sections cut with glass knives were stained with toluidine blue. The pigment epithelium was studied and phagosome counts were made following procedures described previously (6). All measurements of phagosome dimensions, pigment epithelial expanse, rod lengths and retinal areas were made with a Zeiss microscope using 100X oil immersion lens and 10X objectives fitted with a micrometer with 0.93 μm per divisions.

RESULTS

To provide baseline data on the normal phagosome content of the pigment epithelium we have established the phagosome population throughout the day in Rana pipiens maintained in cyclic light. Although our previous work with this animal has documented a dramatic increase in phagosomes due to rod shedding following the onset of the light cycle (5), those data reflect only the time of arrival of newly shed phagosomes and do not follow the change in phagosome numbers as this membrane material is degraded during the remainder of the day. To obtain the latter information we analyzed the phagosome population in eyes from frogs killed at three hour intervals over a 27 hr period. All dense bodies with dimensions above 3 μm within a 990 μm linear expanse of pigment epithelium were counted in eyes from nine frogs recovered at each sample time. We eliminated the smaller phagosomes from these counts since their numbers changed relatively little throughout the day in the adult, as has previously been reported for the tadpole (6). As can be seen in this data (Fig. 1), prior to the onset of the light cycle (7 AM sample) the phagosome population is relatively small. Following the onset of the light cycle there is an increase in phagosome number. Thereafter phagosomes decrease in number, reaching a level virtually identical to that observed at the beginning of the previous morning. These data on the daily changes in the phagosome population during a normal day in cyclic light will be used for comparison with the phagosome population in animals recovered after 1, 7 and 20 days of constant light and constant darkness presented below.

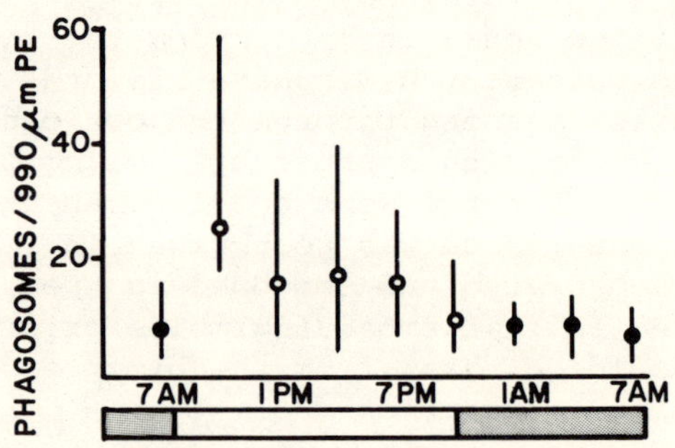

Figure 1: Phagosome counts in the pigment epithelium during a normal day of 14 hrs light followed by 10 hrs darkness (light on 8 AM, off 10 PM). The bar below the abscissa depicts the lighting changes. Each point represents the mean phagosome count in the pigment epithelium from nine frogs sampled at each 3 hr interval. The bars represent the maximum and minimum numbers of phagosomes present in the samples. Note that the phagosome content increases following the onset of the light cycle and decreases thereafter.

Shedding in Constant Light

The lighting conditions of our 1 day constant light animals were as follows: the first sample was taken at 9 AM, 2 hrs after the normal onset of light in the morning and the remainder at 3 hr intervals thereafter. The animals remained in light through the night and into the following morning rather than undergoing the usual transition to darkness at 10 PM. The last sample was taken at 12 Noon, 27 hrs after the first recovery time. Phagosome counts from this material are presented graphically in Fig. 2. As can be seen in this data, the phagosome population is high in the 9 AM samples as is expected for animals which had undergone the transition from darkness to light that morning. Phagosome numbers remain high during the next two sample times, decreasing to lower levels later in the day. The last two samples (9 AM and 12 Noon) on the twenty-first and twenty-seventh hour of constant light treatment were taken at a time when the pigment epithelium in cyclic light animals would have contained large numbers of phagosome. However, the phagosome content in the pigment epithelium remained low through this period, indicating that rod shedding was inhibited by constant light.

The phagosome population after 7 days of constant light treatment (Fig. 3) was well below the level seen during a normal day. Though two samples show a mean phagosome count of around 11-12 phagosome/990 μm PE, all the phagosomes present were around 3-5 μm

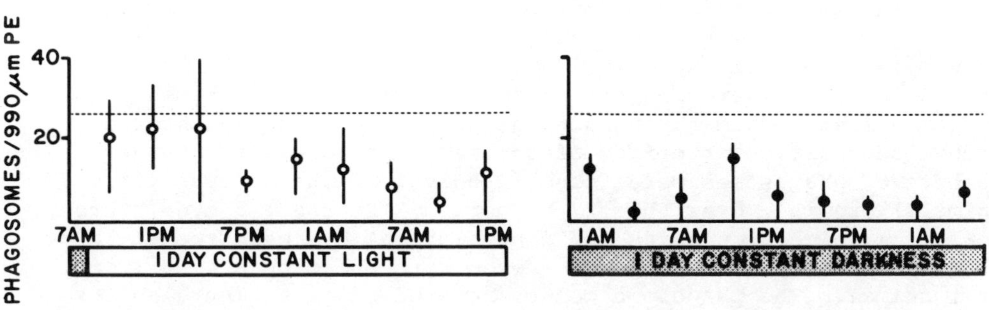

Figure 2: Phagosome counts in the pigment epithelium from frogs fixed during one day of constant light (left) or constant dark treatment (right). Each point represents the phagosome number in the pigment epithelium from 3 animals killed at each sample time. The dotted line across the graph in this and in Figures 3 and 4 represents the mean phagosome content of the pigment epithelium at the peak response time during a normal day of cyclic light as shown in Figure 1. This is used as reference in discerning whether shedding is below or above the level present in frogs maintained in cyclic light. See text for further explanation.

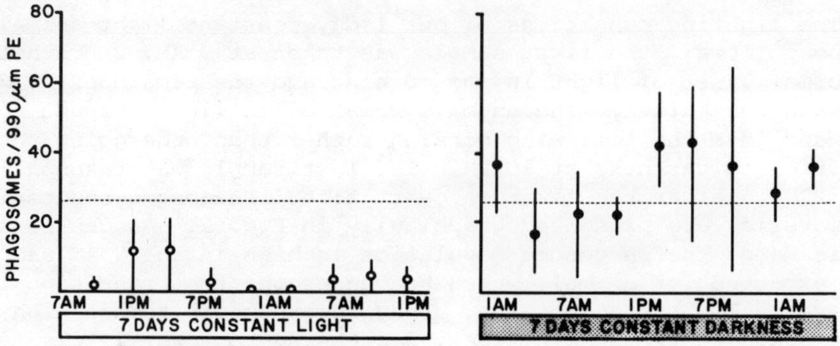

Figure 3: Phagosome content of the pigment epithelium in frogs sampled throughout the seventh day of constant light (left) or constant dark treatment (right). Conventions followed in displaying this data are the same as described for Fig. 2. See text for further explanation.

in diameter, characteristic of phagosomes present in the pigment epithelium for some time and not due to a recent rod shedding event.

In animals maintained for 20 days in constant light (Fig. 4) the phagosome content of the pigment epithelium was either low (5 sample times) or exceedingly high (4 sample times). The high phagosome counts were present in samples recovered at times randomly distributed throughout the day and were not clustered around any particular period of the sampling schedule. Large numbers of phagosomes were found in only 5 of the 54 eyes (from 27 animals) examined over this 27 hr period. Of these 5 eyes, three were from animals in which the contralateral eye was free of phagosomes whereas the other two eyes with phagosomes were from the same animal. The phagosome counts and rod outer segment lengths from these four animals are presented in Table 1 and representative photomicrographs are shown in Fig. 5. As is evident in this data the eyes in which large numbers of phagosomes were observed also possessed rod lengths in the contralateral eye containing only a few phagosomes (Frogs A-C). In the remaining animal, phagosomes were present in both eyes and though the rod lengths differed between these two eyes, the differences were less than those present in the remaining frogs. These observations indicate that spontaneous shedding had occurred unilaterally in three of the constant light frogs but had taken place in both eyes in the remaining animal.

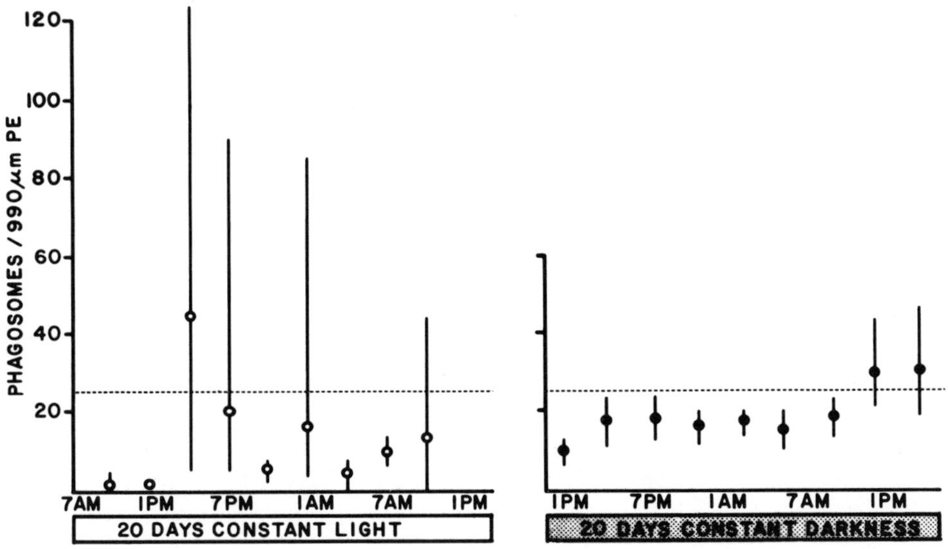

Figure 4: Phagosome content of the pigment epithelium in frogs sampled throughout day 20 of constant light (left) or constant dark treatment (right). Note the elevated phagosome content of the pigment epithelium in 4 sample times on day 20 of constant light. On day 20 of constant darkness note that the beginning of the sampling period is different from day 1 and day 7 of darkness. At the end of the sampling period in the constant dark maintained frogs the phagosome content is above the number present at the peak response time during a normal day. See text for further explanation.

Shedding in Constant Darkness

We began sampling the 1 day constant dark frogs three hours after the normal onset of darkness in the evening. Eyes were recovered at 3 hr intervals through the following morning and for three sampling periods into the next day. Phagosomes counts from this sampling group are presented in Fig. 2. As can be seen in this data, phagosomes are few in number at the beginning of our sampling period. There is a slight increase in the 10 AM samples, suggesting a small amount of shedding, though this is well below the phagosome number present in animals which undergo normal transitions from darkness to light prior to this sampling time. For the remainder of the first day in darkness, the phagosomes population remains at a very low level. It is obvious from this data that shedding did not take place to any appreciable extent during the first day of constant darkness.

On the seventh day of darkness, phagosomes are present in numbers exceeding those observed during the maximum response time

TABLE I

PHAGOSOME COUNTS AND ROD LENGTHS IN EYES FROM SELECTED FROGS TAKEN ON DAY 20 OF CONSTANT LIGHT TREATMENT

FROG	FIXATION TIME	EYE	PHAGOSOMES per 990 μm PE	RED ROS LENGTHS MEAN (μm) ± SEM
A	4 AM	I	123	23.4 ± 0.2
		II	5	36.7 ± 0.5
B	7 PM	I	90	30.1 ± 0.2
		II	14	39.2 ± 0.5
C	10 AM	I	85	30.2 ± 0.9
		II	4	41.7 ± 0.5
D	7 AM	I	44	34.6 ± 0.5
		II	24	39.0 ± 0.7

Table I: Red rod lengths (mean ± SD based on 20 measurements per eye) and phagosome counts from animals sampled on day 20 of constant light treatment in which large numbers of phagosomes were present in the pigment epithelium. Note that in frogs A-C the high phagosome content in one eye (I) is accompanied by short rods in the subjacent retina as compared to the other eye (II) which contains only small numbers of phagosomes. (Though our tissue identification procedure allowed us to identify retinas which came from each eye we do not know whether the retinas are from left or right eyes). In frog D phagosomes are present in both eyes and rod lengths are not as drastically different between the two eyes as in frogs (A-C).

of a normal day (Fig. 3). Large numbers of phagosomes were present in the 1 PM sample.

On the twentieth day of darkness, during the first portion of the sample period phagosome numbers were below those present at the peak response time on a normal day (Fig. 4). The counts remained low through 7 AM but in the 1 PM sample phagosomes were present in excess of the quantity present on a normal day.

DISCUSSION

The observations presented in this study extend our knowledge of the changing characteristics of rod shedding in constant light

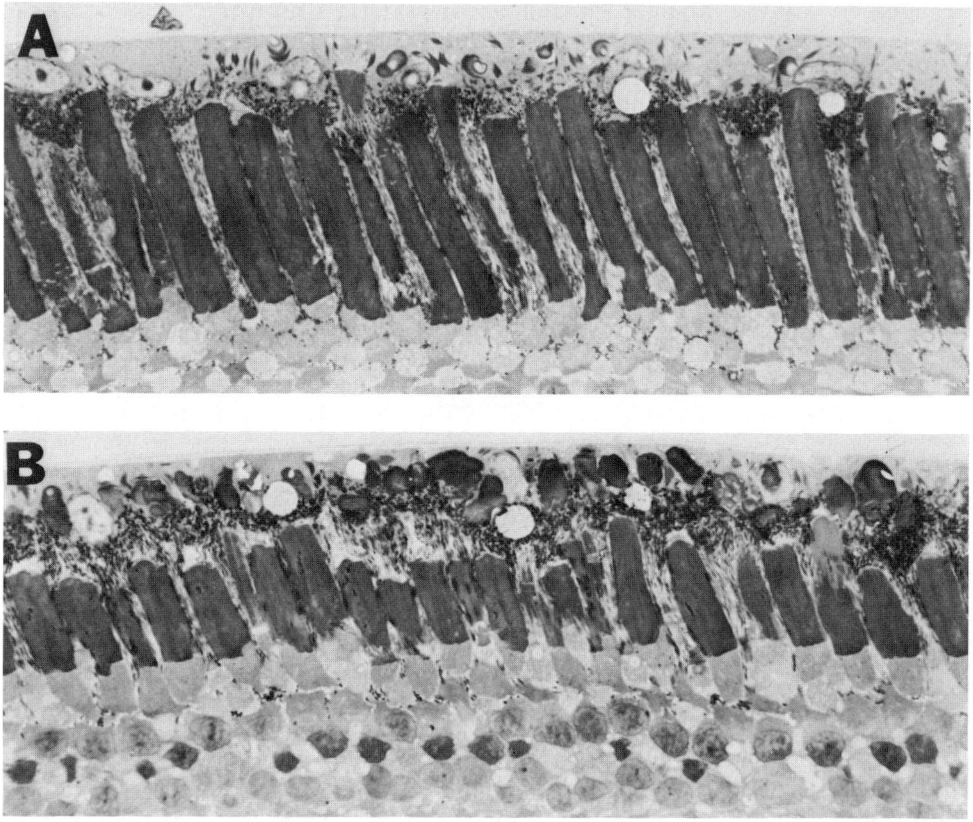

Figure 5: Photoreceptor pigment epithelial interface from both eyes in frog A (see Table I) in which unilateral rod shedding was observed on day 20 of constant light treatment. Note long rods in (A) and a pigment epithelium containing myeloid bodies but free of phagosomes. In (B), rod lengths are reduced as compared to those in (A) and the pigment epithelium contains numerous phagosomes.

and darkness in the frog. In our previous studies of the pigment epithelium during the first few days of darkness we did not observe recently ingested phagosomes, indicating that rod shedding was inhibited by this treatment (5). The one day constant dark data reported herein are consistent with this previous observation. In contrast to the 1 day data, phagosome counts from frogs maintained for 7 and 20 days in darkness indicate that large numbers of phagosomes accumulate within the pigment epithelium. It is clear from these phagosome counts that shedding does occur in darkness after prolonged treatment. The number of phagosomes which have accumulated within the pigment epithelium throughout day 7 are in excess of those present at the peak response time on a normal day of cyclic light. Though the exceedingly high numbers of phagosomes at particular times in the 7 and 20 day constant dark material may be due to the existence of an underlying endogenous rhythm of shedding in darkness it is not possible to reach this conclusion at the present time. An analysis of the portion of phagosomes which are newly shed is in progress and will be dealt with in a forthcoming report (Basinger and Hollyfield, and others, in preparation).

In the rat (4) and *Xenopus* (7) entrained to cyclic lighting, circadian shedding occurs during the first day of maintenance in darkness. Though it is not readily apparent why shedding does not occur in *Rana* maintained under these same conditions, it may be related to a slower rate of membrane addition in this animal. Autoradiographic studies of membrane addition in red rods of *Rana pipiens* maintained at 23°C reveal a rate of ^{3}H-band displacement following ^{3}H-leucine injection of less than 1 μm/day (6) whereas in *Xenopus* (7) and the rat (17, 18) displacement rates above 2 μm/day have been observed. Slower rates of membrane addition in *Rana* rods coupled with the fact that amphibians require daily periods of light exposure to maintain a normal rate of membrane synthesis and/or assembly (7, 14) suggest that the failure of shedding during the first few days in darkness may be related to the slower rate of membrane addition in this animal. Within this regard it should be pointed out that even in *Xenopus* with a rate of membrane addition more than twice that of *Rana* the shedding which occurs on the first and second day of darkness involve only one-half the rods that undergo daily shedding in cyclic light maintained animals. It is not until the third day in darkness that the phagosome population in *Xenopus* reaches normal levels. It is clear from our data that shedding in *Rana* does not occur in the rod population on the first day of darkness but does occur after dark treatment for longer intervals.

Our previous studies on constant light maintained frogs have reported an inhibition of shedding for up to 10 days with this treatment (13). The data presented in this study for 1 and 7 days confirms this previous report. However, the 20 day constant light shedding data reveal a pattern of membrane loss not previously observed. The data indicates that shedding cannot be inhibited indefinitely by constant light. Furthermore, when spontaneous shedding occurs in constant light it can proceed independently in either eye suggesting that the control of shedding is not under central control by a gland or organ elsewhere in the body as was previously suggested (4), but lies within the eye itself. These observations on independent shedding in left and right eyes prompted our directed experiments of eye patching which firmly established that the rod outer segment shedding phenomena is under intraocular initiation (19).

We do not know how light and darkness influence the rod outer segment shedding process. Several lines of evidence suggest that it may be through its effect on serotonin-melatonin metabolism (4). This idea is consistent with our observed cyclic pattern of photoreceptor shedding following prolonged dark treatment and the lack of rhythmicity in constant light. Pineal gland output and blood levels of melatonin occur as free running rhythms in darkness which are eliminated by constant light treatment (20, 21). Though we have ruled out the possibility that rod shedding is mediated by extraocular factors (13, 19), melatonin is also found in the retina and its concentration in this tissue increases in darkness (22). Other possible candidates for a role in photoreceptor shedding are cyclic-nucleotides. The retinal levels of cGMP are high in the dark and are reduced following light exposure (23).

Though the direct involvement of these or other metabolites in the rod shedding process must await further study, it is clear from the pattern of rod shedding after prolonged maintenance in constant light and in our unilateral shedding results following monocular eye patching (19), that the agents which mediate membrane loss are produced within the eye and perhaps within the retina itself.

ACKNOWLEDGEMENT

Supported by grants from the Retina Research Foundation, Houston, Texas, and the National Eye Institute (NIH). Joe G. Hollyfield and Scott F. Basinger are recipients of Research Career Development Awards from the National Eye Institute (NIH).

REFERENCES

1. Bairati, A., Jr., and Orzalesi, N. (1963). J. Ultrastruct. Res. 9, 484-496.
2. Young, R. W. and Bok, D. (1969) J. Cell Biol. 42, 392-403.
3. LaVail, M. M. (1976) Exp. Eye Res. 23, 277-280.
4. LaVail, M. M. (1976) Science (Wash. D.C.) 194, 1071-1074.
5. Basinger, S., Hoffman, R. and Matthes, M. (1976) Science (Wash. D.C.) 194, 1074-1076.
6. Hollyfield, J. G., Besharse, J. C. and Rayborn, M. E. (1977) J. Cell Biol. 75, 490-506.
7. Besharse, J. C., Hollyfield, J. G. and Rayborn, M. E. (1977) J. Cell Biol. 75, 507-527.
8. O'Day, W. T. and Young, R. W. (1978) J. Cell Biol. 76, 593-604.
9. Bunt, A. (1978) Invest. Ophthalmol. Vis. Sci. 17, 90-104.
10. Tabor, G. A., Fisher, S. K. and Anderson, D. H. (1978) Invest. Ophthalmol. Vis. Sci. 17, 135 (Suppl.).
11. Young, R. W. (1977) J. Ultrastruct. Res. 61, 172-185.
12. Young, R. W. (1978) Invest. Ophthalmol. Vis. Sci. 17, 105-116.

13. Currie, J. R., Hollyfield, J. G. and Rayborn, M. E. (1978) *Vis. Res.* 18, 995-1003.
14. Besharse, J. C., Hollyfield, J. G. and Rayborn, M. E. (1977) *Science* (Wash. D.C.) 196, 536-538.
15. Hollyfield, J. G. (1979) *Invest. Ophthalmol. Vis. Sci.* 18, (in press).
16. Hollyfield, J. G. and Basinger, S. (1979) *Invest. Ophthalmol. Vis. Sci.* 18, (Suppl).
17. LaVail, M. M. (1973) *J. Cell Biol.* 58, 650-661.
18. Besharse, J. C., and Hollyfield, J. G. (1979) *Invest. Ophthalmol. Vis. Sci.* 18, (in press).
19. Hollyfield, J. G., and Basinger, S. (1978). *Nature* 274, 794-796.
20. Ralph, C. L., Mull, D., Lynch, H. J. and Hedlund, L. (1971) *Endocrinology* 89, 1361-1366.
21. Kline, D. C. (1974) in *Neurosciences Third Study Program*, F. O. Schmitt and F. G. Worden, eds., The MIT Press Cambridge , Mass., pp. 509-515.
22. Bubenik, G. A., Purtill, R. A., Brown, G. M. and Grota, L. J. (1978) *Exp. Eye Res.* 27, 323-333.
23. Goridis, C., Virmaux, N., Cailla, H. L. and Delaage, M. A. (1974) *FEBS* (*Fed. Eur. Biochem. Soc.*) *Lett.* 49, 167-169.

DAMAGING EFFECTS OF CONSTANT LIGHT AND DARKNESS ON THE RETINA OF THE FROG

Joe G. Hollyfield, Mary E. Rayborn and Donna Medford

Cullen Eye Institute
Baylor College of Medicine
Houston, Texas

Since the discovery of Noell (1) that continuous light of moderate intensity causes severe damage to the albino rat retina, the effects of constant light on retinal structure and function has been the subject of numerous investigations in a variety of animals including man (recently reviewed by Lanum, 2). Most of these studies have dealt with the temporal aspects of light damage to the photoreceptor layer, the cell type most severely affected by this treatment. Though light damage to the retina in pigmented animals has also been reported, in general higher intensity and/or longer duration of exposure are required and the severity of the damage is not as pronounced as in albino strains. Studies in several laboratories indicate that monochromatic light causes specific destruction of selected cone types in pigmented animals and suggests that light damage may be mediated by photopigments (3, 4).

We recently examined the retinas in a large number of frogs maintained in constant light or darkness for 1, 7 or 20 days for studies on rod shedding and changes in retinal metabolism under these conditions. During the analysis of this material we were struck by the frequent observations of pyknotic retinal cells. We systematically analyzed this material for cell death and degenerative retinal changes. The results of this study indicate that both constant light and darkness result in some cell death in all retinal layers. Our quantitative data indicate that constant darkness produces more cell loss than constant light at identical treatment times.

METHODS

The conditions for maintenance of adult *Rana pipiens* used in this study are described in detail in the preceding paper (5). All quantitative data were obtained with light microscopy using an 100X planapochromat oil immersion lens with 10X ocular fitted with a micrometer.

RESULTS

The following criteria were used to distinguish cell death in the retina: The most common feature was a pyknotic nucleus, characterized by the intense affinity for the toluidine blue that was used to stain all of our retinal sections (Fig. 1). In addition to these

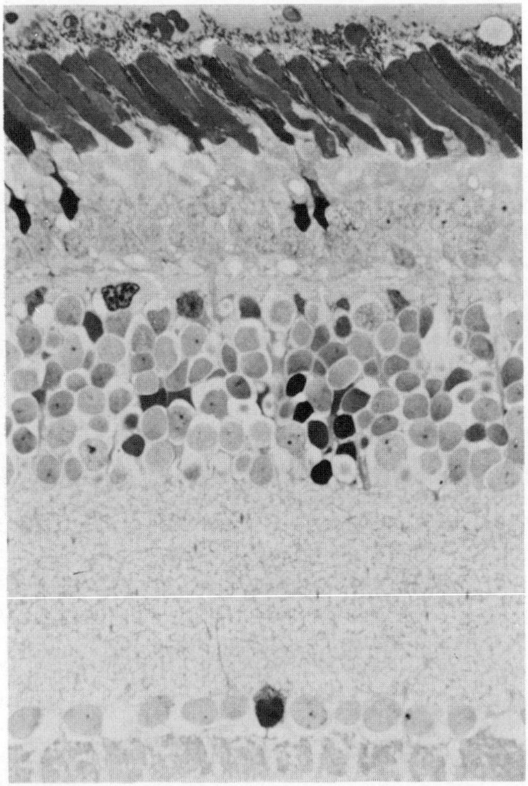

Figure 1. Retina from a frog fixed after 7 days of constant light. Pyknotic cells are present in the ganglion cell, inner nuclear and outer nuclear layer.

nuclear changes, in the photoreceptor layer this intense basophilia was observed throughout the length of the rod from the tip of the outer segment to the synaptic base (Fig. 2). We did not observe these changes in cone photoreceptors. Pyknosis was also observed in retinas from cyclic light maintained animals though infrequently.

To assess the relative differences in the degree of degenerative changes brought about by constant light and darkness, we quantitated these changes by establishing the frequency of dead cells observed at increasing intervals of treatment. The normal frequency of cell death in the retina in cyclic light maintained animals was also determined.

As can be seen in this quantitative data (Table 1), approximately half of the cyclic light maintained frogs sampled throughout a normal day contain some pyknotic cells. This however represented from 0.1-0.4% of the total cells surveyed. In contrast, after 7 to 20 days in constant light or darkness, nearly all the retinas examined showed evidence of cell death. Also the frequency of pyknotic cells from the dark maintained retinas show 3-4 times as many dead cells as compared to retinas from animals maintained for identical periods in the light. These comparisons strongly indicate that constant light under the relative low levels of illumination is detrimental to retinal cell viability but that constant dark is more damaging to retinal cells than constant light.

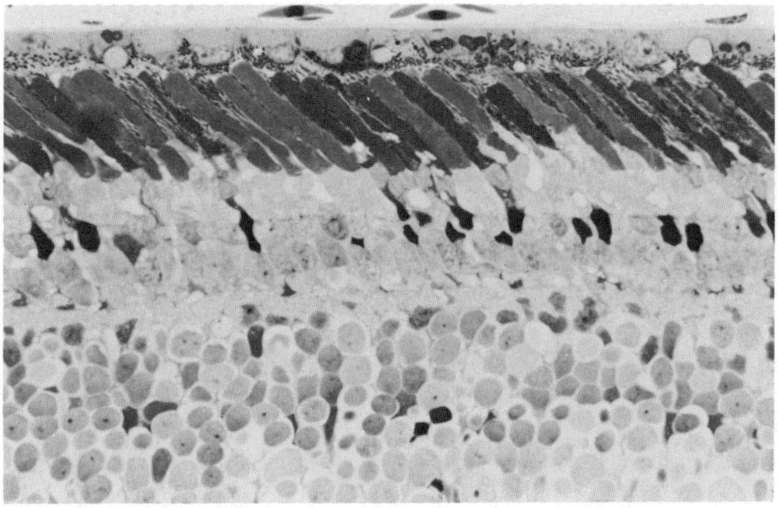

Figure 2. Retina from a frog fixed after 20 days of constant darkness. Note that the intense basophilia along the complete length of numerous rod photoreceptors.

TABLE I.

Distribution of pyknotic cells in retinas from animals maintained under cyclic lighting (14 hrs light; 10 hrs dark) constant light (LL) and constant darkness (DD). Eyes were recovered at 3 hr intervals throughout a 24-27 hr period under the indicated lighting conditions. Three animals were killed at each sample time. The percentage calculations are based on an average of 207 ± 19 photoreceptor cells (rods and cones), 507 ± 62 nuclei in inner nuclear layer, and 66 ± 9 ganglion cell nuclei in a 990 μm expanse of retina (the linear retinal expanse surveyed per eye). Note that there is a progressive increase in the number of pyknotic cells with increasing periods of LL or DD and that maintenance in DD at equivalent treatment times results in more retinal cell death than LL.

CELL DEATH IN THE RETINA OF THE FROG *RANA PIPIENS* IN RESPONSE TO MAINTENANCE IN CONSTANT LIGHT AND DARKNESS

Lighting Conditions	Number of Animals Examined	Animals With Pyknotic Retinal Cells*	Pyknotic Cells Present (Per cent of total)		
			Photoreceptor Layer	Inner Nuclear Layer	Ganglion Layer
Cyclic Light	27	14 (1-4)	3 (0.1%)	17 (0.1%)	7 (0.4%)
7 day LL	27	23 (1-43)	15 (0.3%)	231 (1.7%)	65 (3.7%)
20 days LL	30	28 (1-60)	36 (0.6%)	275 (1.8%)	75 (3.8%)
7 days DD	27	26 (1-54)	63 (1.1%)	376 (2.8%)	104 (5.8%)
20 days DD	30	29 (1-83)	175 (2.8%)	808 (5.3%)	122 (6.2%)

*(Low-high/990μm retina)

In addition to cell death, we also observed depigmentation of the pigment epithelium in five animals maintained for 20 days in constant darkness. This was characterized by the absence of melanin granules in large expanses of the pigment epithelium. In adjacent ocular regions, cells, ladened with melanin granules (probably macrophages) were commonly observed. They were present in choroidal blood vessels, in the space between the retina and pigment epithelium throughout all levels of the retina, and in the vitreous (Fig. 3).

DISCUSSION

Most of the experimental studies evaluating the deleterious effect of constant light on the retina have used albino animals as experimental subjects. These studies demonstrate that severe photoreceptor damage occurs within several hours to a few days treatment. In general, pigmented animals are more resistant to light damage than are albinos of the same species. The results presented in this study indicate that in the frog some cell death results from constant light treatment but more degenerative changes are present following maintenance in darkness.

Early studies by Dowling and Sidman (6) on albino strains of the RCS rat with retinal dystrophy demonstrated that the progression of photoreceptor degeneration was slowed when animals were raised in the darkness. These and other findings prompted the clinical trials of eye patching as a possible means of retarding vision loss in humans with early stages of retinitis pigmentosa (7). Our findings of photoreceptor damage in dark maintained frogs indicate that prolonged dark treatment may be detrimental to photoreceptor viability.

The Ozark cave salamander, *Typhlotriton spelaeus* shows a progressive loss of photoreceptor cells near the end of the larval period. Besharse and Brandon (8) evaluated the effects of light and darkness on the progression of retinal degeneration in this animal. Their studies demonstrate that the changes occurring in continuous darkness were more extensive than in constant light. Their observations are consistent with our findings in *Rana pipiens* that prolonged darkness caused substantial degenerative changes in the retina.

In summary, our observations of retinas from adult frogs subjected to increasing periods in constant light and darkness indicate that both these treatments caused the death of retinal cells. Darkness, however, is more deleterious to retinal viability than is constant light after equivalent treatment times. Our data suggests that a daily cycle of light alternating with darkness is essential for optimum retinal viability in this amphibian.

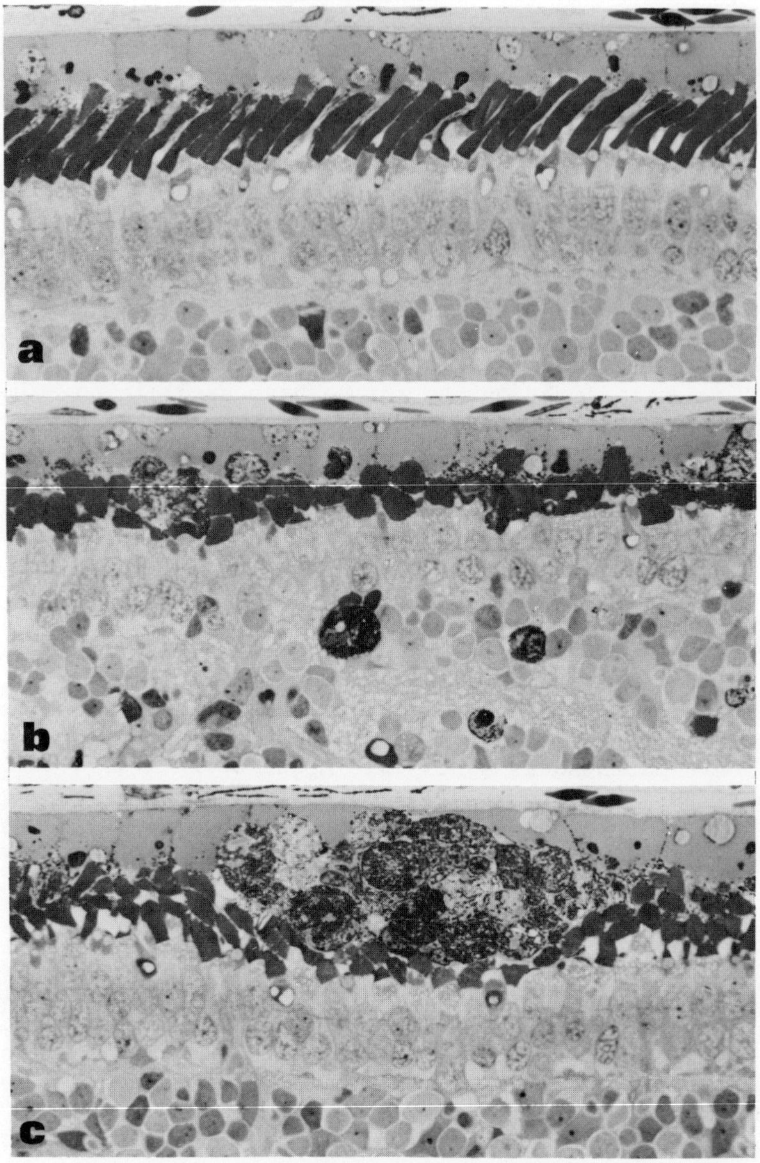

Figure 3. Sections from a retina of an animal fixed after 20 days of continuous darkness. The pigment epithelium is nearly free of melanin. (A) Melanin granules are absent or greatly reduced in this region of the pigment epithelium. (B) Melanin laden cells are present between the retina and pigment epithelium and at the level of the inner plexiform layers (Arrows). (C) Several melanin laden cells are present between the retina and pigment epithelium. These are possible macrophages which have engulfed melanin discarded by the pigment epithelial cells.

This study was supported by a grant from the Retina Research Foundation, Houston and by grants from the National Eye Institute. Joe G. Hollyfield is a receipient of a Research Career Development Award from the National Institutes of Health.

REFERENCES

1. Noell, W. K., Walker, V. S., Kang, B. S. and Berman, S. (1966) Invest. Ophthalmol. 5, 450-473.
2. Lanum, J. (1978) Survey Ophthalmol. 22, 221-249.
3. Marshall, J., Mellerio, J. and Palmer, D. A. (1972) Exp. Eye Res. 14, 164-169.
4. Sperling, H. (1978) in Frontiers of Visual Science, (S. J. Cool and E. L. Smith, III, eds.) pp. 138-153, Springer-Verlag, New York.
5. Hollyfield, J. G., Rayborn, M. E. and Basinger, S. F. (1979) (Preceding paper in this symposium).
6. Dowling, J. E. and Sidman, R. L. (1962) J. Cell Biol. 14, 73-107.
7. Berson, E. L. (1971) Arch. Ophthalmol. 85, 521-529.
8. Besharse, J. C. and Brandon, R. A. (1976) J. Morph. 149, 527-546.

LIGHT AND MEMBRANE BIOGENESIS IN ROD PHOTORECEPTORS OF VERTEBRATES

Joseph C. Besharse

Department of Anatomy
Emory University School of Medicine
Atlanta, Georgia 30322

Throughout the life of an individual, retinal visual cells carry on an elaborate process of photosensitive membrane turnover which is at once prodigious and exquisitely adapted to maintenance of optimal function. The magnitude of the phenomenon, first defined in detail by the work of R. W. Young (1), can best be appreciated by realizing that rod outer segments of rodents are renewed entirely within a period of 10 to 12 days. During this period, outer segment dimensions in adult animals remain stable through the establishment of a balance between synthetic and degradative functions (2). Synthesis of photosensitive membrane precursors occurs in the cell body (inner segment) after which they are transported to the base of the outer segment for incorporation into membrane discs (3, 4). Ultimately, discs are degraded within the lysosomal system of retinal pigment epithelial cells (5, 6).

Outer segment turnover may be regarded as analogous to plasma membrane turnover as it occurs in a variety of other cell types. It is different, however, in that it is highly polarized and involves interactions with pigment epithelial cells. In addition, rod outer segment (ROS) discs along with their major integral membrane component, visual pigment, are degraded in direct relationship to molecular age (4). In most cells plasma membrane turnover appears to be a global process which involves the entire surface structure. Macromolecules are incorporated throughout the membrane and degradation occurs randomly with regard to molecular age, predominantly in the cells own lysosomal system (7, 8). Although photoreceptors are capable of autophagic activity (9), degradation of photosensitive membrane appears to be the exclusive function of pigment epithelial cells (6).

The analogy between plasma membrane turnover and ROS turnover is based on the idea that ROS discs are derived from ROS or ciliary plasma membrane (10, 11). Such a process is thought to involve either infolding of ROS surface membrane (11) or outpocketing of ciliary membrane (12). In either case, the general idea is consistent with a wide variety of observations demonstrating patency of open discs (13-15), the presence of rhodopsin in the ROS plasma membrane (16, 17), and a similarity in the size frequency distributions of intramembranous particles in freeze-fractured ROS plasma membrane, open discs and closed discs (18). Continuation of open disc formation at a rate sufficient to maintain outer segment structure, however, requires plasma membrane expansion. That such an expansion process is not trivial is illustrated by observations on *Xenopus laevis* which indicate that some 3600 μm^2 of membrane are incorporated into discs each day, an amount more than 5 times greater than the total ROS and ciliary surface area (19). At the present time the mechanism by which the ciliary or ROS plasma membrane is maintained in support of disc assembly represents a major unsolved problem.

Insight into the nature of the control mechanism of outer segment turnover has recently been attained with the discovery that photoperiod plays a fundamental role in regulating both disc shedding (20) and disc assembly (21). Shedding of large packages of discs from ROS tips occurs predominantly soon after light onset each day in rats (22), mice (23), frogs (24, 25), and chickens (26). In contrast, shedding of discs from cone outer segments (12) occurs predominantly after light offset (26, 27). In frogs photoperiod related shedding accounts for about 80% of disc loss from the ROS, the remaining 20% being accounted for in a photoperiod independent shedding phenomenon involving small packages of discs (28). Evidence also suggests that cyclic shedding is not strictly a light induced phenomenon. For example, disc shedding is greatly reduced in frogs in constant light (19, 29), but continues in a rhythmic fashion at a reduced level in constant darkness (19, 22). Observations of an endogenous rhythm along with observations suggesting that disc shedding is controlled within the eye (30-32) raise the possibility that an endogenous timing mechanism intrinsic to the eye regulates rhythmic shedding.

Disc assembly normally balances disc loss resulting in maintenance within narrow limits of a standard ROS length (2). In amphibians light accelerates the rate of disc assembly to a variable extent depending on species, ambient temperature and prior state of dark-adaptation (19, 21). Light acceleration of disc assembly is potentially important because it provides an experimental approach and suggests a number of new ideas related to the control of outer segment assembly. The purpose of the present paper is to review the available data relevant to light effects on ROS disc assembly among vertebrates and to present new data which indicates that light stimulates disc assembly *in vitro*. A major impediment to further understanding of control mechanisms at the present time is a lack of knowledge of the essential features of the mechanism of disc assembly. The paper ends, therefore, with consideration of morphological evidence for a carrier vesicle mechanism of plasma membrane expansion during ROS disc assembly.

PHOTOPERIOD AND DISC ASSEMBLY IN ROD PHOTORECEPTORS

Effects of light on ROS disc assembly have been measured by two different techniques. First, displacement of a radioactive band (^{3}H-band) within the ROS has been measured under conditions of controlled temperature in different lighting regimes (19, 21). This approach, particularly useful for measurement of average disc assembly rates over long time intervals, has the virtue of providing data directly related to the total amount of disc membrane produced. Such a relationship is established through a combination of autoradiography and electron microscopy in which the number and dimensions of discs per unit of ROS length are measured (19). A second approach, valuable for monitoring light effects over short time intervals, utilizes electron microscopic observations of basal open discs. The latter method, applicable to date only to *Xenopus laevis*, is based on the observation that during periods of accelerated ^{3}H-band displacement there is a net accumulation of open discs at the ROS base (19). Although the total open disc population can be converted to an estimate of membrane surface area, values thus obtained potentially represent a substantial underestimation of total membrane incorporated into discs because a parallel process involving transformation of open discs into closed discs occurs simultaneously. For example, *in vivo* experiments at 28°C in cyclic light indicate that during the first 8 hours of light 83% of the open discs formed are transformed into closed discs. Nonetheless, the approach provides valuable data particularly in the *in vitro* analysis of disc assembly where measurement over short time intervals is essential. Although it is now apparent that outer segment turnover in cones occurs by a process analogous to that of rods (12, 26, 27), direct measurements of the rate of turnover or of photoperiod effects on COS assembly have not been made.

Effects of Light on Amphibian Photoreceptors

In the frogs *R. pipiens* and *X. laevis* light increases the average rate of ^{3}H-band displacement over that occurring in darkness (DD) by amounts ranging from 14 to 117% depending on species, temperature and total duration of light exposure each day (19, 21). An analogous effect of light has been observed in developing *X. laevis* photoreceptors (33), and a small but significant effect of light has been observed in the Ozark cave salamander (34), where light exposure decreases the extent of postmetamorphic photoreceptor degeneration (35). In general, the effect of light is greater in X. laevis than in R. pipiens, and is most effective when delivered continuously (Table 1). Although stimulation of ^{3}H-band displacement is greater in constant light (LL) than in cyclic light (LD = 12L:12D), the relationship between total light exposure each day and average rate of ^{3}H-band displacement appears to be nonlinear for tadpoles of both species. For example, in X. laevis kept in LD at 28°C, ^{3}H-band displacement is doubled compared to that in DD, but is only increased an additional 8% in LL (Table 1). Likewise, exposure of R. pipiens tadpoles to only 2 hours of light per day at 23°C is sufficient to produce 44% of the effect in constant light as opposed to the 8% expected if the effect were precisely additive (21). These observations suggest that light exposure immediately after dark-adaptation has a "triggering effect", a suggestion consistent with observations on accelerated disc assembly after both prolonged and overnight dark-adaptation (21).

The effect of light on ^{3}H-band displacement is potentiated by prolonged dark-adaptation (21). Such an effect was originally suggested by studies on visual pigment content of R. pipiens eyes where exposure to light after 6 days of darkness resulted in a precipitous decline in visual pigment concentration and ROS length which paralleled ROS disc shedding, but was followed by a rapid recovery of both ROS length and visual pigment levels (36). Subsequently, the suggestion was confirmed by the finding that the rate of ^{3}H-band displacement during 24 hours of LL following 6 days in DD was as much as 4 times that during the previous period in DD (21). As in the case of average ^{3}H-band displacement, the extent of disc assembly after prior dark-adaptation is related to temperature. In both R. pipiens and X. laevis the extent of the effect, evaluated as an acceleration factor (Table 1), is reduced in tadpoles maintained at a higher temperature, and is reduced in R. pipiens adults as well.

The acceleration factor is useful because it expresses light stimulation relative to the prior period in darkness. In R. pipiens, however, the accelerated rate was lower at 28°C in contrast to the reverse situation in X. laevis (Table 1). The significance of this difference between the two species is not immediately obvious. Another difference between the two species is in the pattern of the effects of temperature on ^{3}H-band displacement in different lighting regimes (Table 1). In R. pipiens the temperature coefficient (Q_{10}) is 3.6 - 3.7[1] in each lighting regime. In contrast, a comparable

---

[1]These Q_{10} values may be calculated from data in Table 1 by extrapolation over the range of 18-28°C. Comparable Q_{10} values in the three lighting regimes over the temperature range 16-26°C have been observed (Hollyfield, personal communication). Above 28°C temperature increases ^{3}H-band displacement to a lesser extent (28).

Q_{10} value is found in X. laevis in LD, but the Q_{10} is lower (2.4-2.7) in both LL and DD.

The above observations suggest a complex interaction between temperature and lighting regime. A partial explanation for the reduced acceleration factor and the relatively small difference between LL and LD in X. laevis at 28°C may be that extensive acceleration of disc assembly is only possible at a temperature where the ongoing rate of disc assembly is below its physiological maximum. As temperature increases, the rate of disc assembly approaches its maximum precluding the possibility of further extensive stimulation by subsequent light exposure. Thus, at 18°C LL increases ^{3}H-band displacement substantially above that occurring in LD and a large acceleration factor is recorded, but at 28°C the difference between LL and LD is relatively small and the accelerated rate after prior dark-adaptation does not exceed that occurring in LL or LD (Table 1).

Because of the extremely high rate of disc assembly in X. laevis it has been possible to detect acceleration of ^{3}H-band displacement during the first 8 hours after light onset in tadpoles maintained in LD (19, 21). If the progress of ^{3}H-band displacement is followed at 8 hour intervals during the day following injection of ^{3}H-leucine, 75-80% of the total daily displacement occurs during the first 8 hours in light while significant displacement is not detected during the last 8 hours in darkness (Fig. 1). The suggestion that such a daily rhythm of ^{3}H-band displacement reflects a rhythm of disc assembly, is supported by the observation that the number of basal, open discs increases dramatically during the same period that the ^{3}H-band is maximally displaced (Fig. 1). Inasmuch as ROS discs are believed to form from open discs we have interpreted the autoradiographic and electron microscopic data as indicating that during light accelerated disc assembly each day, open discs accumulate at the ROS base because the rate of open disc formation exceeds the rate of closed disc formation. Later in the day when open disc formation declines, continued formation of closed discs results in a decrease in the number of open discs. Autoradiographic observations indicate that some 64 new discs are formed during the first 8 hours in light. Thus, accumulation of open discs would occur during this period if closed discs formed at a rate of one every 9 minutes and an open disc formed every 7.5 minutes.

Light probably has a direct stimulatory effect on disc assembly. This is in contrast to disc shedding where an endogenous rhythm influences disc loss even in the absence of photoperiod cues (19, 22). The evidence supporting the idea that rhythmic disc assembly, as it occurs in X. laevis, is predominantly a light stimulated rhythm is threefold. First, we have failed to find a rhythmic pattern of ^{3}H-band displacement during successive days in DD (19). Furthermore, an increase in the number of open discs in animals entrained to LD and then transferred to DD fails to occur. Finally,

TABLE 1

Effects of lighting regime on ^{3}H-band displacement in rod outer segments. In each experiment animals were placed in modified lighting regimes 24 hours after intraperitoneal injections of ^{3}H-leucine. The rate was calculated with the formula $(B_{t2} - B_{t1})/(t2 - t1)$ where B is distance in μm of the ^{3}H-band from the ROS base, t1 is 24 hours after ^{3}H-leucine injection and t2 is the time in days at fixation. The accelerated rate is the total ^{3}H-band displacement during 24 hours in constant light after long term dark-adaptation. Except where indicated the prior dark-adaptation period was 6 days. Compiled from references 19 and 21 and unpublished data of Besharse.

Experimental Group	Temperature °C	Rate of ^{3}H-band Displacement μm/day				Acceleration[1] Factor
		12L:12D	Constant Light	Dark	Accelerated Rate	
Rana pipiens tadpoles	23	0.83(32)[2]	0.97(54)[2]	0.63	2.60	4.1
	28	1.33(36	1.52(55)	0.98	1.80	1.8
Rana pipiens adults	23	0.84(14)	1.14(54)	0.74	1.50	2.0
Xenopus laevis tadpoles	18	0.65(30)	0.93(86)	0.50	1.90	3.8
	28	2.40(100)	2.60(117)	1.20	2.20[3]	1.8
	28	2.20(100)	----	1.10	2.00	1.8
Mice, C57Bl pigmented o	--	2.30(5)	2.40(9)	2.20	1.60[3]	---

[1]Acceleration factor is the factor by which ^{3}H-band displacement is increased above the rate during the prior period of darkness.

[2]Parenthetical values represent the percent increase over the rate in darkness in the same experiment.

[3]The period of prior dark-adaptation was 3 days.

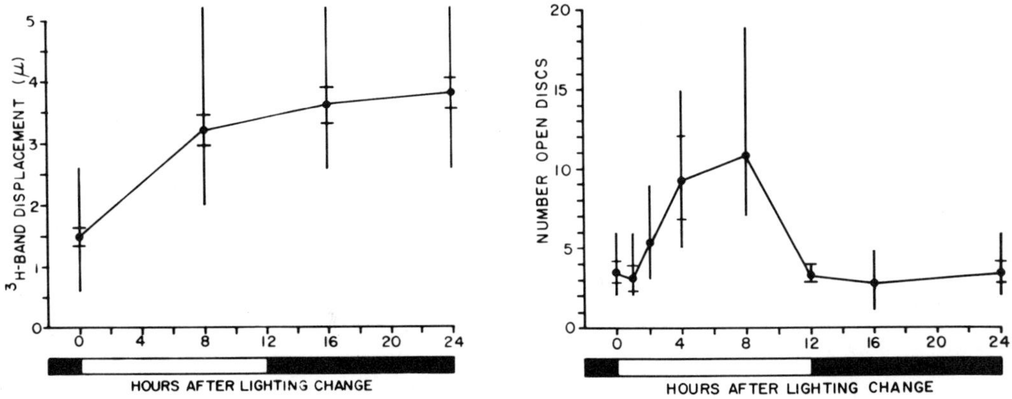

Figure 1. Observations indicating that disc assembly occurs predominantly during the first 8 hours of daylight in *X. laevis* maintained in LD. Left: ^{3}H-band displacement measured at 8 hour intervals beginning 24 hours after injecting animals with ^{3}H-leucine. Note that about 75% of the total daily displacement occurred during the first 8 hours in light. Right: Average number of open discs at the ROS base throughout a day in LD. Vertical bars represent ranges and crossbars delineate a distance corresponding to two standard errors on both sides of the means. From reference 19 with permission of the Rockefeller University Press.

light stimulated disc assembly occurs in retinal explants separated from the pigment epithelium and maintained in culture for 6 to 8 hours (see below).

Perhaps owing to fundamental differences in the control mechanisms of disc assembly and disc shedding, it is possible to dissociate the two processes in modified lighting regimes. For example, in LL disc shedding is decreased to a low level while disc assembly is maintained at a maximal level; the result is a rapid and dramatic increase in ROS length. ROS length increases 21% within 3 days (Table 2) in LL and at the end of 6 days ROS are 43% longer. Thus, ROS increase in length at an average rate of 1.6 μm per day during a period when a ^{3}H-band is displaced 2.6 μm per day. Obviously, disposal continues during this period and must account for about 38% of the discs added. Part of the disc disposal occurs on day one of LL treatment when a normal photoperiod related shedding event occurs (19). This leaves only about 23% of the disposal to be accounted for. Inasmuch as a photoperiod-independent shedding phenomenon, involving small phagosomes, continues unabated in LL (19) and has been shown to normally account for about 20% of disposal (28), it appears that this mechanism may account for the small amount of

TABLE 2

Summary of changes in ROS disc assembly, disc shedding and changes in ROS length among _Xenopus laevis_ tadpoles kept in different lighting regimes for 3 days. Data from reference 19.

	Lighting Regime			
Measurement	12L:12D	Constant Light	2.5L:21.5D	Constant Darkness
^{3}H-band displacement μm	7.2	7.8	4.5	3.6
Renewal rate μm/day	2.4	2.6	1.5	1.2
% of ROS shedding/day[1]	25	12	20	16
ROS length after treatment	23.0 ± 0.5[2]	27.7 ± 0.7	22 ± 1.9	24.4 ± 0.4
% Change in ROS length	0	21	0	0

[1] This value is an estimate of photoperiod related shedding of large packets of discs. The value for LL includes a normal shedding event on day one of treatment with virtual elimination of such shedding on days two and three. Photoperiod independent shedding of small packets of discs continued at about the same level in each lighting regime.

[2] The ROS length for 12L:12D represents the average length throughout the experiment and is the control value to which other treatments should be compared.

disc disposal occurring in LL. Similar observations have been reported for adult *R. pipiens* although the total apparent increase in ROS length exceeds what would be expected on the basis of renewal data (29).

ROS length also increases in DD (24, 36) where both disc assembly and disc shedding are decreased but to a different degree. In addition, it may be possible to decrease ROS length by maintaining animals in a modified lighting regime of 2.5 hours light and 21.5 hours darkness per day (19). Under these conditions rhythmic shedding is maintained at a high level while disc assembly is slowed; the result in three days is a trend toward decreasing ROS length (19).

Photoperiod also influences outer segment elongation of developing *X. laevis* photoreceptors (33). Here the rate of accumulation of volume at 23°C is much greater in LL than DD with LD falling between the two extremes. The rate of ROS assembly in those experiments was about the same in LL and LD whereas outer segment assembly in darkness was reduced by about 30 percent. Differences in outer segment assembly rate, however, did not account entirely for the differences in volume. Differences in disc shedding which began earlier and at a greater rate in LD and DD also contributed to total volume differences. Thus, in developing photoreceptors, as in tadpoles (above), modified lighting regimes influence outer segment dimensions by differentially affecting disc assembly and disc shedding.

The observations suggesting that light has a direct effect in stimulating ROS disc assembly raises the question of the nature of the light effect. Explanations based on local temperature changes in the microenvironment of photoreceptors (1) probably do not apply because the magnitude of the light effect under optimal initial temperature conditions is simply too large (19, 21). As indicated above the endogenous control mechanism which effects rhythmic shedding also seems an unlikely candidate since disc assembly and disc shedding can be uncoupled. We are left, therefore, with the idea that the disc assembly mechanism may be controlled in relationship to some other light induced modification of photoreceptor metabolism. Although, at present, data are not available that would suggest which aspects of photoreceptor metabolism are likely candidates for involvement in the control of disc assembly, light-evoked changes in membrane potential, changes in cyclic nucleotide levels or cytoplasmic calcium represent possible features of the control system. Further studies relating to the nature of the light effect on disc assembly must be considered in relationship to *in vitro* models where appropriate experimental manipulation is possible (*vide infra*).

Photoperiod and Turnover in Mammalian Rods

In contrast to results with amphibians, recent observations on ROS turnover in C57Bl mice (black pigmented) suggest a greater degree of intrinsic regulation of ROS turnover independent of photoperiod (23). Light delivered either as LD or LL has only a small effect on ^{3}H-band displacement or on ROS length (Table 3). The small effects observed, although statistically significant, amount to variations, among the three lighting regimes analyzed, of 7% or less. The observations suggest not only that disc assembly occurs independent of photoperiod, but also that the balance between disc assembly and disc loss is approximately maintained in unusual lighting regimes since ROS dimensions show little change. This conclusion is in keeping with the observation of substantial numbers of phagosomes in the pigment epithelium of animals maintained in both DD and LL (23). The lack of a substantial effect of light over periods of several days is paralleled by the lack of a stimulatory effect of light after a period of three days in darkness (Table 1). These observations raise the question of whether or not control of disc assembly is fundamentally different in amphibian and rodent retinas. Although such a difference must be regarded as a possibility, we have suggested an alternative interpretation based in part on studies of amphibian retinas. As indicated above, the extent of the light effect on disc assembly is related to temperature in amphibia. As the upper limit of the physiological temperature range

TABLE 3

Radioactive band displacement and rod outer segment length in C57Bl mice maintained under different lighting regimes. Values given are the mean plus or minus the standard error. All animals were maintained in 12L:12D for 24 hours after injections and were then transferred to different lighting regimes. Data from reference 23.

Days after Injection	Group No.	Lighting Regime	Distance to ^{3}H-band μm	Rate μm/day	ROS Length μm
1	I	LD	1.8 ± 0.09	---	27.9 ± 0.33
	II	LD	8.8 ± 0.18	2.3	27.6 ± 0.30
4	III	LL	9.3 ± 0.16	2.5	26.3 ± 0.34
	IV	DD	9.0 ± 0.12	2.4	27.1 ± 0.32
	V	LD	11.1 ± 0.14	2.3	27.5 ± 0.37
5	VI	LL	11.2 ± 0.17	2.4	25.4 ± 0.22
	VII	DD	10.3 ± 0.19	2.1	27.1 ± 0.23

is reached the rate of disc assembly approaches its physiological maximum, precluding additional acceleration due to light stimulation alone. In the mouse a similar situation may pertain with the standard disc assembly rate at the normal regulated body temperature ($\sim 37^{o}C$) approaching the physiological maximum. If this were the case, one would expect light to have little effect on disc assembly.

The lack of a substantial light effect on ROS disc assembly in C57Bl mice is similar to that reported for albino rats maintained under conditions of intense illumination (1). The apparent regulation of ROS dimensions in C57Bl mice differs, however, not only from the situation seen in frogs (see above), but also from that observed in albino rats (37, 38). In the latter case, constant light delivered for relatively short time intervals results in degeneration whereas constant darkness results in increased ROS length (up to 32% longer) and an even greater increase in whole eye rhodopsin content (38). Increases in ROS length are apparently due to establishment of a transient "assembly-loss" imbalance in the ROS turnover mechanism, although the nature of the imbalance remains to be characterized. Observations on rhodopsin content of pigmented rat eyes in LD reveal levels of rhodopsin comparable to that seen in albinos after 10 days in DD. Furthermore, maintenance of pigmented rats in DD results in only a small increase in visual pigment (7-9%). These observations have led to the suggestion that apparent differences between C57Bl mice and albino rats may be related primarily to differences in the amount of melanin screening pigment (23).

Light Stimulated Disc Assembly *in Vitro*

One approach to further analysis of light effects on disc assembly is to develop culture conditions under which the relevant phenomena are sustained. Such a culture system would obviously provide ideal conditions for analysis of the cellular mechanism of disc assembly as well as the relevant control mechanisms. Previous studies on frog and rat retinal explants have indicated that disc assembly continues in photoreceptors at least during short periods in culture (39, 40). Explant culture of *R. pipiens* retinas has been used to study aspects of ROS turnover including incorporation of ^{3}H-leucine into rhodopsin (40), metabolism of phosphatidylcholine (41) and distribution of radiolabelled rhodopsin in the ROS plasma membrane (17). On the basis of data from such an explant system it has been suggested that light does not influence incorporation of ^{3}H-leucine into rhodopsin of outer segments (40). Recently, however, we have reexamined the idea of light effects on disc assembly in *X. laevis* retinal explants and found that light stimulates disc assembly under a variety of different culture conditions. Our approach has been to take advantage of both the high inherent rate of disc assembly in *X. laevis* compared to *R. pipiens* as well as the potentiating effect of prior dark-adaptation on light-stimulated

TABLE 4

Observations indicating that light stimulates ROS disc assembly in *X. laevis* retinal explants *in vitro*. Retinas were removed from stage 55-57 tadpoles which had been maintained in DD for 4 (experiments III and IV) or 6 days (experiments I and II). Dissections were carried out under dim red light and retinas were transferred to culture medium. Subsequently groups of explants were transferred to light (tungsten illumination of about 10 lu/m^2) or retained in darkness for 6 (experiments II-IV) or 8 (experiment I) hours before fixation. The dark values for open discs are not significantly different from values at the beginning of the culture period. The estimate of membrane surface area involved in open discs is based on an average disc diameter of 6 μm.

EXPERIMENT	TREATMENT	OPEN DISCS $\bar{X} \pm S\bar{x}$	RANGE	SURFACE AREA μm^2
I	Dark	1.8 ± 0.4	1-3	102
	Light	9.1 ± 1.1	6-15	514
II	Dark	2.0 ± 0.6	0-7	113
	Light	7.0 ± 0.7	3-15	396
III	Dark	1.8 ± 0.2	1-4	102
	Light	7.4 ± 1.8	4-15	424
IV	Dark	0.6 ± 0.1	0-2	34
	Light	7.6 ± 0.5	3-15	430

disc assembly. Furthermore, we have used direct electron microscopic observation of open discs at the ROS base as a criterion for disc assembly. Thus, in a typical experiment animals are dark-adapted for a period of 4-6 days following which retinas are removed under dim red light and transferred to culture medium.[2] Subsequently, retinal explants are incubated in light or darkness and new disc assembly evaluated after varying intervals in culture.

Analysis of open discs in explants maintained in light or darkness indicate that the number of open discs increases dramatically in light compared to darkness (Table 4). In the analysis of open

[2]We have used several different media all of which sustain light stimulated disc assembly but which vary in the extent to which structure of cone outer segments and basal open discs are maintained without vesiculation. At present our standard medium is a slight modification of the amphibian tissue culture medium of Wolf and Quimby (42).

discs we have used criteria like those established in a previous study on open disc formation *in vivo* (19). In addition to confirming our previous observation that basal discs with a wider intradiscal space are open, we have been able to routinely confirm that discs are open at their margins by use of specimen tilting in a goniometer stage. Such observations, in addition to demonstrating light stimulation of ROS disc assembly in the absence of pigment epithelium, also suggest that open discs are formed without incisures initially, and that disc maturation involves formation of peripheral incisures as well as loss of continuity with the plasma membrane.

The quantitative correspondence between the *in vitro* observations (Table 4) and our previous observations on open disc formation *in vivo* (19), suggest that light-evoked membrane assembly detected in culture is of the same order of magnitude as in the intact eye. The measurement, however, is an indirect one since many of the open discs formed are normally transformed into closed discs soon after assembly. Thus, additional data are required to more precisely define the extent of the light effect in culture. Preliminary autoradiographic data, however, suggest that a preformed ^{3}H-band is displaced distally during light exposure *in vitro* which suggests that more discs are formed than are detected as open discs. Thus, the estimate of membrane surface incorporated into discs in Table 4 probably represents a substantial underestimation of the light effect *in vitro*.

Paralleling the changes in open disc formation during culture in light (Table 4) is an increase in incorporation of ^{3}H-leucine into retinal protein. Incorporation of radioactivity into trichloroacetic acid-precipitable protein occurs linearly over a period of 6 to 8 hours *in vitro* and total incorporation is consistently greater in light than darkness. The extent of the increase is variable, however, ranging in different experiments for 25 to 100%. Furthermore, a light-evoked increase in incorporation has only occasionally been detected at shorter time periods (3-4 hours). We have not yet determined whether the light-evoked increase in incorporation reflects increased protein synthesis, nor have we been able to explain the variation in the light effect among different experiments. Nonetheless, autoradiographic examination of retinal explants indicates that ^{3}H-leucine incorporation by both rods and cones is increased substantially in light (Table 5). Although the effect is less extensive, the observations *in vitro* are similar to autoradiographic observations (Besharse and Papermaster, in prep.) which indicate an increase in incorporation of radioactivity by retinal photoreceptors in light among animals receiving intraperitoneal injections of ^{3}H-leucine. In the latter experiments, *X. laevis* tadpoles were dark-adapted 4-6 days and then given injections of ^{3}H-leucine in dim red light. Subsequent light exposure resulted in a 62 to 112% increase in incorporation by photoreceptors.

TABLE 5

Autoradiographic data indicating that light increases incorporation of ^{3}H-leucine into photoreceptors kept for 6 hours in culture medium containing 10μCi/ml of ^{3}H-leucine. Light microscope autoradiographs of paraformaldehyde fixed explants were prepared by methods described previously (19). Analysis of silver grain distributions was made using a Hewlett-Packard Digitizer for determination of surface areas. Data on inner segments represent the myoid and ellipsoid region combined.

	Grains/100 μm^2		
	Light	Dark	% Increase in Light
Rod Inner Segment	37.7 ± 3.0	26.2 ± 1.3	44.2
Cone Inner Segment	27.7 ± 2.5	18.2 ± 2.6	52.3
Outer Nuclear Layer	21.6 ± 1.7	23.5 ± 2.2	----
Outer Plexiform Layer	17.3 ± 2.3	23.4 ± 4.2	----

A natural question arising from these experiments is whether or not the observed increase in ^{3}H-leucine incorporation is necessary for the light-evoked increase in membrane assembly. In an attempt to obtain data related to this question we have analyzed light-stimulated disc assembly after 6 hours in culture in retinal explants incubated in the presence of puromycin and cychloheximide. Both drugs reduce incorporation of ^{3}H-leucine into trichloroacetic acid precipitable protein of retinal explants to about 2% of control levels when provided in a dose of 50 μg/ml of culture medium. With each inhibitor disc assembly, assayed as an increase in the number of open discs in light, failed to occur. Thus, it would appear that ongoing protein synthesis is necessary for light stimulated disc assembly *in vitro*. Whether complete inhibition of disc assembly is accomplished cannot be answered with our present assay system. Nonetheless, light- and dark-treated retinas look virtually identical at the ultrastructural level.

THE MECHANISM OF DISC ASSEMBLY

The nature of the light effect on disc assembly in rod photoreceptors must ultimately be understood in relationship to the mechanism of disc assembly. At the cellular level there are a series of discrete events involved in disc assembly which may

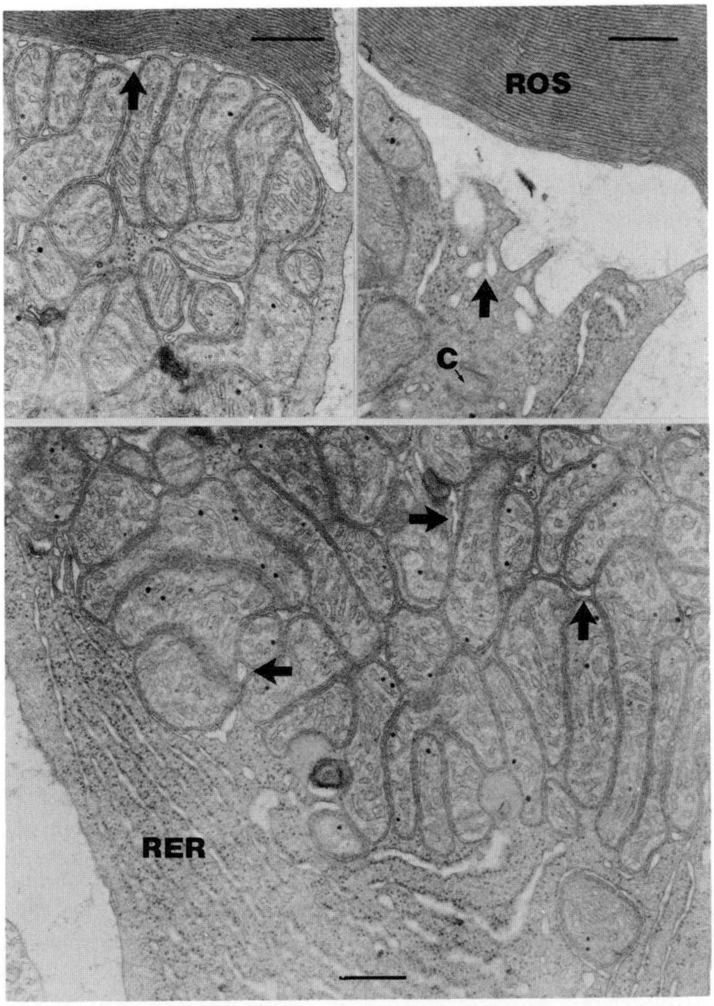

Figure 2. Electron micrographs illustrating vesicles and cisternae in the perimitochondrial and periciliary cytoplasm of the ellipsoid region of rod inner segments of *Xenopus laevis*. Upper left: Micrograph of the inner segment-outer segment junction illustrating vesicles and ROS open discs (Mag. 28,900 X). Upper right: Micrograph illustrating the inner-segment-outer segment junction in the region adjacent to the connecting cilium. The irregular plasma membrane suggesting either endocytotic or exocytotic activity is a consistent feature of this region of the cell. Open discs are visible because of the wider intradiscal space (Mag. 30,000 X). Below: Micrograph illustrating the myoid-ellipsoid junction. Perimitochondrial cisternae are abundant (Mag. 28,900 X). Calibration bars correspond to 0.5 μm. Abbreviations: ROS, rod outer segment; C, centriole; RER, rough endoplasmic reticulum. (Reduced 30% for purposes of reproduction.)

also be regarded as events at which regulatory controls could exert their effect. These events include synthesis and processing of ROS membrane precursors, transport of precursors from the site of synthesis to the ROS and incorporation of precursor into the photosensitive membrane system. Subsequent events involve formation of open discs, maturation of discs (including separation from the plasma membrane as well as formation of incisures), and distal transport of discs within the ROS. As indicated in the introduction, a central feature of any mechanism for ROS turnover must be a mechanism for ROS plasma membrane expansion in support of disc assembly.

A working hypothesis to account for rapid plasma membrane expansion is that preformed membrane packages (i.e., carrier vesicles or "schleppersomes") (43) are added to the cell surface adjacent to the connecting cilium, and that membrane then flows over the cilium into discs. This hypothesis is consistent with a variety of experimental observations on photoreceptors and is in part analogous to that derived from consideration of secretion in pancreatic exocine cells (44, 45). In greater detail the hypothesis predicts that 1) intrinsic membrane components are synthesized in the rough endoplasmic reticulum where they remain associated with the lipid bilayer which 2) flows *via* the Golgi complex to the cell surface. 3) Glycosylation of membrane components occurs in transit beginning at endoplasmic reticulum and saccharide chains of the glycoconjugates remain associated with the luminal leaflet (EF leaflet) of the cytoplasmic membranes. 4) Plasma membrane expansion occurs when carrier vesicles fuse with the surface membrane by a process analogous to exocytosis, thus, exposing saccharides of glycoconjugates to the extracellular space. Considerable published data exists in support of the basic aspects of this scheme. For example, we know the site of synthesis (34) and major sites of glycosylation of visual pigment (46). Cell fractionation studies (43, 47) are consistent with the carrier vesicle hypothesis since opsin is transported to the outer segment in close association with sedimentable membrane. In addition, immunocytochemical observations suggest that immunoreactive opsin passes through the Golgi zone in transit to the ROS (48). The suggested membrane asymmetry with regard to saccharide groups has not been demonstrated in photoreceptors, but the idea is consistent with observations on a variety of other cell types where lectin receptors are restricted to the internal (luminal) leaflet of cytoplasmic membranes and to the outer surface of the plasma membrane (45, 49-51).

Morphological evidence for carrier vesicles is available for both embryonic (52) and mature photoreceptors. In larval and adult *X. laevis* we have analyzed cytoplasmic membranes of both rod and cone inner segments in relationship to lighting regimes which influence disc assembly (19). We have consistently found a population of irregular vesicles in the ellipsoid region immediately adjacent to the connecting cilium which could play a role in membrane assembly

(Fig. 2). In addition to the rather large ellipsoid vesicles, we have found similar vesicles and cisternae among and around the mitochondria of the ellipsoid (Fig. 2), and have also observed a population of smaller vesicles similar in ultrastructure to synaptic vesicles in the same cell. The large periciliary vesicles frequently show continuity with the plasma membrane which may represent sites of plasma membrane expansion. In *X. laevis* such sites (Fig. 2; upper right) are a consistent feature of the plasma membrane adjacent to the cilium. Similar periciliary vesicles have also been found in embryonic *X. laevis* photoreceptors at the time of initial disc assembly where it has been suggested that they represent outer segment membrane precursor (52). Evidence consistent with the idea that the large vesicles and cisternae represent ROS precursor material comes from freeze-fracture (18) and more recently from electron microscope autoradiographic analysis (Besharse and Papermaster, in preparation).

Freeze-fracture analysis of photoreceptor membranes in *X. laevis* has revealed that within the outer segment and connecting cilium, the PF leaflet is characterized by a population of intramembranous particles (IMP) with an average size near 10 nm and a density near 4600 per μm^2 (Table 6). In general, the IMP of the outer segment appear to be uniformly distributed in the connecting cilium, in ROS plasma membrane, in closed discs, and, as far as we are able to judge from a relatively small sample, in open discs as well. In contrast, the inner segment plasma membrane contains smaller IMP with a lower density (Table 6). We have also been able to identify vesicles in replicas of the ellipsoid region of both rods and cones (Fig. 3). The fractured surfaces of the vesicles are characterized by a population of IMP of a size similar to that found in the outer segment (Fig. 3; Table 6) and clearly distinct from that of the inner segment plasma membrane. Although the similarity in average size of IMP in vesicles and outer segment membranes suggests a precursor product relationship, it should be emphasized that the IMP density is much lower in vesicles than in the outer segment. Thus, if the vesicles do represent membrane precursor, a mechanism for increasing the density of IMP relative to lipid bilayer must exist.

Electron microscope autoradiographic observations are also consistent with the hypothesis that ellipsoid vesicles represent membrane precursor material. It has been demonstrated (3, 4) that after injection of radioactive amino acids into frogs, initial incorporation into protein occurs in the rough endoplasmic reticulum of the myoid region of photoreceptors, and that during the following two hours radioactivity is transported to the base of the outer segment for incorporation into discs. During transport it appears that the bulk of the radioactivity passes through the Golgi complex and the perimitochondrial cytoplasm to the region of the connecting cilium (3). Cell fractionation studies during transport of

TABLE 6

Comparison of density and average size of intramembranous particles in the PF leaflet in freeze-fracture replicas of *Xenopus laevis* rods and cones. Retinas were fixed in a glutaraldehyde-paraformaldehyde mixture, treated with glycerol, frozen in Freon 22 and fractured in a Balzer's freeze-fracture apparatus. The average density and size of particles in discs was not significantly different from that in ROS plasma membrane, connecting cilium or open discs. Note the great similarity of IMP density and size in rods and cones, the differences between inner segment (plasma membrane and calycal process) and outer segment (discs) membranes, and the similarity of IMP size in vesicles to that in discs. Note also that the IMP density in vesicles is less than that of other membranes. Compiled from size frequency distributions (Besharse and Pfenninger, in prep.).

	IMP/μm^2		Average Particle Size (nm)	
Membrane	Rod	Cone	Rod	Cone
Inner Segment Plasma Membrane	2989 ± 81	2627 ± 86	7.6 ± 0.12	7.9 ± 0.14
Calycal Processes	3523 ± 161	3092 ± 203	7.5 ± 0.12	7.4 ± 0.11
Discs	4629 ± 63	4613 ± 86	10.2 ± 0.14	10.1 ± 0.15
Ellipsoid Vesicles	1921 ± 121	2050 ± 239	9.9 ± 0.16	9.7 ± 0.15

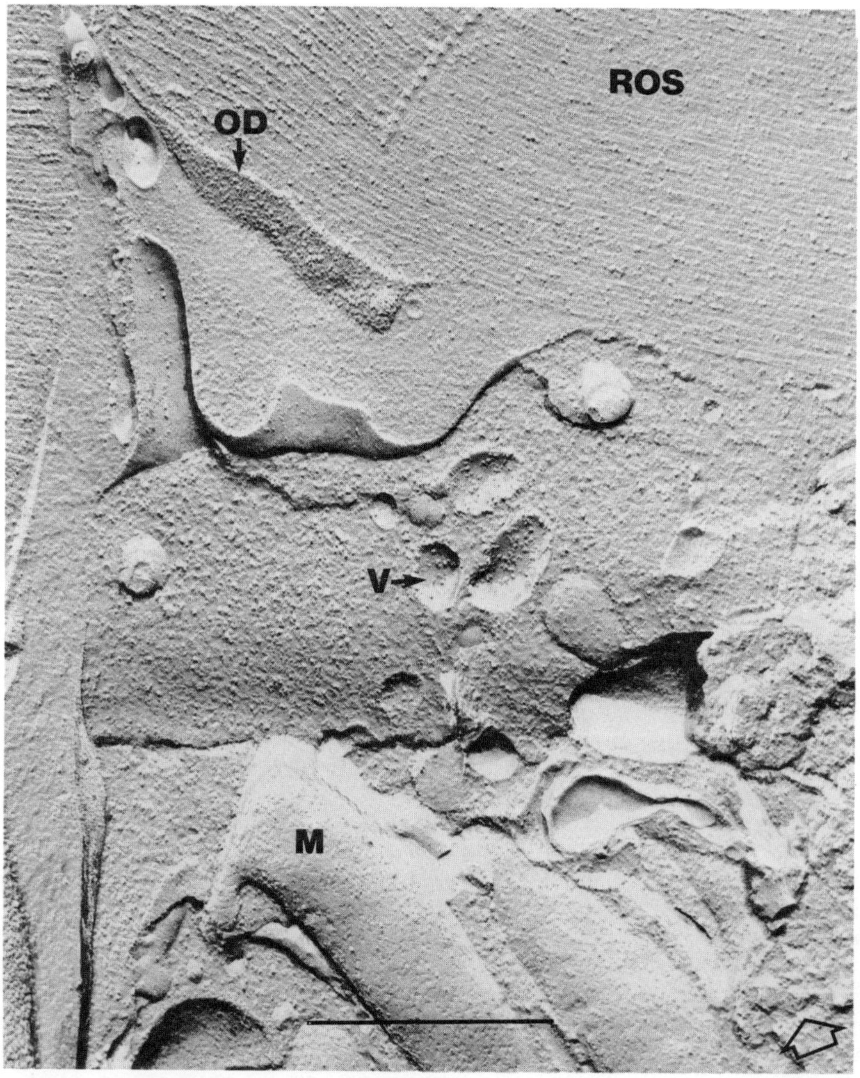

Figure 3. Electron micrograph of a freeze-fracture replica in the region of the inner segment-outer segment junction of an aldehyde fixed rod of *Xenopus laevis* illustrating the vesicles so characteristic of the ellipsoid region. Discs of the ROS are cross-fractured except where the plasma membrane is continuous with a basal disc (OD). Note that the intramembranous particles of the PF leaflet (concave) of vesicles (V) are of a size comparable to that of the outer segment. Abbreviations: ROS, rod outer segment; OD, basal open disc; V, vesicle; M, mitochondrion. Shadowing direction indicated by the arrow in the lower right. Calibration bar corresponds to 0.5 μm. (Mag. 80,750 X). (Reduced 20% for purposes of reproduction.)

radiolabelled opsin have indicated that cytoplasmic membranes act as carriers (43, 47). Recently we have carried out an electron microscope autoradiographic analysis of distribution of radioactive protein during transport to the outer segment in X. laevis (Besharse and Papermaster, in preparation). The objective was to carry out an experiment similar to that of Young and Droz (3) and, in addition, determine to what extent radioactivity is associated with cytoplasmic vesicles. Our work confirms both the general observations and kinetic interpretations of previous studies (3, 4). In addition, however, we have used a quantitative analytical technique for silver-grain localization over underlying cellular organelles which partially compensates for radiation spread from point sources, and completely compensates for differences in relative area of the various organelles in thin sections (53). From this analysis we have been able to demonstrate that within the ellipsoid region radioactivity is associated primarily with the cytoplasmic vesicles described above. The most heavily labelled compartments are the perimitochondrial cytoplasm, which itself is composed primarily of cytoplasmic membranes, and the vesicles in the periciliary region. We interpret the observations as direct evidence for the carrier vesicle hypothesis, since a large fraction of the observed radioactivity is quickly transported to the ROS.

SUMMARY

Light stimulates disc assembly in amphibian photoreceptors to a variable extent depending on species, lighting regime, ambient temperature and prior state of dark-adaptation. Although not so extensive in mammalian photoreceptors, light stimulation of disc assembly may be important because it provides an experimental approach and new ideas relevant to the control of outer segment renewal. Analysis of available data suggests a nonlinear relationship between rate of disc assembly and total light exposure in which light after dark-adaptation has a triggering effect on the disc assembly mechanism. In addition, light stimulation of disc assembly decreases at higher temperature. Since increases in temperature also stimulate disc assembly in amphibia, it is suggested that a light effect is only possible at a temperature at which the ongoing rate of disc assembly is below its physiological maximum. Because rhythmic increases in disc assembly have not been detected in constant darkness and light stimulated disc assembly occurs in constant light which inhibits rhythmic disc shedding, it is suggested that light acts directly on photoreceptor cells.

The latter suggestion is supported by data indicating that light stimulates ROS disc assembly in retinal explants from X. laevis separated from the pigment epithelium and maintained in culture. Light evoked disc assembly *in vitro* is paralleled by increased incorporation of ^{3}H-leucine into photoreceptors, and is blocked by both cycloheximide and puromycin at concentrations which effectively inhibit incorporation of ^{3}H-leucine in the explants. The data suggest, therefore, that ongoing protein synthesis is necessary for light stimulated disc assembly to occur.

Understanding light effects on disc assembly depends ultimately on definition of the underlying cellular processes involved. Quantitative data on light stimulated disc assembly suggest that a major problem may be in expansion of the plasma membrane of the cilium or ROS at a rate sufficient to support rapid disc assembly. A possible mechanism for plasma membrane expansion is that preformed membrane packages (i.e., carrier vesicles or schleppersomes) fuse with the cell surface in the region of the connecting and that the expanded membrane surface is subsequently used for disc assembly. Morphological evidence is reviewed suggesting that a population of vesicles exists in the ellipsoid region of *X. laevis* photoreceptors which may be the structural equivalents of vesicles formerly suggested on the basis of cell fractionation analysis (43). Freeze-fracture observations indicate that intramembranous particles of the vesicles are of a size similar to that of membranes in the ROS, and electron microscope autoradiographic analysis indicates that during transport of radiolabelled protein to the outer segment, vesicles are the major labelled component of the ellipsoid region. The morphological observations are regarded as consistent with the carrier vesicle hypothesis.

ACKNOWLEDGEMENTS

I thank Joe G. Hollyfield for his contributions to the studies reported herein, many of which were either completed or initiated in his laboratory. Rufus O. Terrill for his technical assistance and Vicki Shadix for typing the manuscript. The work on the *in vitro* system and carrier vesicle mechanism was supported by NIH Grant (EY2414).

REFERENCES

1. Young, R. W. (1967) *J. Cell Biol.* 33, 61-72.
2. LaVail, M. M. (1973) *J. Cell Biol.* 58, 650-661.
3. Young, R. W. and Droz, B. (1968) *J. Cell Biol.* 39, 169-184.
4. Hall, M. D., Bok, D. and Bacharach, A. D. E. (1969) *J. Mol. Biol.* 45, 397-406.
5. Bairati, A., Jr. and Orzalesi, N. (1963) *J. Ultrastruct. Res.* 9, 484-496.
6. Young, R. W. and Bok, D. (1969) *J. Cell Biol.* 42, 392-403.
7. Tweto, J. and Doyle, D. (1977) in *The Synthesis Assembly and Turnover of Cell Surface Components* (Poste, G. and Nicholson, eds.) pp. 137-163 Elsevier/North-Holland, Amsterdam.

8. Fambrough, D. M. and Devreotes, P. N. (1976) in Biogenesis and Turnover of Membrane Macromolecules (Cook. J. S., ed.) pp. 121-144, Raven Press, New York.
9. Remé, C. E. (1977) Invest. Ophthalmol. Vis. Sci. 16, 807-814.
10. Sjöstrand, F. S. (1961) in The Structure of the Eye (Smelser, G. K., ed.) pp. 1-28, Academic Press, New York.
11. Nilsson, S. E. G. (1964) J. Ultrastruct. Res. 11, 581-621.
12. Anderson, D. H., Fisher, S. K., and Steinberg, R. H. (1978) Invest. Ophthalmol. Vis. Sci. 17, 117-133.
13. Cohen, A. I. (1968) J. Cell Biol. 37, 424-444.
14. Cohen, A. I. (1970) Vision Res. 10, 445-453.
15. Laties, A. M., Bok, D. and Liebman, P. (1976) Exp. Eye Res. 23, 139-148.
16. Jan, L. Y. and Revel, J.-P. (1974) J. Cell Biol. 62, 257-273.
17. Basinger, S., Bok, D. and Hall, M. (1976) J. Cell Biol. 69, 29-42.
18. Besharse, J. C. and Pfenninger, K. H. (1978) J. Cell Biol. 79, 97a (Abst.).
19. Besharse, J. C., Hollyfield, J. G. and Rayborn, M. E. (1977) J. Cell Biol. 75, 507-527.
20. LaVail, M. M. (1976) Exp. Eye Res. 23, 277-280.
21. Besharse, J. C., Hollyfield, J. G. and Rayborn, M. E. (1977) Science 196, 536-538.
22. LaVail, M. M. (1976) Science 194, 1071-1074.
23. Besharse, J. C. and Hollyfield, J. G. (1979) Invest. Ophthalmol. Vis. Sci. (in press).
24. Basinger, S., Hoffman, R. and Matthes, M. (1976) Science 194, 1974-1976.
25. Hollyfield, J. G., Besharse, J. C. and Rayborn, M. E. (1976) Exp. Eye Res. 23, 623-635.
26. Young, R. W. (1978) Invest. Ophthalmol. Vis. Sci. 17, 105-116.
27. Young, R. W. (1977) J. Ultrastruct. Res. 61, 172-185.
28. Hollyfield, J. G., Besharse, J. C. and Rayborn, M. E. (1977) J. Cell Biol. 75, 490-506.
29. Currie, J. R. and Hollyfield, J. G. (1978) Vision Res. 18, 995-1003.
30. Hollyfield, J. G. and Basinger, S. F. (1978) Nature 174, 794.
31. LaVail, M. M. and Ward, P. A. (1978) Invest. Ophthalmol. Vis. Sci. 17, 1189-1193.
32. Tamai, M., Tiersten, P., Goldman, A., O'Brien, P., and Chader, G. (1978) Invest. Ophthalmol. Vis. Sci. 17, 558-562.
33. Hollyfield, J. G. and Rayborn, M. E. (1979) Invest. Ophthalmol. Vis. Sci. 18, 117-132.
34. Besharse, J. C. and Hollyfield, J. G. (1976) J. Exp. Zool. 198, 287-302.
35. Besharse, J. C. and Brandon, R. A. (1976) J. Morphol. 149, 527-546.
36. Bridges, C. D. B., Hollyfield, J. G., Besharse, J. C., and Rayborn, M. E. (1976) Exp. Eye Res. 23, 637-641.
37. Organisciak, D. T. and Noell, W. K. (1977) Invest. Ophthalmol. Vis. Sci. 16, 188-190.

38. Battelle, B. A. and LaVail, M. M. (1978) Exp. Eye Res. 26, 487-497.
39. O'Brien, P. J., Muellenberg, C. G. and Bunjenberg DeJong, J. J. (1972) Biochem. 11, 64-70.
40. Basinger, S. F. and Hall, M. O. (1973) Biochem. 12, 1996-2003.
41. Basinger, S. and Hoffman, R. (1976) Exp. Eye Res. 23, 117-126.
42. Wolf, K. and Quimby, M. C. (1964) Science 144, 1578-1580.
43. Papermaster, D. S., Converse, C. A. and Zorn, M. (1976) Exp. Eye Res. 23, 105-115.
44. Palade, G. (1975) Science 189, 347-358.
45. Hirano, H., Parkhouse, B., Nicolson, G. L., Lennox, E. S. and Singer, S. J. (1972) Proc. Nat. Acad. Sci. U.S. 69, 2945-2949.
46. Bok. D., Hall, M. O., O'Brien, P. (1977) in International Cell Biology 1976-1977 (Brinkely, B. R. and Porter, K. R., eds.) pp. 608-617, Rockefeller University Press, New York.
47. Papermaster, D. S., Converse, C. A. and Siu, J. (1975) Biochem. 14, 1343-1352.
48. Papermaster, D. S., Schneider, B. G., Zorn, M. A. and Kraehenbuhl, J. P. (1978) J. Cell Biol. 77, 496-510.
49. Nicolson, G. L. and Singer, S. J. (1974) J. Cell Biol. 60, 236-248.
50. Virtanin, I., Miettinen, A. and Wartiovaara, J. (1978) J. Cell Sci. 29, 287-296.
51. Boulan, E. R., Kreibich, G. and Sabatini, D. D. (1978) J. Cell Biol. 78, 874-893.
52. Kinney, M. S. and Fisher, S. K. (1978) Proc. Roy Soc. London B. 201, 149-167.
53. Williams, M. A. (1969) Adv. Opt. Elect. Microsc. 3, 219-272.

ASPECTS OF CHOLINE METABOLISM IN PHOTORECEPTOR CELLS

Richard H. Masland and John W. Mills[†]

Departments of Physiology and Anatomy
Harvard Medical School
and
[†]Departments of Surgery and Medicine
Massachusetts General Hospital
Boston, MA 02114

Elegant experiments by many investigators have described the photoreceptor cell's structural renewal, in which the outer segment is partially shed and replaced in a cyclic daily rhythmn (9, 10, 12, 13, 23, 24, 25). It seems certain that the resynthesis of such a specialized structure requires supporting adaptations by the cell.

Our own interest in this process arose from experiments concerning a different question: we wanted to localize a cholinergic cell known from other evidence to be present in the rabbit retina. Toward that goal we incubated isolated rabbit retinas in solutions containing low concentrations of tritiated choline, expecting that most of the choline would be taken up by the high-affinity mechanism present in cholinergic cells. We found that the retinas did synthesize acetylcholine from the extracellular choline; but they synthesized even more of a major phospholipid precursor, phosphorylcholine (2-5). If the high-affinity uptake system were present only in cholinergic cells, most of the choline taken up at low concentrations would have been acetylated (11). It seemed at the least that the rabbit retina contains cells with an unusually effective mechanism for incorporating extracellular choline into phospholipids.

Autoradiographic studies showed that the photoreceptors synthesize a disproportionate amount of the retina's total phospholipid. This was not surprising. Because of the high amounts of membrane contained in outer segment disks, the maintenance of the photoreceptor cell's steady state demands very rapid membrane synthesis.

If one assumes, from LaVail's findings (12), that the outer segment of a rat rod is renewed every 10 days, then each rat rod must replace something between 200 and 600 μM^2 of membrane per day. (A range is given because the calculation is quite sensitive to the assumed diameter of the disks, and to the shape of the incisures). This estimate is probably a lower limit, since it considers only the replacement of the shed material; if the possibility of an additional renewal of membrane in existing disks, as suggested by the data of Bibb and Young (8), is considered, then the estimate would be higher.

When all of these things were considered together, it appeared that the photoreceptors might possess the high-affinity choline uptake system thought to be present primarily in cholinergic cells. We therefore carried out a kinetic analysis of the retina's overall choline uptake, and studied the choline metabolites formed at each choline concentration. In addition, we directly evaluated the choline uptake of the photoreceptors by dissecting them from sections of freeze-dried tissue.

The main findings of these studies will be reviewed below; the experiments will be described in detail elsewhere (16).

The Metabolism of Choline by Retinal Neurons

Our original goals were to describe the overall metabolism of choline by the retina, and to identify the cells in which the various choline metabolites are formed.

For these studies rabbit retinas were removed from the pigment epithelium and incubated in a medium resembling cerebrospinal fluid. The rabbit retina is quite thin, the distance from the layer of the optic nerve fibers to the tips of the photoreceptors averaging 110 μm. It is essentially avascular, being primarily nurtured *in vivo* by diffusion from the choroidal vessels. Perhaps for this reason, it can be successfully nurtured by diffusion *in vitro*, while maintaining its protein synthesis (1), neurotransmitter release (14), photoreceptor sensitivity (20), and electrophysiological function as judged by ganglion cell receptive fields (15) for more than six hours.

The isolated retinas were exposed for 15 minutes to 0.3 μM ^{3}H-choline. At the end of this time they were found to contain radioactive acetylcholine (30% of the total radioactivity), phosphorylcholine (54%), free choline (8%) and phospholipid (8%). Because several choline metabolites were formed, meaningful autoradiography required a method by which ^{3}H-choline incorporated into acetylcholine could be distinguished from ^{3}H-choline incorporated into compounds of the phospholipid pathway. In attempting this we exploited the differing properties of the two pathways. Newly synthesized acetylcholine is quickly released from the stimulated retina. The choline metabolites of the phospholipid pathway, on the other hand, are stable within the time of an experiment--choline incorporated into membranes has a turnover much slower than that of choline incorporated into the neurotransmitter. This allowed a subtractive identification of acetylcholine, in which retinas that contained both ^{3}H-acetylcholine and ^{3}H-compounds of the phospholipid pathway were compared autoradiographically with retinas that contained radioactivity only in the lipid pathway.

Retinas were pulse-labeled by a 15 minute exposure to ^{3}H-choline. After labeling, different groups of retinas were incubated under chase conditions designed either to retain ^{3}H-acetylcholine within synapses or promote its release. In the former case, the retinas were incubated for one hour in the presence of 20 mM Mg^{++} (to prevent acetylcholine release), 30 μM physostigmine (an anticholinesterase that prevents the hydrolysis of intracellular acetylcholine by the presynaptic esterase) and 1 mM unlabeled choline (to clear free ^{3}H-choline from the tissue). Retinas in the acetylcholine-releasing condition were incubated in medium identical to control except for the presence of unlabeled choline, and were stimulated by flashing light.

At the end of the chase period, the retinas were removed and cut into three pieces. One was analyzed chemically, so that the retina's contents of the different choline metabolites could be established. A second was fixed with 1% osmium tetroxide in bicarbonate buffer. The third was cut into smaller samples, which were plunged into a propane slurry at liquid nitrogen temperature, slowly freeze-dried, lightly fixed with osmium vapor, and embedded in Spurr's resin (18). Sections from the tissue initially fixed with aqueous osmium were autoradiographed by conventional techniques. Those from the freeze-dried sector of the retina were autoradiographed by pressing them onto dry emulsion (7, 16).

Analysis of the retina's radiochemical contents showed that the chase period achieved its objective: the retinas incubated under acetylcholine-releasing conditions contained substantial amounts of radioactivity in the phospholipid pathway but essentially no radioactivity in the form of choline or acetylcholine. The retinas incubated under the phospholipid pathway; but these retinas also contained a substantial amount of radioactive acetylcholine.

Autoradiography of freeze-dried tissue from retinas incubated under the acetylcholine-releasing condition showed an accumulation of silver grains over the photoreceptor cells and faint labeling of all ganglion cells (Fig. 1). Retinas incubated under the acetylcholine-protecting condition--which contained ^{3}H-acetylcholine as well as ^{3}H-compounds of the phospholipid pathway--showed the same

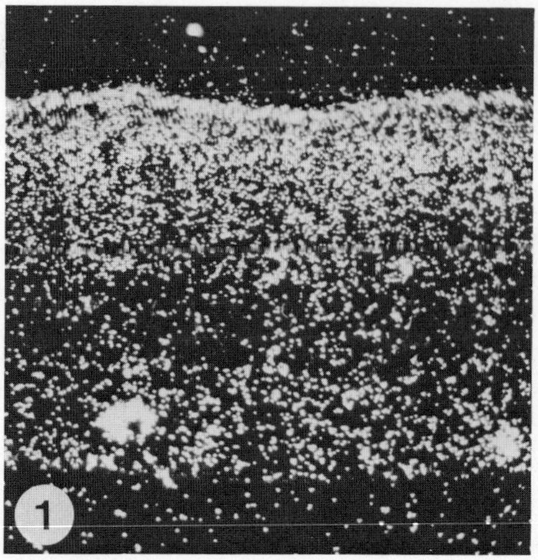

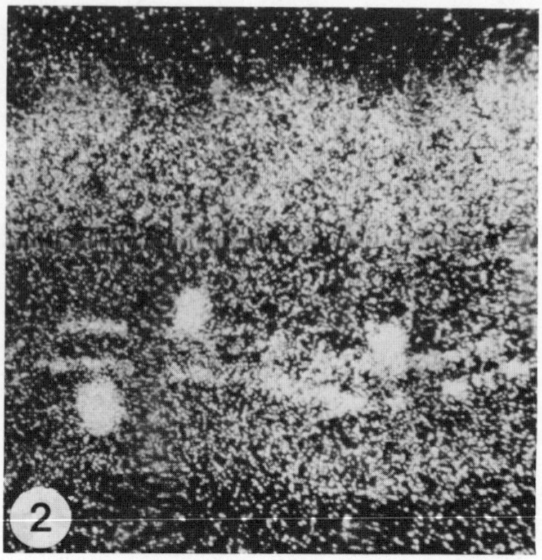

Figure 1. Autoradiograph of a retina pulse-labeled with ^{3}H-choline and then incubated under acetylcholine-releasing conditions. This retina contained substantial amounts of ^{3}H-phosphorylcholine and ^{3}H-phospholipid, but essentially no ^{3}H-acetylcholine. The photoreceptor cells are densely labeled, and the ganglion cells are faintly labeled.

Figure 2. Autoradiograph of a retina pulse-labeled with ^{3}H-choline and then incubated under acetylcholine-protecting conditions. This retina contained the same amount of radioactivity in the phospholipid pathway as the retina shown in Fig. 1, but also contained a substantial amount of ^{3}H-acetylcholine. The photoreceptors and ganglion cells are again labeled, but an additional group of labeled cells at both margins of the inner plexiform now appear.

pattern of silver grains; but they also showed, superimposed on that pattern, a few densely labeled cells at both margins of the inner plexiform layer (Fig. 2).

These findings implied that acetylcholine is synthesized by the densely labeled cells seen near the inner plexiform layer while the photoreceptor cells (and to a lesser extent the ganglion cells) are distinguished by an unusually rapid synthesis of choline-containing phospholipids. This conclusion was confirmed by autoradiographs of tissue initially fixed with aqueous osmium tetroxide, which retains a substantial part of the tissue's phospholipid while eluting the water-soluble choline metabolites. In autoradiographs of these retinas silver grains were only seen over the photoreceptor and ganglion cells (Fig. 3).

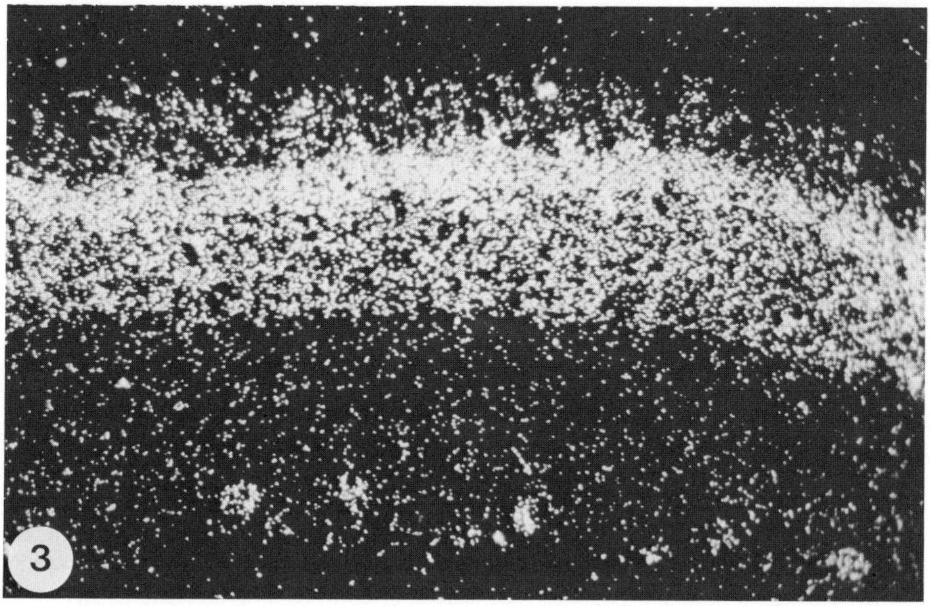

Figure 3. Autoradiograph of tissue initially fixed with aqueous osmium tetroxide. Fixable label is seen in the photoreceptor and ganglion cells.

Finally, the localization seen in autoradiographs was confirmed by direct chemical analysis of the radioactive choline metabolites eluted from sections of the freeze-dried tissue. Because the water-soluble choline metabolites are not fixed by osmium, we found that their localization in autoradiographs was extremely vulnerable to wetting. (In fact, the moisture from a single breath was enough to convert the autoradiographic pattern characteristic of an acetylcholine-containing retina to that of a retina containing radioactivity only in the phospholipid pathway.) This suggested that we might be able to systematically extract the water-soluble choline metabolites for chemical analysis. Immersion of 10 μm sections from freeze-dried tissue in formate-acetate buffer for 48 hours extracted more than 85% of the section's radioactivity, and the compounds recovered faithfully reflected those present in the retina before processing. Sections were divided along the outer plexiform layer, using a razor blade chip attached to the objective of a microscope. The radioactivity of the photoreceptors or of the inner half-retina could then be separately analyzed. The photoreceptors were found to contain substantial amounts of ^{3}H-phosphorylcholine but negligable ^{3}H-acetylcholine.

From these results we concluded that acetylcholine is synthesized by a small number of cells that line both margins of the inner plexiform layer. The photoreceptors, and to some extent the ganglion cells, are distinguished by a high rate of synthesis of choline-containing phospholipid. Of these, the photoreceptors are quantitatively the more important; the autoradiographs and the analysis of microdissected retinas both indicated that about two-thirds of the retina's total synthesis of choline-containing phospholipid occurs distal to the outer plexiform layer.

The Uptake of Choline by Photoreceptor Cells

We were surprised to find, in the experiments described above, that the rabbit retina synthesized a large amount of ^{3}H-phosphorylcholine when exposed to low concentrations of ^{3}H-choline. Studies in other acetylcholine-synthesizing tissues have provided evidence that cholinergic cells possess a specialized choline transport system, characterized by a high affinity for choline. With the exception of scattered data on cultured cells, this carrier has been thought to be present exclusively in cholinergic cells; other neurons are thought to obtain choline via a second transport mechanism, also identifiable in kinetic measurements, that has a much lower affinity for choline (see 11 for review). The rat (19) and chicken (7) retinas have been shown to possess both choline carriers. Since the rabbit retina synthesizes acetylcholine at a rate not strikingly different from that observed in the rat and chicken, it seemed reasonable to use the published kinetic values to compute, as a first approximation, the relative amounts of choline that should enter cells _via_ the high and low affinity carriers.

Such calculations indicated that, at concentrations of choline below about 5 μM, more choline should be transported by the high affinity mechanism than by the low affinity. This was suggestive that some of the phosphorylcholine synthesis observed in our experiments obtained its choline _via_ the high-affinity route. And the fact that the photoreceptors synthesize a large fraction of the rabbit retina's total phosphorylcholine suggested that this might be happening in the photoreceptors.

In order to evaluate these possibilities we needed a kinetic analysis of choline uptake in the rabbit retina, and we needed to know the choline metabolites formed at concentrations of choline where transport by one or the other uptake system predominates.

Retinas were incubated for five minutes in the presence of ^{3}H-choline, at concentrations from 0.7 to 300 μM. At the end of incubation they were briefly rinsed, after which they were homogenized and the total choline uptake and distribution of choline metabolites were evaluated. The kinetics of choline uptake were established by conventional graphic analysis. As shown in Fig. 4, the retina's total choline uptake could be resolved into two components, with apparent kinetics similar to those reported for the rat and chicken retinas (and for many other tissues that contain cholinergic neurons).

The choline metabolites synthesized by these retinas are shown in Fig. 5. Two things are immediately apparent: first, ^{3}H-acetylcholine synthesis and ^{3}H-phosphorylcholine synthesis increase in parallel as the extracellular concentration of choline increased. If phosphorylcholine synthesis and acetylcholine synthesis were obtaining choline via carriers with the kinetics defined by Fig. 4, and if the conventional assumption that transport is limiting is correct, then the rates of synthesis should have diverged as high-affinity uptake saturates and low affinity uptake comes to predominate. Second, at low concentrations of choline--concentrations below saturation for the high-affinity mechanism--more phosphorylcholine than predicted was synthesized. If choline for

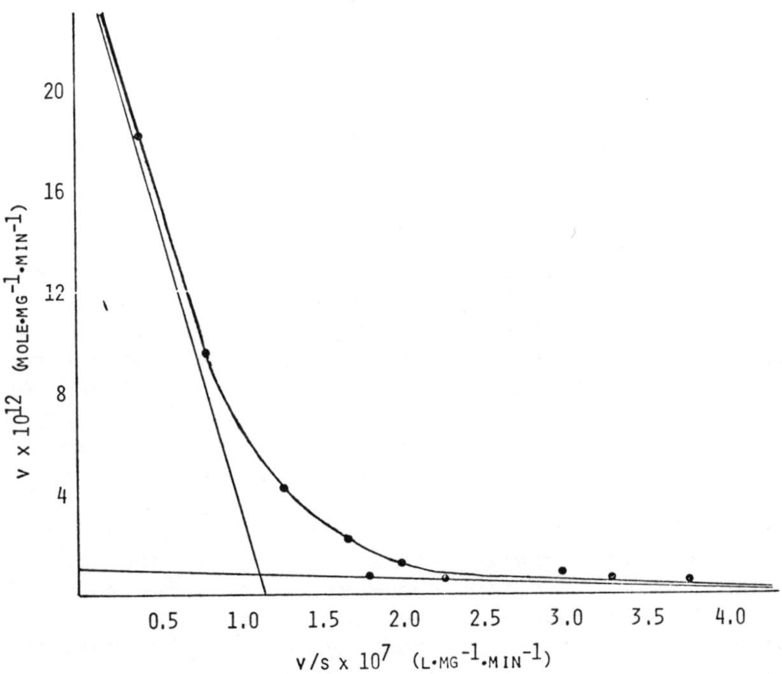

Figure 4. Eadie-Hofstee plot of the retina's total uptake of choline. Two saturable components are evident, with Km values of approximately 2 μM and 230 μM.

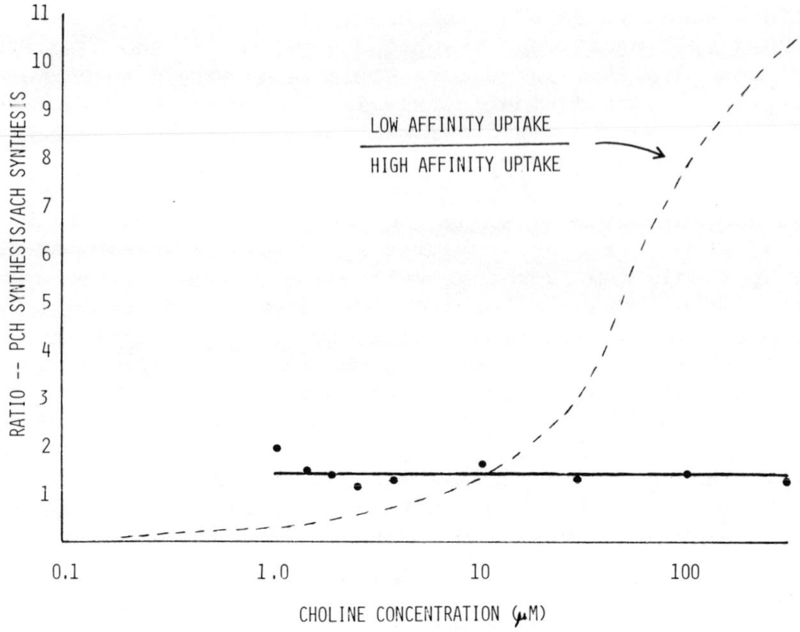

Figure 5. Relative amounts of ^{3}H-phosphorylcholine and ^{3}H-acetylcholine formed at various concentrations of extracellular ^{3}H-choline. The ratio of the two metabolites is essentially constant. For comparison, the dashed line shows the ratio of low-affinity to high-affinity uptake at the same choline concentrations.

phosphorylcholine synthesis were provided exclusively by the low-affinity mechanism, and choline for acetylcholine synthesis were provided exclusively by the high-affinity mechanism, then ^{3}H-phosphorylcholine and ^{3}H-acetylcholine should have been formed in a ratio defined by the overall uptake kinetics. At 1.0 μM extracellular choline the ratio of low affinity to high-affinity uptake was 0.3. The ratio of ^{3}H-phosphorylcholine to ^{3}H-acetylcholine was 2.2.

These findings indicate that phosphorylcholine and acetylcholine synthesis are supplied choline *via* mechanisms with similar overall kinetics. In addition, because of the large amount of phosphorylcholine formed at low choline concentrations, it appears that at least some of the cells that synthesize phosphorylcholine have a high-affinity choline uptake system. Since the majority of the retina's phosphorylcholine synthesis occurs in the photoreceptors, the findings thus argue quite strongly that the photoreceptors have high-affinity choline uptake.

In confirmation of this conclusion, we used the microdissection technique already described. Samples from the retinas incubated at varying choline concentrations were freeze-dried by our usual technique. The photoreceptors were isolated from 10 μm sections by a cut passing parallel to the outer plexiform layer and through the inner row of photoreceptor nuclei. The cut-off area of photoreceptors was photographed, and the accuracy of the dissection was verified in a large print. The image of the photoreceptors was then cut from the print and weighed. From the surface area thus established and the section's known thickness, the volume of photoreceptor tissue in each sample could be ascertained. The radioactivity present in the isolated tissue was extracted into acid, and kinetic analysis was carried out graphically (Fig. 6). The results showed the photoreceptors to have taken up choline by two distinct processes, one of which had a high affinity for choline.

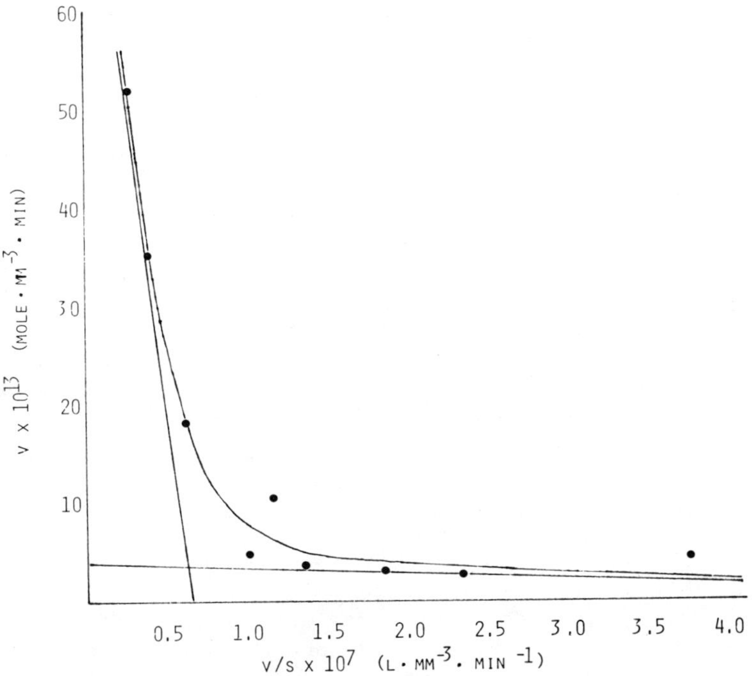

Figure 6. Eadie-Hofstee plot of choline uptake by the isolated photoreceptors. Two saturable components are evident, with Km values of approximately 0.6 μM and 140 μM.

DISCUSSION

In other systems, high-affinity uptake of small molecules has been most studied in the context of their possible release as neurotransmitters, and has in fact been used as a means of neurotransmitter identification. Whether the high-affinity carrier observed in the photoreceptor cells is identical to those associated with neurotransmitter metabolism is not certain; but the present results indicate quite clearly that high-affinity choline uptake, kinetically defined, is not limited to neurons that use acetylcholine as a neurotransmitter. Cultured fibroblasts (6) and glia (17, 21) also appear to have a high-affinity choline uptake that leads only to the synthesis of phosphorylcholine, and Suszkiw and Pilar (22) have suggested that this is concomitant to the fact that such cells are dividing and must synthesize much new membrane. It seems quite possible that high-affinity choline uptake will be found wherever cells have an especially active choline metabolism.

The photoreceptors almost certainly fall into that class; and one of our main points is to emphasize the demand placed on the photoreceptor cell by the requirement for resynthesis of the shed disks. The demand is apparent *a priori*, from the sheer amount of membrane required to maintain the steady state; and is presumably reflected in the fact that the cell has evolved a specialized carrier that would serve to ensure its supply of a crucial phospholipid precursor. The ability to synthesize phospholipids at an unusually high rate seems essential to the viability of the visual cell.

ACKNOWLEDGMENTS

The work was supported by NIH grants EY01075 and HE06664. Richard Masland holds a Research Career Development Award of the National Eye Institute. John Mills is a Junior Investigator of the King Trust.

REFERENCES

1. Ames, A. III, Parks, J. M. and Nesbett, F. B. (1976) *J. Neurochem.* 27, 999-1015.
2. Anderson, R. E. (1970) *Exp. Eye Res.* 10, 339-344.
3. Anderson, R. E., Feldman, L. S. and Feldman, G. L. (1970) *Biochim. et Biophys. Acta* 202, 367-373.
4. Anderson, R. E. and Maude, M. B. (1970) *Biochemistry* 9, 3624-3628.
5. Anderson, R. E. and Risk, M. (1974) *Vision Res.* 14, 129-131.
6. Barald, K. F. and Berg, D. K. (1978) *Devel. Biol.* 65, 90-99.

7. Baughman, R. W. and Bader, C. R. (1977) Brain Res. 138, 469-485.
8. Bibb, C. and Young, R. W. (1974) J. Cell Biol. 62, 378-389.
9. Hall, M. O., Bok, D., Bacharach, A. D. E. (1969) J. Molec. Biol. 45, 397-407.
10. Hollyfield, J. G., Besharse, J. C. and Rayborn, M. (1976) Exp. Eye Res. 23, 623- 629.
11. Kuhar, M. J. and Murrin, L. C. (1978) J. Neurochem. 70, 15-21.
12. LaVail, M. M. (1973) J. Cell Biol. 58, 650-661.
13. LaVail, M. M. (1976) Exp. Eye Res. 23, 277-280.
14. Masland, R. H. and Livingstone, C. J. (1976) J. Neurophysiol. 39, 1210-1219.
15. Masland, R. H. and Ames. A. III (1976) J. Neurophysiol. 39, 1220-1235.
16. Masland, R. H. and Mills, J. W. (1979) J. Cell Biol. (in press).
17. Massarelli, R., Sansenbrenner, M., Ebel, A., and Mandel, P. (1974) Neurobiology 4, 414-418.
18. Mills, J. W., Ernst, S. A., and DiBona, D. R. (1977) J. Cell Biol. 72, 88-111.
19. Neal, M. J. and Gilroy, J. (1975) Brain Res. 93, 548-551.
20. Pepperberg, D. R. and Masland, R. H. (1978) Brain Res. 151, 194-200.
21. Richelson, E. and Thompson, E. J. (1973) Nature, Lond. 241, 201-208.
22. Suszkiw, J. B. and Pilar, G. (1976) J. Neurochem. 26, 1133-1138.
23. Young, R. W. (1967) J. Cell Biol. 33, 61-72.
24. Young, R. W. and Droz, B. (1968) J. Cell Biol. 39, 169-184.
25. Young, R. W. and Bok, D. (1969) J. Cell Biol. 42, 292-403.

LIST OF CONTRIBUTORS

Kenneth V. Anderson
Department of Anatomy
Emory University
Atlanta, Ga. 30322

E. Auerbach
Vision Research Laboratory
Hadassah University Hospital
Jerusalem, Israel

Howard D. Baker
Department of Psychology
Florida State University
Tallahassee, Fla. 32306

Robert B. Barlow, Jr.
Institute for Sensory Research
Syracuse University
Syracuse, New York 13210

Scott F. Basinger
Cullen Eye Institute
Baylor College of Medicine
Houston, Tx. 77030

Joseph C. Besharse
Department of Anatomy
Emory University
Atlanta, Ga. 30322

A. D. Blest
Department of Neurobiology
Research School of Biological Sciences
Australian National University
Canberra City, A.C.T., 2601, Australia

Steven C. Chamberlain
Institute for Sensory Research
Syracuse University
Syracuse, New York 13210

R. S. Crockett
University of Louisville
School of Medicine
Louisville, Ky. 40202

Rosalie K. Crouch
Departments of Ophthalmology and Biochemistry
Medical University of South Carolina
Charleston, S. C. 29403

G. Currier
University of Louisville
School of Medicine
Louisville, Ky. 40202

Dian Gifford
Department of Biology
University of Massachusetts/Boston
Boston, Mass. 02125

DuPont Guerry, III
Department of Biophysics
Virginia Commonwealth University
Richmond, Va. 23298

William T. Ham, Jr.
Department of Biophysics
Virginia Commonwealth University
Richmond, Va. 23298

Joe G. Hollyfield
Cullen Eye Institute
Baylor College of Medicine
Houston, Tx. 77030

M. Kaitz
Vision Research Laboratory
Hadassah University Hospital
Jerusalem, Israel

Thomas K. Kuyk
Department of Psychology
Florida State University
Tallahassee, Fla. 32306

Matthew M. LaVail
School of Medicine
Department of Anatomy
University of California, San Francisco
San Francisco, Calif. 94143

Theodore Lawwill
University of Louisville
School of Medicine
Louisville, Ky. 40202

Vance Lemmon
Department of Physiology and Biophysics
Washington University
St. Louis, Mo. 63130

Ellis R. Loew
Department of Physical Biology
Cornell University
Ithaca, New York 14853

Richard H. Masland
Department of Physiology and Anatomy
Harvard Medical School
Boston, Mass. 02114

Donna Medford
Cullen Eye Institute
Baylor College of Medicine
Houston, Tx. 77030

Norman A. Michaud
Department of Biology
University of Massachusetts/ Boston
Boston, Mass. 02125

John W. Mills
Departments of Surgery and Medicine
Massachusetts General Hospital
Boston, Mass. 02114

H. A. Mueller
Department of Biophysics
Virginia Commonwealth University
Richmond, Va. 23298

Werner K. Noell
Department of Physiology
School of Medicine
State University of New York
Buffalo, New York 14214

W. Keith O'Steen
Bowman Gray School of Medicine
Wake Forest University
Winston-Salem, N. C. 27103

L. M. Rapp
Department of Biological Sciences
Florida State University
Tallahassee, Fla. 32306

Mary E. Rayborn
Cullen Eye Institute
Baylor College of Medicine
Houston, Tx. 77030

John J. Ruffolo, Jr.
Department of Biophysics
Virginia Commonwealth University
Richmond, Va. 23298

Charles R. Shear
Department of Anatomy
School of Medicine
University of Maryland
Baltimore, Md. 21201

Harry G. Sperling
University of Texas
Health Science Center
Sensory Sciences Center
Houston, Tx. 77025

Richard H. White
Department of Biology
University of Massachusetts/Boston
Boston, Mass. 02125

T. P. Williams
Department of Biological Sciences
Florida State University
Tallahassee, Fla. 32306

INDEX